Emerging Wireless Communication and Network Technologies

Karm Veer Arya · Robin Singh Bhadoria
Narendra S. Chaudhari
Editors

Emerging Wireless Communication and Network Technologies

Principle, Paradigm and Performance

 Springer

Editors
Karm Veer Arya
Department of Computer Science
  and Engineering
Institute of Engineering and Technology
Lucknow, Uttar Pradesh
India

Narendra S. Chaudhari
Visvesvaraya National Institute
  of Technology
Nagpur, Maharashtra
India

Robin Singh Bhadoria
Department of Computer Science
  and Engineering
Indian Institute of Information Technology
  (IIIT) Nagpur
Nagpur, Maharashtra
India

ISBN 978-981-13-4405-3   ISBN 978-981-13-0396-8   (eBook)
https://doi.org/10.1007/978-981-13-0396-8

This Springer imprint is published by the registered company Springer Nature Singapore Pte Ltd. part of Springer Nature
The registered company address is: 152 Beach Road, #21-01/04 Gateway East, Singapore 189721, Singapore

Preface

This edited book on "Emerging Wireless Communication and Network Technologies: Principle, Paradigm and Performance" aims to discuss a broader view of all futuristic wireless communication and network technologies being used. Moreover, this book would be helpful for the future research to be done in the field of communication engineering. It also explores the recent progress in several computing technologies and evaluates the performance based on previous development. This book also covers a wide range of topics such as cognitive radio networks, mobile opportunistic and reliable cooperative networks.

This book starts with advancements in wireless communication that includes the emerging trends and research direction for wireless technologies. It also briefs about the importance and need for advancement in technology along with existing basics of analog and digital signals, frequency, amplitude, encodings, channel access methods. It also pays a focus on fast and flexible technology like long-term evolution. This book would also present the detailed survey and case studies for current trends in wireless technology and communications for smart home, secure data access control in vehicular networks. It also focuses on latest methods for detecting and avoiding congestion in wireless communication based on stream engineering.

The book consists of three parts with 17 chapters equally focusing on new trends and explorations, methodologies and implementation, advancement and future scope. Part I provides recent advancements in wireless technologies and networks, cognitive radio networks, emerging trends in vehicular networks, 5G technologies, reliable cooperative networks, and delay-tolerant networks. Part II selects the chapters from generic design in wearable sensor technology, mobile opportunistic networks, long-term evolution, and Internet of things. Part III includes the chapter on security attacks and green generation of wireless communication systems, software-defined networks, spectrum decision mechanism, and state estimation for wireless sensor networks.

This book also deliberates the role of wireless communication technology in day-to-day human life. Some of the features of this book are as follows:

- Detailed survey for the wide variety of wireless and network technologies.
- Concepts and visualization of wireless communications into current trends like Li-Fi technology and intelligent transportation systems.
- Helpful for young researchers and practitioners especially in the area of security attacks for wireless networks.
- Talks about wearable sensor technology, cognitive radio networks, Internet of things, and many more.
- Different case studies for experimental wireless communication system in software-defined networking, state estimation and anomaly detection in wireless sensor network, green generation of wireless communication systems, etc.

We honestly believe that readers of today, as well as the future, would have interest in emerging wireless communication and network technologies. This book would also be useful in building new concept and perception to forthcoming advancements in the modern era of communication. We wish all readers of this book the very best in their journey of wireless communication and network technologies.

Lucknow, India   Karm Veer Arya
Nagpur, India   Robin Singh Bhadoria
Nagpur, India   Narendra S. Chaudhari

Contents

About the Editors

Prof. Karm Veer Arya received his Ph.D. from the Indian Institute of Technology Kanpur (IITK), India, and his Master's from the Indian Institute of Science (IISc), Bangalore, India. His research areas include Image Processing, Biometrics, Information Security, and Wireless Ad hoc Networks. He is currently working as a Professor at the Institute of Engineering and Technology (IET), Lucknow, and as Dean of PG Studies and Research, AKTU, Lucknow, Uttar Pradesh, India. He has published more than 150 papers in international journals and conferences and supervised 6 Ph.D. scholars and more than 100 Master's students. In addition, he has completed several funded research projects.

Robin Singh Bhadoria has worked in various fields including Data Mining, Cloud Computing, Service-Oriented Architectures, Wireless Sensor Networks. He has published more than 60 research articles in the form of chapters, conference, and journal papers and has released 3 edited books. Most recently, he completed his Ph.D. in the discipline of Computer Science and Engineering at the Indian Institute of Technology Indore (IITI), Madhya Pradesh, India.

Prof. Narendra S. Chaudhari has more than 35 years of academic and research experience. He was Professor of Computer Science in the Ministry of Defense (Government of India) M.Sc. DRDO Program from 1990 to 2001 and has been Professor of Computer Science and Engineering at the Indian Institute of Technology Indore (IITI) since 2009. He has also been a member of the Computer Engineering Faculty at Nanyang Technological University (NTU), Singapore, since 2002–2009. He has also been the Director of Visvesvaraya National Institute of Technology (VNIT), Nagpur, Maharashtra, India, since 2013. His research contributions are in the areas of Algorithms and Graph Theory, Network Security and Mobile Computing, Novel Neural Network Models, Context-free Grammar Parsing, and Optimization. He has authored more than 340 research publications.

Part I
Wireless Technology and
Communications—Explorations & Trends

Advancement in Wireless Technologies and Networks

Bathula Siva Kumar Reddy

Abstract This chapter discusses the emerging trends and research direction for advanced wireless technologies. This chapter also presents the necessity of spectral efficiency for next-generation wireless technologies by discussing different spectrum sensing techniques. Moreover, a recent survey reveals that almost 70% of the available spectrum is not utilized efficiently. Therefore, more research is needed to determine whether the spectrum is being used by primary user or not for efficient utilization of spectrum. This chapter analyses the sensing by identifying a few situations, and then these behaviours have been reported to the operator for further action. Generally, spectrum sensing techniques are classified into three such as transmitter detection, receiver detection and interference temperature detection. This chapter mainly focuses on the performance analysis of transmitter-based detection techniques, such as matched detection, energy detection and cyclostationary detection. In addition, the receiver operating characteristics (ROC) for various number of sensing samples are presented in this chapter.

Keywords Cyclostationary detection · Cognitive radio · Energy detection
Matched detection · ROC · Spectrum sensing

1 Introduction

The communications are broadly divided into two categories, namely, wired and wireless communications (see Fig. 1). The wireless systems mainly include satellite systems, cellular systems, paging systems, Bluetooth and wireless LANs [1]. Recent researchers are mainly focussing on performance issues of wireless LANs, i.e. wireless fidelity (Wi-Fi) and worldwide interoperability for microwave access (WiMAX) [2]. Wi-Fi is based on the IEEE standard 802.11 while WIMAX is based on IEEE 802.16. Both standards are designed for the Internet protocol applications. Both

B. Siva Kumar Reddy (✉)
Department of Electrical Engineering, Institute of Infrastructure Technology Research and Management (IITRAM), Ahmadabad 380026, Gujarat, India
e-mail: bsivakumar100@gmail.com

© Springer Nature Singapore Pte Ltd. 2018
K. V. Arya et al. (eds.), *Emerging Wireless Communication and Network Technologies*,
https://doi.org/10.1007/978-981-13-0396-8_1

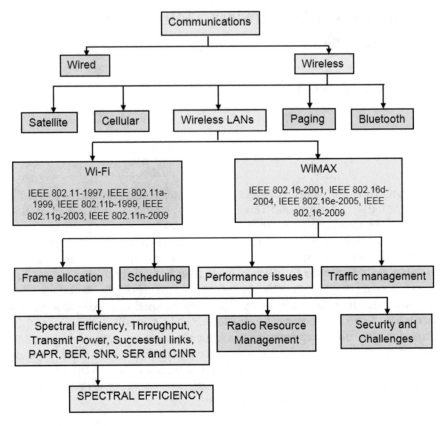

Fig. 1 A detailed review of literature

Wi-Fi and WiMAX wireless networks support real-time applications such as Voice over Internet Protocol (VoIP) and are frequently used for wireless Internet access [3]. The WiMAX standard provides fixed services (IEEE 802.16d-2004) as well as mobility services (IEEE 802.16e-2005). The WiMAX standard provides fixed services (IEEE 802.16d-2004) as well as mobility services (IEEE 802.16e-2005). However, WiMAX has some issues, namely, frame allocation, scheduling, traffic management, security and performance issues [4].

1.1 Need for Advancement in Wireless Technologies

The performance metrics such as packet loss, throughput and delay of WiMAX are measured on the basis of optimal boundary per WiMAX cell under different WiMAX network models. The performance metrics considered are spectral efficiency, throughput, transmit power, percentage of successful links, PAPR, BER,

SNR and CINR. This chapter mainly focusses on spectrum sensing techniques to achieve better spectral efficiency [5].

Recently, there is a lot of demand for tremendous technologies such as 3G, 4G and 5G, where voice-only communications are transitioned into multimedia type applications [6, 7]. These applications may be mobile TV, mobile P2P, streaming multimedia, video games, video monitors, interactive video, 3D services and video sharing. These high data rate applications consume more and more energy to guarantee quality of service [8]. However, the current frequency allocation schemes are unable to handle the requirements of recent higher data rate systems due to the limitations of the frequency spectrum.

Therefore, more efforts are kept on efficient frequency spectrum usage, and then a solution is found by Joseph Mittola [9], in the name of cognitive radio. The basic definition given by him is that cognitive radio (CR) is a type of a transceiver which can intelligently sense or detect unusable communication channel, and instantly allocate those channels to the unlicensed users without disturbing occupied channels [10]. Though there is no formal meaning of cognitive radio, various definitions can be seen in several contexts. A cognitive radio is, as defined by the researchers at Virginia Tech, 'A software defined radio with a cognitive engine brain' [11, 12]. The evolution of SDR in current technologies is provided in Fig. 2. The physical, data link and network layers of OSI model can be implemented by using SDR as shown in Fig. 3. The SDR Forum proposed a multi-tiered definition of SDR by providing the use of open architectures for advanced wireless systems and supports deployment and development [13–15]. An abstraction of the five-tier definition is illustrated in Fig. 4, where the length of the arrow represents the distribution of the software content within the radio [16].

Software-defined radio architecture comprises three sections such as radio frequency (RF), intermediate frequency (IF) and baseband section [17, 18]. It is observed from Fig. 5 that an RF signal received by smart antenna is sent to the hardware (here USRP) in which various components are inbuilt such as daughterboard, ADC/DAC, FPGAs, DSPs and ASICs. This hardware converts RF signal to IF signal and then to low-frequency baseband signal (digitized) and that will be sent to a personal computer (PC) for baseband signal processing in the transmitter (Tx) path. In this experimentation, an open-source software, GNU Radio, is employed as a software to perform baseband processing in which most of the signal processing blocks are inbuilt. All the reverse operations are performed in receiver (Rx) path such that baseband signal is converted to analogue by DAC and then sent into the air by RF hardware.

2 Emerging Trends and Research Direction for Wireless Technologies

A simple and typical dynamic spectrum access (DSA) network consists of a pair of primary user (PU) or licensed user and a pair of secondary user (SU) or unlicensed user and both are operated at the same frequency band. The PU has higher priority

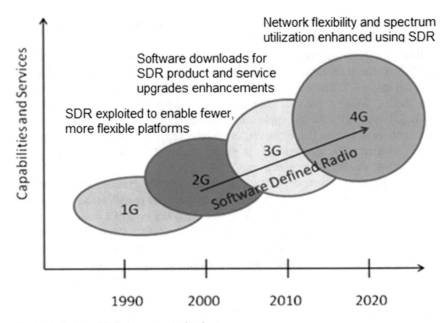

Fig. 2 Evolution of SDR in current technology

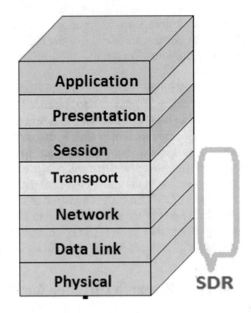

Fig. 3 The open systems interconnection (OSI) reference model

to access the spectrum, as it is a licensed user. Several spectrum sensing techniques are broadly classified into three such as transmitter detection, receiver detection and

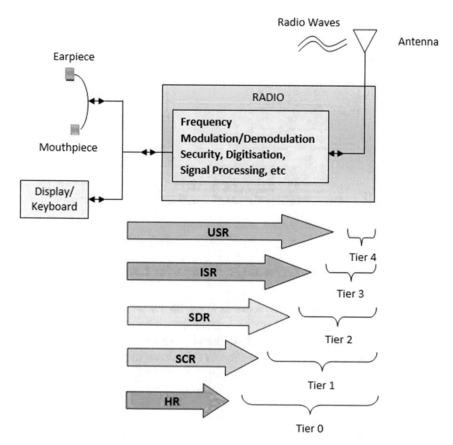

Fig. 4 SDR definition

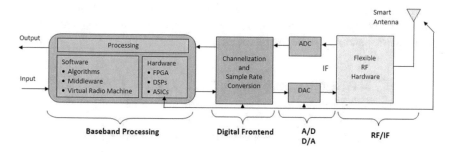

Fig. 5 Signal processing in software defined radio

interference temperature detection as shown in Fig. 6. However, this chapter presents the performance analysis of transmitter sensing techniques such as energy detection, matched filter detection and cyclostationary feature detection.

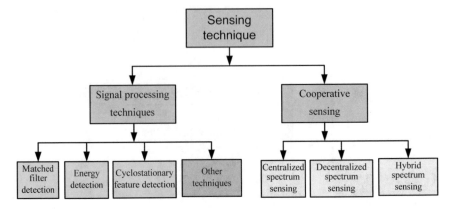

Fig. 6 Further classification of spectrum sensing techniques

2.1 Advancement in Technology Along with Existing Works

Signal processing techniques are proposed in the literature [11, 12] based on man-made signals that present periodicity in their statistics. First of all, we need primary user waveform on which we can apply different spectrum sensing techniques [13]. Radial basis function network based on energy detection is implemented in [19]. Sun et al. proposed a blind OFDM signal detection on cyclostationary sensing [20].

3   Results and Discussion

Generally, energy detection performance is measured in terms of probability of false alarm P_{fa} (detection algorithm falsely decides that PU is present when it actually is absent) and probability of detection P_d (correctly detecting the PU signal). Mathematically, P_{fa} and P_d can be expressed as [16]:

$$P_{fa} = P_r(\text{signal is detected}|H_0 = P_r(u > \lambda|H_0) = \int_{\lambda}^{\infty} f(u|H_0)du \qquad (1)$$

$$P_d = P_r(\text{signal is detected}|H_1 = P_r(u > \lambda|H_1) = \int_{\lambda}^{\infty} f(u|H_1)du \qquad (2)$$

where $f(u|H_i)$ denotes the probability density function (pdf) of test statistic under hypothesis H_i with $i = 0, 1$.

Thus, we target at maximizing P_d while minimizing P_{fa}. P_d versus P_{fa} plot depicts receiver operating characteristics (ROC) and is considered as an important

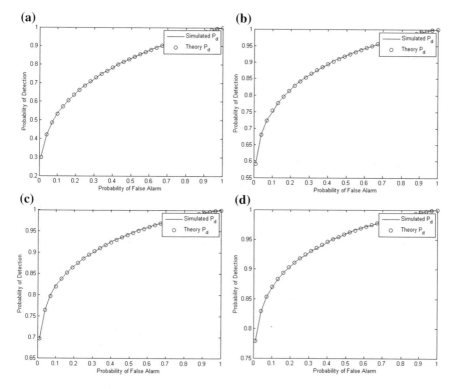

Fig. 7 ROC curve for number of sensing samples: **a** 10, **b** 50, **c** 100 and **d** 200

performance indicator. The receiver operating characteristics (ROC) for various number of sensing samples, such as 10, 50, 100 and 200, are presented in Fig. 7a, b, c and d, respectively [16]. It can be observed from Fig. 7 that the probability of detection (p_d) is increased with the number of sensing samples. In our simulations, some assumptions are made such as the primary signal is deterministic, and noise is real Gaussian with mean 0 and variance 1 [17]. The probability of detection for Rayleigh channel is calculated by the averaging the probability of detection for AWGN channel [18].

4 Conclusion

This chapter mainly focused on the fundamental issues, challenges and semantics for advancement in wireless technology and network. This chapter also analysed the performance analysis of transmitter-based spectrum sensing techniques, such as matched detection, energy detection and cyclostationary detection. A new research methodology, software-defined radio (SDR), is clearly explained with a detailed literature survey. It is concluded that cyclostationary feature detection is the best method, when there is no prior knowledge about primary user's waveform. It is also concluded that modulation technique can be estimated by counting the number of peaks.

References

1. Ghasempour, Yasaman, and Edward W. Knightly. "Decoupling Beam Steering and User Selection for Scaling Multi-User 60 GHz WLANs." Proceedings of the 18th ACM International Symposium on Mobile Ad Hoc Networking and Computing. ACM, 2017.
2. Varma, Ruchi, Jayanta Ghosh, and Rajarshi Bhattacharya. "A compact dual frequency double U-slot rectangular microstrip patch antenna for WiFi/WiMAX." Microwave and Optical Technology Letters 59.9 (2017): 2174–2179.
3. Badr, Ahmed, et al. "FEC for VoIP using dual-delay streaming codes." INFOCOM 2017-IEEE Conference on Computer Communications, IEEE. IEEE, 2017.
4. Chen, Olga, Catherine A. Meadows, and Gautam Trivedi. "Stealthy Protocols: Metrics and Open Problems." Concurrency, Security, and Puzzles. 2017.
5. Weng, Zu-Kai, et al. "60-Gbit/s QAM-OFDM Direct-Encoded Colorless Laser Diode Uniform Transmitter for DWDM-PON Channels." CLEO: Science and Innovations. Optical Society of America, 2017.
6. Alkhazaali, Naseer Hwaidi, et al. "Mobile Communication through 5G Technology (Challenges and Requirements)." International Journal of Communications, Network and System Sciences 10.05 (2017): 202.
7. Thuemmler, Christoph, et al. "Information Technology–Next Generation: The Impact of 5G on the Evolution of Health and Care Services." Information Technology-New Generations. Springer, Cham, 2018. 811–817.
8. Vannithamby, Rath, and Shilpa Talwar, eds. Towards 5G: Applications, Requirements and Candidate Technologies. John Wiley & Sons, 2017.
9. I. Mitola, J. and J. Maguire, G. Q., "Cognitive radio: making software radios more personal," IEEE Personal Commun. Mag., vol. 6, no. 4, pp. 13–18, Aug. 1999.
10. K. Chowdhury, R. Doost-Mohammady, W. Meleis, M. Di Felice, and L. Bononi "Cooperation and Communication in Cognitive Radio Networks based on TV Spectrum Experiments," 2011 IEEE International Symposium on a World of Wireless, Mobile and Multimedia Networks, pp. 1–9, 2011. IEEE https://doi.org/10.1109/wowmom.2011.5986378.
11. Yucek, Tevfik, and Huseyin Arslan. "A survey of spectrum sensing algorithms for cognitive radio applications." IEEE communications surveys & tutorials 11.1 (2009): 116–130.
12. Gueguen L, Sayrac B (2009) Sensing in cognitive radio channels: a theoretical perspective. IEEE Trans Wirel Commun 8(3):1194–1198.
13. D. Cabric, S. M. Mishra, and R. W. Brodersen, "Implementation issues in spectrum sensing for cognitive radio," Proc. Asilomar Conf. on Signals, Syst., and Comput., vol. 1, pp. 772–776, Nov. 2004.
14. Waleed Ejaz, Dr. Shoab A. Khan, "Spectrum Sensing In Cognitive Radio Networks", PhD Thesis, National University of Sciences and Technology, 2008.
15. Haykin S, Thomson DJ, Reed JH (2009) Spectrum sensing for cognitive radio. Proc IEEE 97(5):849–877.
16. Umar, Raza, Asrar UH Sheikh, and Mohamed Deriche. "Unveiling the hidden assumptions of energy detector based spectrum sensing for cognitive radios," IEEE Communications Surveys & Tutorials, 16.2 (2014): 713–728.
17. S. Kyperountas, N. Correal, Q. Shi and Zhuan Ye, "Performance analysis of cooperative spectrum sensing in Suzuki fading channels," in Proc. of IEEE Intern. Con. on Cognitive Radio Oriented Wireless Networks and Communications (CrownCom'07), pp. 428–432, June 2008.

18. Zhang, Wei, Ranjan K. Mallik, and Khaled Ben Letaief. "Optimization of cooperative spectrum sensing with energy detection in cognitive radio networks." IEEE Transactions on wireless Communications 8.12 (2009): 5761–5766.
19. Dey, Barnali, et al. "Function approximation based energy detection in cognitive radio using radial basis function network." Intelligent Automation & Soft Computing 23.3 (2017): 393–403.
20. Sun, Xiang, et al. "A Blind OFDM Signal Detection Method Based on Cyclostationarity Analysis." Wireless Personal Communications 94.3 (2017): 393–413.

Cognitive Radio Network Technologies and Applications

Rajorshi Biswas and Jie Wu

Abstract Mobile devices are advancing every day, creating a need for higher bandwidth. Because both the bandwidth and spectrums are limited, maximizing the utilization of a spectrum is a target for next-generation technologies. Government agencies lease different spectrums to different mobile operators, resulting in the underutilization of spectrums in some areas. For some operators, limited licensed spectrums are insufficient, and using others' unused spectrums becomes necessary. The unlicensed usage of others' spectrums is possible if the licensed users are not using the spectrum, and this gives rise to the idea of cognitive radio networks (CRNs). In CRN architecture, each user must determine the status of a spectrum before using it. In this chapter, we present the complete architecture of CRN, and we additionally discuss other scenarios including the applications of the CRN. After the Federal Communications Commission (FCC) declared the 5 GHz band unlicensed, Wi-Fi, LTE, and other wireless technologies became willing to access the band, leading to a competition for the spectrums. Because of this, ensuring that the spectrum is fairly shared among different technologies is quite challenging. While other works on DSRC and Wi-Fi sharing exist, in this chapter, we discuss LTE and Wi-Fi sharing specifically.

1 Introduction

Currently, governmental agencies assign wireless spectrums to license holders in large areas for long terms. For this kind of static spectrum allocation, licensed users of any spectrum cannot use others' licensed spectrums. This increase in data transmission results in a spectrum crisis for the mobile users. One method that can help in this situation is dynamic spectrum allocation. Users use their spectrum in an opportunistic manner. This way, if others' spectrums are free, then any licensed user can use their

R. Biswas (✉) · J. Wu (✉)
Temple University, Philadelphia, PA 19122, USA
e-mail: rajorshi@temple.edu

J. Wu
e-mail: jiewu@temple.edu

© Springer Nature Singapore Pte Ltd. 2018
K. V. Arya et al. (eds.), *Emerging Wireless Communication and Network Technologies*,
https://doi.org/10.1007/978-981-13-0396-8_2

licensed spectrum. Users in this kind of network architecture need to sense channels to find a free channel. If they find more than one free channel, they need to choose the best channel for their transmissions. Generally, the number of users is greater than the number of free channels and users need to share a channel. While using a free, unlicensed channel, users must be cautious about licensed transmissions because if any licensed transmission is detected, the user must vacate the channel. Therefore, the operations of users can be divided into four major steps: spectrum sensing, spectrum decision, spectrum sharing, and spectrum mobility. From these four major operations in a CRN, we can conclude that there are two kinds of transmissions. One is transmissions in a licensed band; we call this the *primary transmissions* and we call the user transmitting in the licensed band a *primary user* (PU). The other type is transmissions in the unlicensed band, which we call the *secondary transmissions*; the user transmitting in the unlicensed band is an *SU* (SU).

Transmissions in an unlicensed channel depend on the sensing information of CR users. There are various methods for detecting transmissions in a spectrum. Primary transmitter detection, primary receiver detection, and cooperative sensing are the most common techniques. Cognitive radio (CR) users must be able to decide the best channel out of all the available channels. This notion is called *spectrum decision*. Spectrum decision depends on the channel characteristics and operation of PUs. *Spectrum sharing* deals with sharing the same channel with multiple CR users. Many users can detect that the same channels are free and their channel choice decisions can be the same. Because of this, the channel must sometimes be shared between different CR users. While a CR user is transmitting in a secondary channel, a PU may need to use the channel. In this situation, the SU vacates the channel to the PU, but a secondary transmission cannot be stopped. The SU must find another channel and resume transmissions in that channel.

At the end of this chapter, we discuss some coexistence scenarios in the 5 GHz band which is currently an unlicensed band. Currently, some Wi-Fi standards (802.11ac and 802.11ax) are operating in the 5 GHz band. Dedicated Short Range Communication (DSRC) also operates in the 5 GHz band. LTE shareholders are now trying to operate in that band. We discuss two coexistence scenarios: the coexistence between LTE and Wi-Fi and the coexistence between Wi-Fi and DSRC.

2   Network Architecture of Cognitive Radio Networks

This subsection describes the network architecture and components of a CRN. Figure 1 depicts the whole network system. User devices, primary base stations, and CR base stations are the components of a basic CRN. In Fig. 1, there are two channels: channel 1 and channel 2. One primary base station operates in channel 1 and another in channel 2. Transmissions with the primary base station are done through licensed channels by mobile users, and the transmissions are called primary transmissions (denoted by solid lines). Transmissions with the CR base station can be done through either licensed or unlicensed channels and these transmissions are

Fig. 1 Network architecture of a CRN

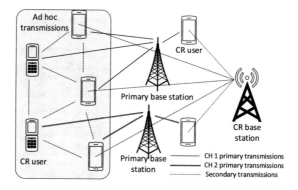

called secondary transmissions (marked by dotted lines). There is also another kind of transmission in which any user device can transmit directly to another user device. Therefore, transmissions in a CRN can be grouped into three classes:

- **Primary transmissions**: Primary transmissions are most prioritized transmissions and cannot be compromised by other transmissions. These transmissions are done in a licensed channel between primary base stations and PUs. Primary transmissions are denoted by solid lines in Fig. 1.
- **Secondary transmissions**: Secondary transmissions are done in the absence of primary transmissions. Transmissions between the CR base station and the CR user are usually secondary transmissions.
- **Secondary ad hoc transmissions**: User-to-user communications are called ad hoc transmissions. These transmissions can continue without base stations or other components of the network architecture. Users create their own network topology and adapt any routing protocols of ad hoc networks. Users in the gray area form an ad hoc network in Fig. 1. There are a lot of routing protocols for mobile ad hoc networks. For example, the proposed routing algorithm in [1], which ensures a fair amount of communications among nodes and improves the load concentration problem, can be used in secondary ad hoc networks. The on-demand cluster-based hybrid routing protocol proposed in [2] is also applicable here.

3 Spectrum Sensing

Secondary transmissions depend on spectrum sensing information, so this step should be done very accurately. Inaccurate sensing detection can lead to interferences with the PU that are highly unexpected. Though false alarms (in which channel is not occupied, but is detected as occupied) do not create interferences with the primary transmissions, it makes the CR user choose a channel from a narrower range of channels. As a result, a channel must be shared with many CR users and there would be increased competition among CR users to access the channel. The authors

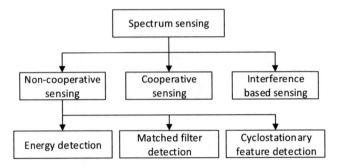

Fig. 2 Classification of spectrum sensing technologies

of [3] present a classification of spectrum sensing techniques. First, they classify sensing techniques into three groups: noncooperative sensing, cooperative sensing, and interference-based sensing. Noncooperative sensing is again classified into three groups: energy detection, matched filter detection, and cyclostationary feature detection. The classification is depicted in Fig. 2.

3.1   Noncooperative Sensing

In noncooperative sensing, CR users do not share sensing information with one another. A CR user makes a decision about the PU's presence using its own sensing information. We discuss primary transmitter detection and primary receiver detection, which are presented in [4, 5], in the following subsection.

3.1.1   Primary Transmitter Detection

Transmitter detection techniques emphasize detecting low power signals from any PU. Low power signals mix with noise from the environment and make it hard for the CR user to detect primary signals. A low signal-to-noise ratio, multipath fading effects, and time depression make primary transmissions detection very difficult for the CR user. We discuss some primary transmitter detection techniques including energy detection, coherent detection, and matched filter detection.

Energy Detection

This technique does not require CR users to have knowledge of PU signal characteristics, and it is easy to implement. Because of this, it is widely used to detect primary transmissions. Let us assume $S(n)$ is the signal received by the CR user, $W(n)$ is white Gaussian noise, and $P(n)$ is the original signal from the PU.

$$H_0 : S(n) = W(n) \tag{1}$$

$$H_1 : S(n) = W(n) + hP(n) \tag{2}$$

Hypothesis H_0 indicates the absence of a PU and hypothesis H_1 indicates the presence of PU transmissions. h denotes the channel gain between the primary and secondary transmissions. Then, the average energy S of N samples is

$$S = 1/N \sum_{n=1}^{N} S(n)^2 \tag{3}$$

The CR user collects N samples, calculates the average energy, and compares it with a threshold λ. If the average energy is greater than the threshold, λ, then the CR user concludes that primary transmissions are present. To measure the performance, we denote the probability of the false positive (CR detects the presence of PU transmissions when there is no PU transmission) as P_f and probability of the detection as P_d.

$$P_f = P(S > \lambda | H_0) \tag{4}$$

$$P_d = P(S > \lambda | H_1) \tag{5}$$

To improve the performance, we need to keep the PU's transmission secured. Therefore, the false positive probability should be less than 0.1 and the detection probability should be greater than 0.9.

Coherent Detection

When characteristics like signal patterns, pilot tones, and preambles of primary signals are known, coherent detection techniques can be used. These techniques work better than energy detection in an environment with noise level uncertainties. To describe this technique, we define the binary hypothesis slightly differently than energy detection.

$$H_0 : S(n) = W(n) \tag{6}$$

$$H_1 : S(n) = \sqrt{\epsilon} P_{pt}(n) + \sqrt{1 - \epsilon} P(n) + W(n) \tag{7}$$

Here, pilot signal energy is denoted by P_{pt}, and ϵ is the fraction of energy allocated to the pilot tone. Pilot signals are a special kind of signals used to send control signals. Hypothesis H_0 indicates the absence of primary transmissions, and hypothesis H_1 indicates the presence of primary transmissions.

If the CR user collects N samples and $\hat{X}_p$ is the unit vector in the direction of the pilot tone, then the average energy, S, is

$$S = 1/N \sum_{n=1}^{N} S(n)^2 \times \hat{X}_p(n) \tag{8}$$

Problems of Transmitter Detection

There are some situations where this detection technique does not work. We discuss two such situations: the hidden terminal problem and shadowing and multipath effects. Figure 3 depicts a scenario where a CR user remains outside of a base station's coverage area and it detects that the channel is free. Because it thinks the channel is free, it transmits in the channel and interference occurs at the other PU remaining in the coverage of the base station and the CR user. Figure 4 depicts the shadowing effect. The CR user behind the wall cannot detect primary transmissions. So, the same problem occurs.

Fig. 3 Hidden terminal problem

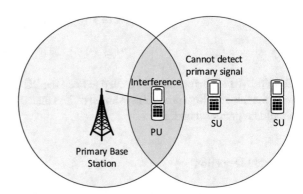

Fig. 4 Shadowing effect

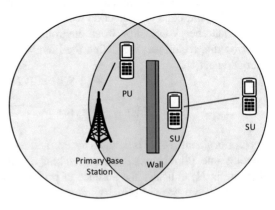

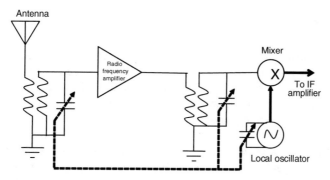

Fig. 5 Simple RF receiver circuit

3.1.2 Primary Receiver Detection

The most effective way to detect PU transmissions is to detect the primary receivers who are receiving from the primary channel. The circuit in Fig. 5 shows a simple RF receiver. It has a local oscillator that emits a very low power signal for its leakage current in the circuit. A CR user can detect the leakage signals from the RF receiver circuit and identify the presence of primary transmissions. This detection technique solves both the hidden terminal and shadowing effect problems. Since the signal power is very low, it is very challenging and costly to implement the circuit for primary receiver detection.

3.1.3 Matched Filter Detection

When primary signal features like modulation type, pulse shape, operating frequency, packet format, noise statistics, etc., are known, matched filter detection can be an optimal detection technique. If these parameters are known, the CR user only needs to calculate a small number of samples. As the signal-to-noise ratio decreases, the CR user needs to calculate a greater number of samples. The disadvantages of this technique are the complexities in low signal-to-noise ratio, the high cost of implementation, and the very poor performance if the features are incorrect.

3.1.4 Cyclostationary Feature Detection

In a broader sense, a signal can be called a cyclostationary process if its statistical properties vary cyclically with time. In [6], the authors presented a signal classification procedure that extracts cyclic frequency domain profiles and classifies them by comparing their log-likelihood with the signal type in the database. This technique can work very well in a low SNR. The drawback of this technique is that it needs a huge amount of computation and thus, a high-speed sensing is hard to achieve [7].

3.2   Cooperative Sensing

Cooperative sensing deals with sharing CR users' sensing information and making decisions by combining this information. A CR user collects sensing information from other CR users (in a distributed system) or from the base station (in a centralized system). Then, it analyzes the sensing information and makes a decision about whether the primary transmission is ongoing or not. Though this detection technique overcomes the hidden terminal problem and the shadowing and multi-path problems, it is more complex than previously mentioned detection techniques. Though its implementation is costly and its time complexity is higher, this technique has the best sensing accuracy and very few false alarms.

3.2.1   Data Aggregation Center of Cooperative Sensing

This system can be located either in user devices (distributed system) or in base stations (centralized system). A *Data aggregation center* is responsible for the collection and combination of sensing information. The system runs some aggregation functions over the collected data continuously and emits results about the primary transmission status. There are different methodologies to combine and calculate, but we must discuss the hard combining and the soft combining methodologies.

Hard Combining

CR users send their sensing results to the data aggregation center. This is just one bit information: 1 for the presence of primary transmissions and 0 for the absence of primary transmissions. After receiving the sensing information, the data aggregation center calculates the final result. The final result can be calculated based on AND, OR, or MAJORITY voting.

Soft Combining

Unlike the hard combining methodology, CR users send their raw sensing information (energy level w.r. to time, signal power, SNR, etc.) to the data aggregation center. Then, the data aggregation center decides the presence of primary transmissions.

3.3   Interference-Based Sensing

One user's transmissions can interfere with another user's transmissions at the receivers. The FCC introduced a new model to measure interference. According to the model, a receiver can tolerate up to a certain level of interference. This limit is called the *interference-temperature limit*. As long as CR users do not exceed this limit, they can use any spectrum. In this sensing method, the PU calculates the noise level and sends the information to the CR users. The CR users use the information to control their transmissions to avoid exceeding the interference-temperature limit for PUs. The authors in [8] present interference-based sensing and a technique to calculate the interference at a PU.

3.4   Predicting Channel to Sense

Due to limitations in the hardware, the CR users cannot sense a wide range of channels at a time. In addition, sensing a wide range of channels would raise the CR users' power consumption. Instead of sensing a huge number of channels, a CR user can predict which channel to sense. The authors in [9] model the prediction as a *multi-armed-bandit* problem. In the multi-arm-bandit problem of probability theory, a gambler tries to maximize his reward by playing different slot machines. The gambler has to decide which slot machine to play, how many times to play each machine, and in which order to play. The main objective of the gambler is to learn through every play and to predict which machine to play next so that the cumulative reward is maximized.

Let us assume there are N SUs and K channels, and SUs are trying to learn from their past history to predict the next channel to sense. Every CR user keeps a log of the transmitting channel in an array of length K. We denote the array by B_n where $n \in \{1, \ldots, N\}$.

$$B_n[k] = \begin{cases} 1, & \text{if CR user } n \text{ transmitted in channel } k \\ 0, & \text{Otherwise} \end{cases} \tag{9}$$

CR users share their B_n with other CR users. CR users preserve B_n with the time of arrival t_{B_n}. Then, CR users apply ϵ-GREEDY methods to predict the channel for sensing [9].

ϵ-GREEDY Method

This is the simplest solution to the multi-arm-bandit problem. The next channel is selected randomly with a probability of ϵ. The rest of the time, the maximum average valued channel is selected. The average value of channel k is denoted by A_k.

$$A_k = \frac{1}{N} \sum_{n=1}^{N} B_n[k] \tag{10}$$

Another approach that considers forgetting factor β while averaging the channel values works better. Let transmission logs $B_{n_1}, B_{n_2}, \ldots, B_{n_z}$ come to a CR user at $t_1, t_2, \ldots, t_z$. The forgetting factors for $t_1, t_2, \ldots, t_z$ are $\beta_{t_1}, \beta_{t_2}, \ldots, \beta_{t_z}$, respectively. The average value of channel k is denoted by A_{k_β}.

$$B_{n_\beta}[k] = \sum_{z=1}^{Z} \beta_{t_z} \times B_{n_z}[k] \tag{11}$$

$$A_{k_\beta} = \frac{1}{N} \sum_{n=1}^{N} B_{n_\beta}[k] \tag{12}$$

$B_{n_\beta}[k]$ denotes the effective value of channel k for CR user n. The effective values of different CR users for a channel are averaged to find the average effective value of the channel. The channel with the maximum average effective value is selected to sense next.

4 Spectrum Decision

CR users get a list of free channels after completion of the sensing process. A CR user can transmit in only one channel at a time. Therefore, the CR user must choose one channel among all the free channels. It is likely that any rational CR would choose the best channel. A channel can be characterized as "good" or "bad" according to some channel properties. Channel choice not only depends on channel characteristics but also on other CR users' activities. For example, if a channel is crowded by many CR users, despite being a good channel, a CR may not choose that channel. Normally, the spectrum decision process is done in two steps. We discuss some characteristics of channel in the following.

Interference

Interference in a channel reflects the channel's capacity. If interference is high, its capacity is low. A CR user should choose a channel with low interference. The permissible power of a CR can be calculated from the interference at the receiver.

Path Loss

Path loss is the reduction in power density of an electromagnetic wave as it propagates through space. It is related to both distance and frequency. If the carrier frequency is high, the path loss is also high. To reduce path loss, a CR can increase the transmission power. Interference with other users also increases with the increase of transmission power. Usually, a CR user chooses a channel with low path loss. If the distance between the sender and the receiver is short enough that the path loss is ineffective, then the CR user can ignore the path loss effects.

Wireless Link Error

Errors are more likely to happen in wireless than in wired connections. The error rate also depends on modulation techniques. These errors are handled by transport layer protocols. Therefore, CR users choose channels with low link error rates.

Transmission Delay

Different channels have different interference levels, packet loss rates, wireless link errors, and path loss effects. As a result, different types of link layer protocols are appropriate for different channels. For this heterogeneity, different transmission delays are observed in different channels. A CR user might choose a channel with few transmission delays.

PU Activity

If PU transmissions are very likely in the channel, then the CR cannot continue transmission for a long time in that channel. In this sense, the CR should choose the channel with the lowest user activity.

Contagious Frequency Channel

If a CR user can find some channels with contagious frequencies, it can extend the channel's bandwidth by combining channels. However, if PU activities are seen in any of the channels, it cannot yield one particular channel. As a result, the CR segregates the channels and takes different channels for transmission. Since the probability of PU activity increases by the number of combined channels, combining channels may not be a good spectrum decision in situations with high user activity channels. In addition, channel aggregation and segregation take time and can increase the latency of a transmission.

5 Spectrum Sharing

Usually, the number of available free channels is less than the number of CR users. Therefore, CR users must share channels. CR users can be competitive or cooperative with each other. A scenario where CR users are competitive can be modeled as a static game where each CR user tries to maximize their reward by transmitting in the shared channel. There are three paradigms (underlay, overlay, and interleave) that are used to facilitate the spectrum sharing.

Underlay

In this paradigm, secondary and primary transmissions are done simultaneously. CR users transmit in very low power that appears as noise to the primary receiver. Secondary transmission power can be determined by the interference-temperature limit. If the secondary transmission power does not exceed the interference-temperature limit, then it does not hamper the primary transmission. The biggest advantage is that CR users do not need to sense PU transmissions, so, secondary transmissions can be operated regardless of PUs' activities. SUs suffer from packet loss due to primary transmissions. The authors in [10] propose an energy-efficient algorithm to minimize the loss rate of SUs. The algorithm also maximizes energy efficiency in information bits per Joule.

Overlay

In this paradigm, CR users utilize the unused portion of the primary spectrum. Using a portion of the spectrum reduces interference with a PU who uses the whole spectrum. Unlike the underlay, there is no transmission power limit; an SU can transmit in its maximum power. SUs must have knowledge (codebook, message format, frequency, etc.) about the primary spectrum. CR users can get this knowledge from the broadcasting of the PU or from a uniform standard. Since the CR user knows the codebook, it can divide its power between its own message transmissions and relay the primary message [4].

Interweave

This is the original proposal for CRN. In this paradigm, the SU can only transmit if there is no PU activity. This requires that one sense the primary channel. SUs use their detection techniques to detect primary transmissions, and if a channel is not occupied by a PU, then the SU starts transmitting.

5.1 Spectrum Allocation in Centralized Interweave Cognitive Radio Network

In a centralized system, channel allocations to SUs are done by a base station, and secondary transmissions can occur in the absence of PUs in a interweave system. We assume a heterogeneous network with M number of PUs $m \in \{1, \ldots, M\}$. An N number of SUs $n \in \{1, \ldots, N\}$ compete with each other to get access to any $k \in \{1, \ldots, K\}$ channel among K free channels. $P_p(x)$ and $P_s(y)$ are the transmission powers of the PU x and the SU y, respectively. $g_p(x)$ denotes the gain of the signal of PU x in the channel and $g_s(y)$ denotes the gain of the signal of SU y. So, the total noise in any channel, k, at any SU, s, is:

$$Total\,Noise = \sum_{n=1}^{N} g_s(n) P_s(n) + \sum_{m=1}^{M} g_p(m) P_p(m) + N_k \qquad (13)$$

N_k denotes white Gaussian noise from external sources. The first part of the total noise equation is caused by the signals of other secondary transmissions and the second part is caused by the signals of all primary transmissions. Therefore, the signal-to-noise ratio at the SU y in channel k is:

$$SNR_k(y) = \frac{g_s(y) P_s(y)}{Total\,Noise} \qquad (14)$$

Figure 6 shows the bipartite graph made of the SUs and the free channels. An N number of SUs form a disjoint set, and a K number of channels form another disjoint set of the bipartite graph. Edges in this graph represent an allocation of the channel to an SU. The graph in Fig. 6 is a weighted graph whose edge weight represents the allocation cost of the channel in terms of the decrease in the total signal-to-noise ratio. Let us assume that after allocating channel k to SU n, the total signal-to-noise ratio decreases from SNR_p to SNR_n. Then, the cost of allocation is:

Fig. 6 Channel allocation algorithm

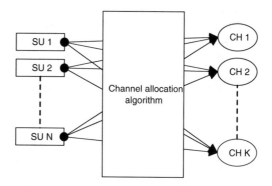

$$C(n, k) = \frac{SNR_p - SNR_n}{SNR_p} \tag{15}$$

We can also call $C(n, k)$ the weight of the edge (n, k) in the bipartite graph constructed by N SUs and K free channels. Now, the problem comes down to minimizing the matching cost in a bipartite graph. The Hungarian algorithm provides the solution to the maximum weight matching [11], which can be adapted to minimize the cost of the matching by inverting the cost.

6 Spectrum Mobility

After the spectrum choice, any CR user can access a secondary spectrum, but a PU is sensed to be present when a CR user is transmitting. The CR user must vacate the channel for the PU. The transmissions of the CR user cannot be stopped and may be continued in another channel. This process is referred to as spectrum mobility. The main function of spectrum mobility is to do a spectrum handoff. The spectrum handoff process consists of two phases: the evaluation phase and the link maintenance phase [12]. The evaluation phase deals with observing the environment to find handoff-triggering events, like primary transmission detection or channel condition degradation. After the handoff-triggering event, SUs decide to handoff and go to the link maintenance phase. In the link maintenance phase, SUs stop the ongoing transmission and resume transmissions in another free channel. After completing this phase, SUs return to the evaluation phase. In [13], authors present different handoff strategies:

- **Non-handoff Strategy**: In this strategy, the CR user remains idle while the primary transmissions continue. The CR user expects to transmit in the same channel. This strategy is inefficient if the primary transmissions continue for a long time. Long waiting times cause the QoS to degrade. This strategy is preferable when CR users know the channel statistics and short time primary transmissions are likely in the channel.
- **Pure Reactive Handoff Strategy**: In this strategy, the CR user hands off the channel after detection of a primary transmission in the current channel. The CR must choose another free channel to continue the transmission. Finding the next free channel can take time, which is not acceptable for the smooth data connection. Since the CR user finds the next channel after the handoff-triggering events, the majority of the time is spent finding the free channel.
- **Pure Proactive Handoff Strategy**: In the proactive handoff strategy [14], the CR user finds the next free channel before the detection of the primary channel; the free channel can work as a backup channel. The CR user can predict the time of the PU's presence and handoff channel before handoff-triggering events occur. This strategy needs hardware support to sense and transmit simultaneously. Still, there is the possibility of the presence of a PU in the backup channel that could lead to

transmission delays. Predicting the time of the presence of primary transmissions requires a lot of machine learning and can lead to a high power consumption by CR users due to computation complexities.

- **Hybrid Handoff Strategy**: This strategy is a combination of the pure reactive and pure proactive handoff strategies. In the hybrid handoff strategy, finding the free channel is done before the handoff-triggering event (like in the proactive handoff strategy) but the channel handoff is done after the triggering event (like reactive handoff strategy). This strategy can achieve a faster channel handoff, but the possibility that the backup channel will be obsolete is still a concern.

Multiple strategies for selecting the next channel exist. The hidden Markov model is used to predict channel behavior in [15–17]. However, prediction-based channel selection can be harmful when predictions are wrong. Delays in selecting the next channel can exacerbate the QoS. Therefore, we consider a search-based approach to select the next channel. Let us consider a 2-D search space of time and frequency. We consider all slots as nodes in a graph. An edge between one node to another represents the channel switching cost, which is either zero or one. Figure 7 shows the formation of the graph. Let us denote a slot by (T, CH), where T represents the time and CH represents the channel. For example, the switching cost from $(T1, CH3)$ to $(T2, CH3)$ is 0 because the CR user has actually continued transmission in the same channel. The switching cost from $(T1, CH1)$ to $(T2, CH1)$, $(T2, CH2)$, and $(T2, CH4)$ is 1. The weight of an edge can be found by adding the switching cost and the channel density. In the figure, darker slots have more channel density. Now, we get a directed weighted graph. Graph traversal algorithms like Dijkstra can be applied to this graph to find the best slot. However, spectrum mobility is challenging. When a CR user switches its channel, the routing breaks, and the routing table needs to be updated. Routing recalculation is a costly and time-consuming process. Therefore, routing calculation becomes a part of the channel handoff process. Instead of recalculating the routing before the handoff, a CR user can prepare a backup channel. The CR user needs to maintain the backup channel periodically so that it can transfer communication links immediately to the backup channel after a handoff-triggering event.

Fig. 7 Spectrum search space

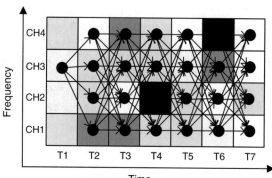

7 Security Issues in Cognitive Radio Networks

In the last few years, CRNs have become a promising technology in solving the spectrum scarcity. However, this network technology has a lot of security challenges. Authors in [18–20] describe a lot of security issues and solutions in CRNs. In a primary user emulation attack (PUEA), an attacker mimics a PU's signal to fool SUs so that they refrain from using the channel. As a result, network congestion and a denial of service happen. To solve the PUEA, classification methods are used to classify whether a signal is from a PU or an attacker. Location verification-based solutions are proposed in [21, 22]. Though a PUEA mimics PU signals, it is very hard to mimic the signal's energy distribution of a PU. Based on this principle, [23, 24] propose solutions which classify signals based on power mean, power variance, and the Wald's sequential probability ratio. Usually, all SUs know all PUs' locations. Based on the received signals' power, an SU can determine whether the power level is feasible if the signal comes from the PU's location [25].

The spectrum sensing data falsifying (SSDF) attack is applicable in a cooperative spectrum sensing system with a data aggregation center or fusion center. Malicious SUs (MSU) send false information to data aggregation centers so that other SUs get wrong decisions. In the worst case, MSUs and the benign SUs with the wrong sensing information can win the election. So, it is important to detect the MSUs and exclude them from the voting. Solutions for SSDF attacks are based on clustering the benign SUs to a group and reputation-trust based. Authors in [26, 27], propose two different clustering algorithms based on the hamming distance between the sensing results of different time slots of different SUs. Associative rule mining-based classification is proposed in [28]. Authors propose an apriori algorithm to get a frequent subset of the sensing results from all the SUs. They assume that the MSU will remain in the frequent subset of the sensing result. Based on the probability of PUs' presence, they classified the SUs into benign SUs and MSUs. A trust-based spectrum sensing scheme against SSDF attacks is proposed in [29]. In the proposed method, data aggregation center selects some of the SUs to take local decisions and combines the detection results based on their reliability. Authors in [30] propose a distributed spectrum sensing method. The reputation computed based on the deviation from the majority's decision.

A PU emulation-based testing scheme, FastProbe, is proposed in [31]. FastProbe creates PU signals to test whether the SUs are reporting honestly or not. This detection technique is now ineffective because there are a lot of mechanisms that detect PU emulation signals [21, 25, 32–34]. These mechanisms are based on distribution, mean, variance of energy, and transmitter localization. So, any MSU can detect the PU-emulated signal from the data aggregation center and report correct results in that time slot to get a high reputation and keep reporting false results in other time slots.

8 Applications of Cognitive Radio Networks in LTE and Wi-Fi

Wi-Fi and LTE are the most prominent wireless access technologies nowadays. The migration from PCs to mobile devices leads to an exponential increase of data usage in wireless technologies. The 84.5 MHz of unlicensed spectrum in the 2.4 GHz band which is allocated for Wi-Fi has been saturated and is unusable for new wireless applications [35]. Therefore, Wi-Fi stakeholders have been showing interest in using the 5 GHz bands. There is up to 750 MHz of unlicensed spectrums in the 5 GHz band that falls under the Unlicensed National Information Infrastructure (U-NII) rules of the FCC.

Some LTE stakeholders, including Qualcomm (an American multinational semiconductor and telecommunications equipment manufacturer), are also keenly interested in the 5 GHz bands. They proposed extending the deployment of LTE-Advanced (LTE-A) to the 5 GHz band using channel aggregation (CA) and supplemental downlink technologies (SDL). Carrier aggregation in LTE-A enables using multiple carriers to provide high data rates. A supplemental downlink is a multi-carrier scheme for enhancing the downlink capacity in Evolved High Speed Packet Access(HSPA+). Some Wi-Fi systems, such as 802.11a and 802.11n, are already operating in 5 GHz bands. However, Wi-Fi stakeholders have been lobbying the government for access to more spectrums within the 5 GHz bands. In response, the FCC issued a Notice of Proposed Rule Making (NPRM) 13–22 in 2013 [36] that recommends adding 195 MHz of additional spectrums for use by unlicensed devices. The Wi-Fi Innovation Act was introduced in the U.S. Senate and House [37] recently. This act directs the FCC to conduct tests to assess the feasibility of opening the upper 5 GHz band, including the Intelligent Transportation System (ITS) band, for unlicensed use. ITS stakeholders are very concerned about sharing spectrums with Wi-Fi. They fear that coexistence with Wi-Fi may severely degrade the performance of ITS applications, especially safety applications that are sensitive to communication latency. When the ITS band was first allocated in 1999, the FCC's original intention was for this band to support Dedicated Short Range Communications (DSRC) for ITS exclusively. As a result, ITS protocol stacks and the relevant applications are not designed to coexist with unlicensed devices. Access to the 5 GHz spectrum has become a cause of contention between the LTE, Wi-Fi, and DSRC stakeholders, but more importantly, the 5 GHz bands have become a proving ground for spectrum sharing between three heterogeneous wireless access technologies: LTE-U, Wi-Fi (802.11ac/802.11ax), and DSRC. Recognizing the importance of this problem, a research opportunity focusing on two coexistence scenarios has opened: the coexistence between LTE-U and Wi-Fi and the coexistence between DSRC and Wi-Fi (Fig. 8).

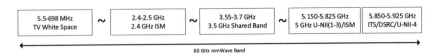

Fig. 8 Different bands for wireless applications

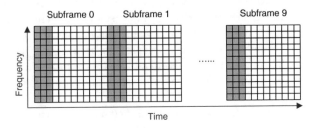

Fig. 9 LTE subframe and resource allocation to users

8.1 LTE and Wi-Fi Coexistence

Enabling harmonious coexistence between LTE and Wi-Fi in 5 GHz bands is particularly challenging for two main reasons. First, Wi-Fi networks are contention-based, whereas LTE communications are schedule-based. Figure 9 shows the basic resource block of LTE. Each user is assigned to a slot in the time and frequency domain of the spectrum. LTE HeNB (LTE base station) does not sense before transmission. On the other hand, Wi-Fi is a CSMA/CA-based protocol, which means a Wi-Fi device senses before transmission and if the channel is occupied, it does not transmit. As a result, LTE always shows eminent behavior while coexisting with Wi-Fi. In fact, experiments done by Nokia Research [38] show that in coexistence scenarios, the Wi-Fi network is heavily influenced by LTE-U interference. Specifically, the Wi-Fi APs stay on LISTEN mode more than 96% of the time, which causes severe degradation to their throughput. So, the challenge facing LTE and Wi-Fi coexistence is ensuring a fair share between them. There are several studies on ensuring a fair share and coexistence between LTE and Wi-Fi. We discuss some of the approaches next.

8.1.1 Self-interference Suppression Technology with LTE and Wi-Fi

Wi-Fi and LTE coexistence can be achieved using self-interference suppression (SIS) and Full Duplex (FD) capabilities. The SIS is a technique to remove interference induced by its own transmission. The self-interference cancellation circuits [39] can be used to achieve full duplex capabilities. We consider a coexistence scenario that consists of one or more Wi-Fi networks along with several LTE operators. Each Wi-Fi network is comprised of an 802.11 AP and several wireless users (WUs). Figure 10 shows the LTE and Wi-Fi coexistence scenario. At the beginning, LTE-U starts the transmission only in the licensed spectrum (without CA). After a while,

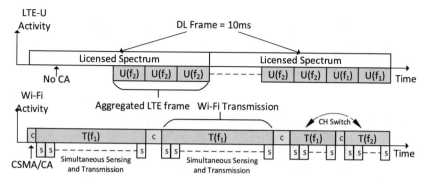

Fig. 10 LTE and Wi-Fi activity and channel switching

LTE aggregates an unlicensed channel, f_2, with the licensed spectrum and continues transmission for some time. After that, LTE releases f_2 because of channel quality degradation, and it aggregates another unlicensed channel, f_1. Wi-Fi starts with CSMA/CA activity (sensing the channel) and starts transmission in channel f_1. At the same time, it also starts sensing f_1. When LTE switches channels from f_2 to f_1, the Wi-Fi sensing detects the transmission and releases the channel f_1. Then, Wi-Fi again does CSMA/CA on another channel (f_2) and continues transmitting if the channel is free.

8.1.2 Backward Compatibility Approaches

The previous approach focuses on interactions between Wi-Fi and LTE-U and advocates the design of new FD/SIS-based interference mitigation techniques without consideration of the issue of fairness. In this section, we study fair spectrum sharing in heterogeneous systems.

Mechanism 1: Indirect Coordination Using Carrier Sensing

In this scenario, there will be no information flow between LTE and Wi-Fi. Wi-Fi and LTE-U may not even know what their fair portions of the spectrum should be. LTE senses the environment and detects the preambles of Wi-Fi and an LTE HeNB determines how many Wi-Fi APs are within its transmission range. Then, the LTE HeNB determines its fair portion of the spectrum based on the number of coexistent Wi-Fi networks. If the LTE-U's spectrum usage so far is larger than its fair portion, then the LTE network would slow down. That is, LTE turns off its transmission for a longer time (i.e., reducing its duty cycle) and lets Wi-Fi transmit more. On the other hand, if the LTE-U's spectrum usage so far is less than its fair portion, it may increase its transmission.

				3 reserved bits			2 reserved bits	
L-STF	L-LTF	L-SIG	VHT-SIGA	VHT-STF	VHT-LTF	VHT-SIGB	Data
2-symbols	2-symbols	1-symbols	2-symbols	1-symbols	≥ 1-symbols	1-symbols	N symbols

Fig. 11 IEEE 802.11ac packet format

Mechanism 2: Embedding Wi-Fi Information in Preamble

Fair coexistence between LTE-U and Wi-Fi can be achieved by modifying the Wi-Fi preamble. Only the reserved bits in the preamble can be used for this purpose. Useful information can be embedded in them, and LTE can adjust its operations according to the information; this enables fair usage. There are some options for embedding Wi-Fi usage information in the preamble. There are at least 5 reserved bits that may be used to embed Wi-Fi information. Figure 11 shows the Wi-Fi (IEEE 802.11ac) packet format. Based on the information embedded in the Wi-Fi preamble, LTE can adjust its operations accordingly to achieve fairness between the two systems, i.e., the LTE may reduce (or increase) its duty cycle if the Wi-Fi has been using less (or more) bandwidth than the fair portion.

Mechanism 3: Indirect Communication Between LTE and Wi-Fi

This mechanism can be applied to the scenario where there are some service providers that provide both LTE and Wi-Fi networks. Those service providers' HeNB supports both LTE and Wi-Fi. Suppose provider 1 has both LTE and Wi-Fi networks and provider 2 has only Wi-Fi networks. There can be two kinds of communication. Firstly, indirect communication between provider 1's LTE and provider 1's Wi-Fi. Secondly, communication between provider 1's Wi-Fi AP and provider 2's Wi-Fi AP. The indirect communication may be used for various purposes, such as time synchronization between LTE and Wi-Fi, exchanging spectrum usage information (e.g., aggregated bandwidth, aggregated throughput, or the total air time so far), or for other signaling information to achieve fair spectrum sharing between the two systems (Fig. 12).

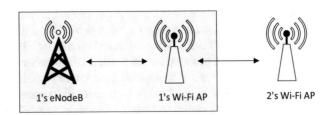

Fig. 12 Indirect communication

8.2 Wi-Fi and DSRC Coexistence

The FCC allocates 75 MHz spectrums in the 5.9 GHz band to DSRC which is used for vehicle-to-vehicle communications. The DSRC is based on the IEEE 802.11p standard. After the FCC declared the 5.9 GHz unlicensed band (U-NII-4), Wi-Fi could operate in that band. The DSRC remains as a PU and others act as SUs. DSRC's inter-frame space (IFS) parameters are two times longer than 802.11ac/802.11ax. Similarly, the slot time used in the MAC backoff is longer for DSRC (13 μs) than for 802.11ac/802.11ax (9 μs). These differences give Wi-Fi effective priority over DSRC when accessing the channel. One solution is changing the IFS values of 802.11ac or 802.11ax so that DSRC gets a higher access priority. For instance, we can increase the values of the Wi-Fi's IFS parameters by adding a DSRC priority time offset value, giving priority to DSRC. Only the IFS adjustment cannot guarantee DSRC's protection. Channelization of U-NII-4 by 802.11ac standards affects DSRC transmission significantly. Experiments by the authors of [40] show that if the 802.11ac primary channel remains in the same band as DSRC, then adjusting the DIFS ensures the DSRC's protection. On the other hand, if the primary channel remains in another band and some of the secondary channels remain in the same band as DSRC, adjusting DIFS does not protect the DSRC transmissions.

9 Conclusion

Spectrum is a valuable resource in wireless communication systems. The CRN is an excellent method of wireless communication in which underutilized channels can be fairly used. The implementation of a CRN includes PU detection, channel choice, channel sharing, channel handoff, and routing reestablishment. Though current mobile devices have hardware that support operation in 2.4, 5 GHz, GSM, WCDMA, and LTE bands, simultaneous sensing and transmitting are still lacking in them. The promising thing is that most of the physical layers of communication are software managed. Therefore, changing the software may adapt some functionality of CRN more easily than changing hardware. A lot of changes in the base stations are also required to implement a CRN. Commercial issues inducing usage policies and charges to SUs or cognitive radio operators are also not defined. Therefore, the implementation of a CRN is very complex and expensive. In this chapter, we present the full architecture of CRN at a high level. We discuss applications of a CRN in the 5 GHz band for the coexisting Wi-Fi, LTE, and DSRC. We present different mechanisms for ensuring the fair sharing of spectrums among different technologies in the 5 GHz band. Thereby, a CRN may become an excellent means of wireless communication in which underutilized channels are fairly used.

Acknowledgements This research was supported in part by NSF grants CNS 1629746, CNS 1564128, CNS 1449860, CNS 1461932, CNS 1460971, and CNS 1439672. We would also like to express our gratitude to the people who provided support, comments, information, proofreading, and formatting.

References

1. M. Yoshimachi and Y. Manabe, "A new AODV route discovery protocol to achieve fair routing for mobile ad hoc networks," in *2016 6th International Conference on Information Communication and Management (ICICM)*, Oct 2016, pp. 222–226.
2. M. Zareei, E. M. Mohamed, M. H. Anisi, C. V. Rosales, K. Tsukamoto, and M. K. Khan, "On-demand hybrid routing for cognitive radio ad-hoc network," *IEEE Access*, vol. 4, pp. 8294–8302, 2016.
3. R. Abdelrassoul, E. Fathy, and M. S. Zaghloul, "Comparative study of spectrum sensing for cognitive radio system using energy detection over different channels," in *2016 World Symposium on Computer Applications Research (WSCAR)*, Mar 2016.
4. I. F. Akyildiz, W. y. Lee, M. C. Vuran, and S. Mohanty, "A survey on spectrum management in cognitive radio networks," *IEEE Communications Magazine*, vol. 46, no. 4, pp. 40–48, Apr 2008.
5. A. Wisniewska, "Spectrum sharing in cognitive radio networks: A survey," 2014.
6. K. Kim, I. A. Akbar, K. K. Bae, J. S. Um, C. M. Spooner, and J. H. Reed, "Cyclostationary approaches to signal detection and classification in cognitive radio," in *2007 2nd IEEE International Symposium on New Frontiers in Dynamic Spectrum Access Networks*, Apr 2007.
7. N. Muchandi and R. Khanai, "Cognitive radio spectrum sensing: A survey," in *2016 International Conference on Electrical, Electronics, and Optimization Techniques (ICEEOT)*, Mar 2016.
8. S. Sinha, S. Mehfiiz, and S. Urooj, "Energy detection of unknown signals over rayleigh fading channel," in *2014 International Conference on Issues and Challenges in Intelligent Computing Techniques (ICICT)*, Feb 2014.
9. S. K. Rashed, R. Shahbazian, and S. A. Ghorashi, "Spectrum decision in cognitive radio networks using multi-armed bandit," in *2015 5th International Conference on Computer and Knowledge Engineering (ICCKE)*, Oct 2015.
10. J. Martyna, "Green power control in underlaying cognitive radio networks," in *2016 8th IFIP International Conference on New Technologies, Mobility and Security (NTMS)*, Nov 2016.
11. H. W. Kuhn, "The hungarian method for the assignment problem," *Naval Research Logistics Quarterly*, vol. 2, 1955.
12. Y. Zhang, "Spectrum handoff in cognitive radio networks: Opportunistic and negotiated situations," in *2009 IEEE International Conference on Communications*, Jun 2009.
13. D. M. Alias and R. G. K, "Cognitive radio networks: A survey," in *2016 International Conference on Wireless Communications, Signal Processing and Networking (WiSPNET)*, Mar 2016.
14. I. Christian, S. Moh, I. Chung, and J. Lee, "Spectrum mobility in cognitive radio networks," *IEEE Communications Magazine*, vol. 50, no. 6, Jun 2012.
15. I. A. Akbar and W. H. Tranter, "Dynamic spectrum allocation in cognitive radio using hidden Markov models: Poisson distributed case," in *Proceedings 2007 IEEE SoutheastCon*, March 2007.
16. C. H. Park, S. W. Kim, S. M. Lim, and M. S. Song, "Hmm based channel status predictor for cognitive radio," in *2007 Asia-Pacific Microwave Conference*, Dec 2007.
17. Z. Chen and R. C. Qiu, "Prediction of channel state for cognitive radio using higher-order hidden Markov model," in *Proceedings of the IEEE SoutheastCon 2010 (SoutheastCon)*, Mar 2010.

18. J. Li, Z. Feng, Z. Feng, and P. Zhang, "A survey of security issues in cognitive radio networks," *China Communications*, vol. 12, no. 3, Mar 2015.
19. G. Baldini, T. Sturman, A. R. Biswas, R. Leschhorn, G. Godor, and M. Street, "Security aspects in software defined radio and cognitive radio networks: A survey and a way ahead," *IEEE Communications Surveys Tutorials*, vol. 14, no. 2, Second 2012.
20. R. K. Sharma and D. B. Rawat, "Advances on security threats and countermeasures for cognitive radio networks: A survey," *IEEE Communications Surveys Tutorials*, vol. 17, no. 2, Secondquarter 2015.
21. D. Salam, A. Taggu, and N. Marchang, "An effective emitter-source localisation-based puea detection mechanism in cognitive radio networks," in *2016 International Conference on Advances in Computing, Communications and Informatics (ICACCI)*, Sept 2016.
22. Z. Ma, W. Chen, K. B. Letaief, and Z. Cao, "A semi range-based iterative localization algorithm for cognitive radio networks," *IEEE Transactions on Vehicular Technology*, vol. 59, no. 2, Feb 2010.
23. Z. Jin, S. Anand, and K. P. Subbalakshmi, "Detecting primary user emulation attacks in dynamic spectrum access networks," in *2009 IEEE International Conference on Communications*, June 2009.
24. Z. Chen, T. Cooklev, C. Chen, and C. Pomalaza-Rez, "Modeling primary user emulation attacks and defenses in cognitive radio networks," in *2009 IEEE 28th International Performance Computing and Communications Conference*, Dec 2009.
25. R. Chen, J. M. Park, and J. H. Reed, "Defense against primary user emulation attacks in cognitive radio networks," *IEEE Journal on Selected Areas in Communications*, vol. 26, no. 1, Jan 2008.
26. S. Nath, N. Marchang, and A. Taggu, "Mitigating SSDF attack using k-medoids clustering in cognitive radio networks," in *2015 IEEE 11th International Conference on Wireless and Mobile Computing, Networking and Communications (WiMob)*, Oct 2015.
27. K. Rina, S. Nath, N. Marchang, and A. Taggu, "Can clustering be used to detect intrusion during spectrum sensing in cognitive radio networks?" *IEEE Systems Journal*, vol. PP, no. 99, 2017.
28. S. Bhattacharjee, R. Keitangnao, and N. Marchang, "Association rule mining for detection of colluding SSDF attack in cognitive radio networks," in *2016 International Conference on Computer Communication and Informatics (ICCCI)*, Jan 2016.
29. F. Zeng, J. Li, J. Xu, and J. Zhong, "A trust-based cooperative spectrum sensing scheme against SSDF attack in CRNS," in *2016 IEEE Trustcom/BigDataSE/ISPA*, Aug 2016.
30. R. Chen, J. M. Park, and K. Bian, "Robust distributed spectrum sensing in cognitive radio networks," in *IEEE INFOCOM 2008 - The 27th Conference on Computer Communications*, Apr 2008.
31. T. Bansal, B. Chen, and P. Sinha, "Fastprobe: Malicious user detection in cognitive radio networks through active transmissions," in *IEEE INFOCOM 2014 - IEEE Conference on Computer Communications*, Apr 2014.
32. S. Anand, Z. Jin, and K. P. Subbalakshmi, "An analytical model for primary user emulation attacks in cognitive radio networks," in *2008 3rd IEEE Symposium on New Frontiers in Dynamic Spectrum Access Networks*, Oct 2008.
33. Y. Liu, P. Ning, and H. Dai, "Authenticating primary users' signals in cognitive radio networks via integrated cryptographic and wireless link signatures," in *2010 IEEE Symposium on Security and Privacy*, May 2010.
34. C. Chen, H. Cheng, and Y. D. Yao, "Cooperative spectrum sensing in cognitive radio networks in the presence of the primary user emulation attack," *IEEE Transactions on Wireless Communications*, vol. 10, no. 7, July 2011.
35. S. R. Group, "The prospect of LTE and Wi-Fi sharing unlicensed spectrum: Good fences make good neighbors," February 2015.
36. FCC, "Revision of part 15 of the commissions rules to permit unlicensed national information infrastructure (U-NII) devices in the 5 GHz band," February 2013.

37. H.R.821, "Wi-Fi innovation act," https://www.congress.gov/bill/114th-congress/house-bill/821.
38. A. M. Cavalcante, E. Almeida, R. D. Vieira, S. Choudhury, E. Tuomaala, K. Doppler, F. Chaves, R. C. D. Paiva, and F. Abinader, "Performance evaluation of LTE and Wi-Fi coexistence in unlicensed bands," in *2013 IEEE 77th Vehicular Technology Conference (VTC Spring)*, Jun 2013.
39. D. Bharadia, E. McMilin, and S. Katti, "Full duplex radios," in *Proceedings of the ACM SIGCOMM 2013 Conference on SIGCOMM*.New York, NY, USA: ACM, 2013.
40. J.-M. J. P. Gaurang Naik, Jinshan Liu, "Coexistence of dedicated short range communications (DSRC) and Wi-Fi: Implications to Wi-Fi performance," in *IEEE International Conference on Computer Communications*, June 2017.

Emerging Trends in Vehicular Communication Networks

Marco Giordani, Andrea Zanella, Takamasa Higuchi, Onur Altintas
and Michele Zorzi

Abstract The potential of connected and autonomous vehicles can be greatly magnified by the synergistic exploitation of a variety of upcoming communication technologies that may be embedded in next-generation vehicles, and by the adoption of context-aware approaches at both the communication and the application levels. In this chapter, we discuss the emerging trends, potential issues, and most promising research directions in the area of intelligent vehicular communication networks, with special attention to the use of different types of data for multi-objective optimizations, including extremely large capacity and reliable information dissemination among automotive nodes.

1 Introduction: Vehicular Communication Networks

In recent years, Vehicle-to-Everything (V2X) communications have been investigated as a means to support emerging automotive applications ranging from safety services to infotainment [14]. However, next-generation automotive systems, which will include advanced services based on sophisticated sensors to support enhanced automated driving applications [32], are expected to require very high data rates (in the order of terabytes per driving hour), that cannot be provided by current V2X

M. Giordani (✉) · A. Zanella · M. Zorzi
Department of Information Engineering, University of Padova, Padua, Italy
e-mail: giordani@dei.unipd.it

A. Zanella
e-mail: zanella@dei.unipd.it

M. Zorzi
e-mail: zorzi@dei.unipd.it

T. Higuchi · O. Altintas
Network Division, TOYOTA InfoTechnology Center, U.S.A., Inc., Mountain View, CA, USA
e-mail: ta-higuchi@us.toyota-itc.com

O. Altintas
e-mail: onur@us.toyota-itc.com

© Springer Nature Singapore Pte Ltd. 2018
K. V. Arya et al. (eds.), *Emerging Wireless Communication and Network Technologies*,
https://doi.org/10.1007/978-981-13-0396-8_3

technologies. A possible answer to this growing demand for ultrahigh transmission speeds can be found in next-generation Radio Technologies (RTs) and interfaces, such as the millimeter wave (mmWave) bands [27] between 10 and 300 GHz[1] or the Visible Light Communication (VLC) bands [3] from 400 to 790 THz. On the one hand, the extremely large bandwidths available at those frequencies can support very high data rates. On the other hand, the increased carrier frequency makes the propagation conditions more challenging, as blockage becomes an important issue since signals do not penetrate most solid materials and are subject to high signal attenuation [12]. The space and time variability of the channel quality and the need for beam alignment between moving nodes have an impact not only on the design of the physical (PHY) and Medium Access Control (MAC) layers but also on the upper-layer protocols, an aspect that has been mostly overlooked in the literature so far.

A possible way to improve the performance of vehicular communications is to make a clever use of the different type of data (e.g., road structure and positions of connected vehicles) to optimize network control. In addition, vehicles may exploit multiple RTs as a fallback in the case of short outages of the high-frequency links or limitations of existing interfaces. Furthermore, the network and transport protocols need to be adapted to fulfill the strict performance requirements of V2X communications.

In this chapter, we aim at identifying potential issues and research directions in the area of intelligent V2X networks that will make use of data for multi-objective optimizations, including extremely high-capacity and reliable information dissemination among the nodes. We start our journey into the next-generation automotive world from the description, in Sect. 2, of the automotive services and applications that require V2X communication. In Sect. 3, we focus on the enabling technologies that can support V2X data exchange and on their possible shortcomings in relation with the target application requirements. In addition to the transmission technologies, a key role in a vehicular communication system is also played by the networking protocols. Therefore, in Sect. 4, we provide a brief survey of existing MAC, network, and transport protocols for vehicular networks, discussing their possible evolutions to better support next-generation automotive scenarios. In Sect. 5, we discuss how different types of data can be obtained from the existing systems and exchanged among the vehicles and the infrastructure to better support the final applications. Finally, in Sect. 6, we address the key aspects related to the security/privacy issues in V2X communication systems and discuss the emerging protocols for privacy management and secure data dissemination. Section 7 concludes the chapter by summarizing the discussion and suggesting promising research directions in this context.

---

[1] Although strictly speaking mmWave frequencies are between 30 and 300 GHz, industry has loosely defined as mmWave bands frequencies above 10 GHz.

2 Requirements for Next-Generation Vehicular Communication Networks

Next-generation automotive systems are expected to provide multiple services with diverse goals and requirements. However, providing an exhaustive list of all applications that can possibly be offered through vehicular communication systems is rather difficult, considering their large number and wide variety. In the following, therefore, we focus on four "macro-applications" that, for their generality, complementarity and significance we believe are good representatives of the main types of next-generation automotive services. Although the requirements of such services are not yet fully specified, some qualifying characteristics can be outlined as follows.

- **Infotainment** generically refers to a set of services that deliver a combination of information and entertainment. Infotainment requires low latency and stable throughput (especially for streaming of high-quality video contents) and the dynamic maintenance of a multicast communication (e.g., for gaming), which can be an issue. Reliability requirements are typically loose for these services.
- **Basic Safety** services are typically characterized by very strict requirements. While the size of the exchanged safety messages is typically small (up to a few hundreds of bytes), latency must be very small to ensure prompt reactions to unpredictable events. V2X connections must also be very reliable and stable, due to the sensitive nature of the exchanged information and the potential consequences of a communication failure.
- **Cooperative Perception** services deal with the enhancement of the sensing capabilities of a vehicle by sharing information with neighboring vehicles and infrastructures, with the final goal of extending the perception range of the driver beyond the line-of-sight or field-of-view of one single vehicle. This operation usually requires stable, reliable, and high throughput connections, due to the detailed nature of the shared contents, while some latency could be tolerated depending on the type of data contents exchanged among vehicles.
- **Platooning** refers to the services that make it possible for a group of vehicles that follow the same trajectory to travel in close proximity to one another, nose-to-tail, at highway speeds. A significant amount of information needs to be shared by V2X communications. In addition to the strict latency requirement, the connection reliability and stability are also very critical.

Figure 1 provides a visual and qualitative comparison of the different requirements of the above-listed target applications. As we can see, strict reliability constraints and small latency values are common to all such applications, while stable communications should be guaranteed especially for safety-related services.

From the above discussion, it is apparent that advanced vehicular services are expected to challenge the capabilities of current communication technologies, calling for innovative solutions. In the following, in particular, we discuss some specific requirements for the V2X communication system, called Communication Key Performance Indices (CKPIs) that go beyond the classical Quality-of-Service (QoS)

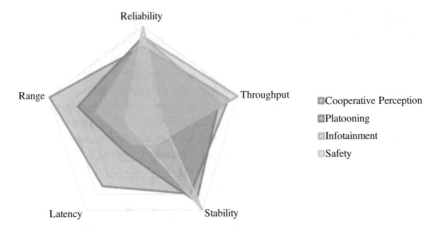

Fig. 1 Visual representation of the features and requirements of the vehicular applications and services presented in this section

metrics such as minimum required bitrate, maximum end-to-end latency and jitter, and maximum packet loss probability.

- **Range**. In a vehicular scenario, the classical QoS requirements have to be associated with a spatial range. In fact, most services will set tighter communication requirements toward nearby vehicles, so that the resulting aggregate CKPI requirements are likely to be stricter in close proximity of the transmitter, and progressively more relaxed with distance.
- **Speed**. The CKPIs should also account for the speed of the vehicles. In fact, stricter requirements (e.g., in terms of latency and connection stability) are usually associated to faster nodes, whose communication quality levels should therefore be monitored more closely than for slower vehicles.
- **Directionality**. CKPIs may also depend on the communication direction. For example, safety services are likely to require higher transmission capacities toward the vehicles located within the range of the transmitting beam, to avoid accidents. Therefore, the space-dependent characterization of the CKPIs described above can actually be anisotropic in space, with some directions that are more demanding than others in terms of communication requirements.
- **Nodes Density**. The CKPIs can also be affected by the density of nodes. On the one hand, a higher density may require a higher bitrate to maintain coordination among the cars. On the other hand, the bitrate and reliability requirements for broadcasting information may be relaxed in the presence of multiple vehicles that share the same data content.
- **Broadcast and Multicast**. A significant portion of vehicular applications requires broadcast-type V2V communications (typically enabled by omnidirectional radios). While the supplemental use of directional radios (e.g., as in mmWave) could help improve the communication performance, they typically do not provide native support for broadcast. Some other services, such as platooning, may

require multicast links instead. The support of both broadcast and multicast connectivity, hence, represents another CKPI for vehicular systems.

3 Enabling Radio Technologies for Next-Generation Vehicular Networks

In this section, we present features and limitations of target RTs that are expected to play a key role in next-generation automotive applications.

3.1 Dedicated Short-Range Communications (DSRC)

The IEEE 802.11p standard supports the PHY and MAC layers of the Dedicated Short-Range Communications (DSRC) transmission service. It can operate without a network infrastructure, removing the need for prior exchange of control information and thus bringing a significant advantage in terms of latency [1]. However, the throughput and delay performance can degrade as the network load increases (e.g., due to high user density), mainly because of the limited bandwidth and the "hidden node" problem. Furthermore, some V2X applications may require reliable transmissions beyond the communication range of IEEE 802.11p, which is typically limited to hundreds of meters. Moreover, the maximum data rate supported by DSRC, between 6 and 27 Mbps for each channel, may not be sufficient to sustain the transmission rates required by some next-generation automotive applications. For instance, high-resolution sensors may require more than 50 Mbps, while rates produced by cameras range from around 10 Mbps for low-resolution compressed images up to around 500 Mbps for high-resolution images [5].

3.2 Long-Term Evolution (LTE) Cellular

LTE offers ubiquitous coverage and collision-free packet transmission, but the support of vehicular communication services may still be limited. For example, access and transmission latency increase with the number of users in the cell, thus raising scalability issues. Despite the almost ubiquitous coverage of LTE, still the connection may not be always available, or good enough to satisfy the stringent reliability requirements under weak coverage (e.g., in tunnels, underground parking lots, rural areas, mountains). Finally, the maximum data rate of 4G-LTE systems is limited to around 100 Mbps for high mobility (though much lower rates are typical), which may not be sufficient to handle the potential gigabit rates that can be generated by next-generation vehicles [1].

3.3 Wi-Fi

Wireless networking based on the IEEE 802.11 standard, i.e., Wi-Fi technology, is popular and broadly available at low cost for home networks [19]. Raw data rates from 10 to 300 Mbps have proven to scale to several hundreds of concurrently active users when properly designed. However, Wi-Fi is mainly used by stationary or slowly moving indoor and outdoor users while, in a vehicular context, high mobility and link instability must be considered. Moreover, despite the high data rate, still the transmit speed may be insufficient to fully satisfy the requirements of some next-generation automotive applications. Finally, despite the huge popularity of Wi-Fi, the availability of access points that can be potentially used for V2X communications (e.g., at street corners, co-located with traffic lights, in parking lots, gas stations, cars, and so on) is still scarce, and the resulting intermittent connectivity will affect both the data rate and the latency of many vehicular services.

3.4 Millimeter Wave Bands

Communication in the millimeter wave (mmWave) bands [27] between 10 and 300 GHz is a promising candidate to support high data rates, in the order of Gbps, in line with the requirements of the next-generation cellular communication standard (5G). Moreover, the small wavelengths at mmWave frequencies make it practical to build very large antenna arrays (e.g., with 32 or more elements) to provide spatial isolation, reduce interference, increase security/privacy, and support multiplexing [12]. However, there are many concerns about the transmission characteristics at these frequencies [13]. The path loss is indeed very large and the communication range is quite limited. Moreover, mmWave signals do not pass through most solid materials, and movements of obstacles and reflectors, or even changes in the orientation of a handset, cause the channel to rapidly appear and disappear [35]. Additionally, mmWave links are typically directional to benefit from the resulting beamforming gain, requiring the fine alignment of transmitter and receiver beams and, consequently, a large overhead. Finally, dense deployments of short-range cells, as foreseen in future cellular networks operating at mmWaves, may increase the rate of handovers and reassociation events between adjacent cells, with consequent throughput degradation [21][2].

---

[2] A preliminary performance comparison between the mm Wave technology and the LTE and the DSRC standards (currently employed for V2I and V2V communications, respectively) has been recently provided in [10] and [11], in relation with future automotive applications' requirements.

3.5 Visible Light Bands

Visible Light Communication (VLC), whose bandwidth extends from 400 THz up to 790 THz, is an optical wireless communication technology that uses low-power Light Emitting Diodes (LEDs) not only to provide light but also to transmit data (e.g., brake signaling from car's taillights). The large and unregulated available bandwidth (390 THz) provides attractive opportunities for many automotive applications, due to the huge achievable data rates. Furthermore, other radios can be used simultaneously with VLC, without interference. However, VLC coverage is restricted to small areas and to Line-of-Sight (LoS) links. Moreover, the limited modulation bandwidth of today's inexpensive LEDs and the inter-symbol interference due to multipath propagation represent data transmission bottlenecks [3].

3.6 Satellite Communication

Satellite communication guarantees huge coverage areas, reaching zones that are not serviced by either landline or cellular networks. Moreover, since the cost of satellite broadcasting is basically independent of the number of receivers, the system scales very well with the number of served vehicles. Nevertheless, from a network perspective, satellite communication suffers from long delays, packet losses, intermittent connectivity, and link disruptions. Transmission also requires LoS conditions with the vehicle, limiting the accessibility to the service, in particular in dense urban areas. Finally, satellite channels are mostly broadcast and downlink, making this technology unsuitable for services that require unicast and uplink communications.

3.7 Low-Power Wide Area Networks (LPWANs)

Low-Power Wide Area Network (LPWAN) technologies may provide low power and low data rate connectivity over tens of kilometers. Furthermore, LPWAN base stations can connect a large number of devices, thus making it possible to cover wide geographical areas with a small number of base stations, significantly reducing the costs for infrastructure deployment. However, this technology offers very low data rates (in the order of tens of kilobits per second) with high latency (in the order of seconds or even minutes), thus restricting the possible employment of these technologies to noncritical vehicular services [26].

Table 1 provides a schematic and qualitative comparison of the requirements of the above-listed target RTs. As we can see, most radio interfaces ensure wide range, but with relatively high latency and small/medium throughput, while only a few technologies (i.e., mmWave and VLC systems) can provide high throughput.

Table 1 Features and requirements of the radio technologies presented in Sect. 3

RT	Throughput	Latency	Stability	Reliability	Range
DSRC	6 and 27 Mbps	Low	High	High	Medium ~1 km
4G/LTE	Medium < 100 Mbps	Medium	High	High	Medium ~1 km
Wi-Fi	Medium < 100 Mbps	Small	Low	Medium	Short < 100 m
mmWaves	High > 10 Gbps	Small	Low	Low	Short < 100 m
VLC	High > 10 Gbps	Small	Very low	Very low	Very short < 50 m
Satellite	Low ~Mbps	High > 250 ms	High	High	Very large ~ km
LPWAN	Very low ~Kbps	Small	High	High	Very large ~40 km

4   Next-Generation Vehicular Architecture

The demanding features of future vehicular networks, together with the limits of current and future radio access technologies and the peculiarities of upcoming wireless systems, have driven the redesign of the communication stack. In this section, we propose some guidelines for the design of next-generation MAC, network and transport layers specifically tailored to high-frequency vehicular communication systems.

4.1   Medium Access Control (MAC) Protocol Design

Medium Access Control (MAC) layer design has been extensively studied in the context of DSRC and 4G-LTE, while only a limited amount of literature has investigated solutions for other types of radios that are expected to be available in next-generation automotive systems. Conventional MAC solutions are suitable for situations in which the velocity/position of the vehicles can be accurately predicted. However, this may not be the case for V2X communication systems operating at high frequencies, mainly due to the intrinsic variability of the channel. Moreover, most recent solutions lack consideration of some important KPIs like reliability and delay. In particular, mmWave radio links require new schemes to enable vehicles and infrastructures to quickly determine the best directions to establish directional links. This functionality can be hardly supported by traditional communication protocols, which are often significantly affected by the high speed of the nodes and by the presence of frequent blockages on the propagation path.

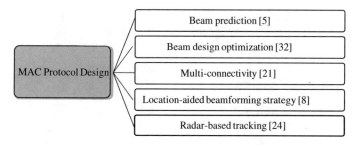

Fig. 2 Proposed solutions for MAC protocol design for vehicular networks

The above discussion makes apparent the need for innovative MAC protocol design, specifically tailored to future vehicular networks, as represented in Fig. 2. This objective can be achieved by enabling *multi-connectivity,* thus coupling a high-frequency data plane with a lower frequency control plane, to support the required rates, while increasing the robustness of the communication [9].

The authors in [5] present a *beam prediction* technique based on periodical speed and position information exchanged among network nodes through DSRC messages. Using the acquired information, the system is then able to estimate the vehicle's trajectory and derive the optimal beam orientation accordingly.

Beam design optimization is also being considered as a solution to maximize the data rate [32]. Results are consistent with the intuition that narrower beams should be used for users near the cell edge, where coverage is weaker.

In [8], a *location-aided beamforming strategy* is proposed to achieve ultrafast connectivity between nodes. In particular, adaptive channel estimation based on location information allows the estimation time to be substantially reduced.

Efficient beam alignment schemes can also be designed by extracting information from radar signals [24]. Simulations confirm that radars can be a useful source of side information and can help configure the mmWave V2I links.

In conclusion, although mmWave communication is a viable approach to provide high-bandwidth connectivity to future intelligent vehicles, innovative MAC-layer solutions should be engineered to overcome the limitations that prevent the direct employment of traditional communication protocols on high-frequency links.

4.2 Network and Routing Protocol Design

While the literature on network protocols[3] for legacy vehicular scenarios is quite rich, little work exists regarding the communication performance of the network layer (especially routing) in a next-generation V2X context. More specifically, traditional

---

[3]The network layer primarily aims at maximizing the throughput while minimizing the packet loss and limiting the overhead. A comprehensive taxonomy of the current routing protocols for vehicular communication systems can be found in [28].

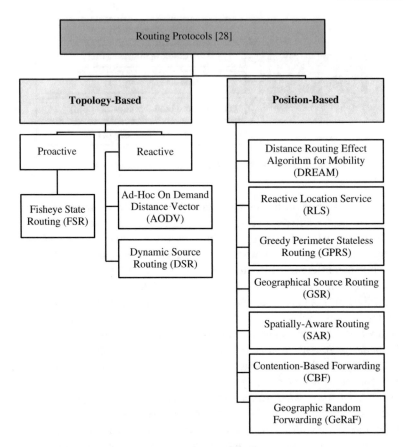

Fig. 3 Review of network and routing solutions for vehicular networks

routing solutions can be classified into two categories, as reported in Fig. 3: (i) topology-based routing protocols, and (ii) position-based routing protocols.

Topology-Based Routing Protocols. These schemes use link information within the network to send the data packets from the source to the destination. In particular, *proactive routing protocols* continuously maintain up-to-date routes for all the valid destinations, thus guaranteeing low-latency packet forwarding but suffering from scalability issues. *Reactive routing protocols*, instead, establish the path to follow for packet delivery only when a message needs to be actually exchanged, thus saving precious bandwidth resources but increasing the latency to find a reliable route.

Position-Based Routing Protocols. These schemes do not require routing tables, but only use the position information of neighboring nodes to determine the next forwarding hop to the destination. Since those protocols are based only on local knowledge, they are considered more scalable and robust against topological changes. However, they exclusively rely on position information that may be inaccurate or unavailable (e.g., in tunnels or where the satellite signal is absent) [23], and may suffer large overheads or additional delays caused by collision and contention of the underlying MAC protocols.

In this context, the propagation characteristics and the directional nature of mmWave links bring both challenges and opportunities for routing protocol design. For instance, due to the presence of communication blockages, the shortest path connecting two network nodes (in terms of geographical or topological distance) is not necessarily the best, and may actually yield lower throughput and higher packet loss than a longer path. It is thus important to make a judicious selection of relaying nodes, for example trying to keep the number of hops to a minimum when using multi-hop communications to overcome an impaired direct path.

Recently, some works tried to design network layer protocols specifically tailored to multi-hop systems with directional antennas. In [4], the authors proposed an *Optimal Geographic Routing Protocol (OGRP)* that selects the appropriate multi-hop relays considering the specific features of mmWave propagation. Other solutions implement some sort of *multipath routing* that allows a vehicular node to establish multiple connections through different access technologies, besides using device-to-device (D2D) transmissions.

In [25], a multi-hop concurrent transmission scheme is proposed and, by properly breaking one single-hop low-rate link into multiple shorter high-rate links and allowing non-interfering nodes to transmit concurrently, the network resources can be efficiently used to improve the network throughput.

4.3   Transport Protocol Design

A relevant issue in vehicular networks is the performance analysis of transport-layer protocols, especially congestion control using the *Transmission Control Protocol* (TCP). In fact, traditional implementations are not suitable for high-frequency vehicular systems. First, standard slow start mechanisms can take several RTTs to achieve the full throughput offered by the mmWave physical layer, increasing the latency of the communication. Second, sudden drops in the data rate, which are likely to occur in LoS-NLoS transitions, can result in very large queuing at the nodes, dramatically increasing the packet drop probability. Third, after a retransmission timeout, even aggressive TCP protocols (e.g., Cubic) can take inordinately long to recover the full data rate [36].

As summarized in Fig. 4, one possible way to design mmWave-aware transport layer protocols is to dynamically adapt the TCP flow according to the instantaneous channel propagation conditions of the surrounding nodes, thereby reducing the congestion window size in case the path between the endpoints is obstructed [37].

The hybrid and joint use of V2V and V2I communications can also ensure better QoS and transmission efficiency, especially when delivering large data contents [18]. The message may indeed be divided into several segments and delivered to multiple vehicles (i.e., to remove possible points of failure on the propagation paths), or

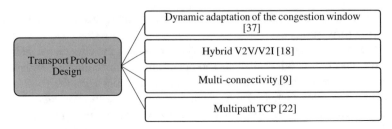

Fig. 4 Proposed solutions for transport protocol design for vehicular networks

shared among multiple infrastructure nodes, which are less affected by forwarding constraints.

Multi-connectivity can also be used to increase the overall throughput performance. In fact, while mmWaves can be exploited in the data plane to achieve the multi-Gbps rates required by next-generation automotive systems, legacy frequencies add robustness to the network thanks to the higher obstacles penetration capabilities and their inherent stability [9].

Finally, multipath-TCP, a standard that makes it possible to multiplex a TCP connection over multiple end-to-end paths, is another promising approach to improve the reliability of high-capacity networks. However, there are several issues with the traditional congestion control algorithms, in particular when coupling mmWave and LTE links [22].

To sum up, the first step toward the design of efficient transport protocols specifically tailored to next-generation vehicular systems operating at high frequencies involves the identification of the challenges that are specific to a high-mobility highly dynamic automotive environment and the potential performance pitfalls of existing V2X strategies. However, the definition of innovative transport-layer schemes is just in its infancy and therefore represents a wide-open research area.

5 Knowledge Acquisition and Distribution in Vehicular Communication Networks

New sensor technologies and advances in automotive electronics enable enhanced control systems and vehicle safety, and ease the driver's workload [6]. Besides some "endogenous" sources of information, the driver's experience can be further enhanced by exploiting inter-vehicle communication, which has the potential to dramatically expand the driver's perception of the surrounding environment (e.g., eliminating blind spots and enlarging the field of vision). How to best utilize the acquired information to improve the communication capabilities of the system and better support the automotive services is still an open challenge. In this section, we discuss different aspects of this problem. First, we consider the most promising techniques for knowledge acquisition in vehicular communication networks. More specifically, we list the main type of information that can be exchanged among the network nodes. Second, we measure the usefulness of such acquired information for automotive-related applications. The results of such analysis can be used to drive the design of vehicular networking protocols. More specifically, identifying the correlations among different data, and measure the importance of each type of signal, can be of help toward a preliminary understanding of which piece of data can be more critical in providing actionable information toward network and application optimization. Third, we discuss the most promising techniques for knowledge distribution in V2X networks, to enable multi-objective optimization. In particular, we consider the synergistic exploitation of multiple RTs (either selectively, in parallel, or hierarchically) guarantees reliable, efficient, and stable exchange of data among nodes.

5.1 Knowledge Acquisition

In this section, we list some typical examples of information that can be collected by connected cars and utilized to optimize the vehicular communication networks and the supported services.

Global Positioning System (GPS) Data. Vehicles equipped with GPS[4] can collect a variety of information, including position, velocity, and acceleration, that can be exploited to improve V2X communications, e.g., by differentiating the data to be exchanged based on the vehicles position, or by supporting beam tracking for directional communications. However, GPS accuracy has a great impact on the overall communication performance, especially when directional communication is used (i.e., narrow beams are much more sensitive to position inaccuracy).

Cameras, LIDARs, and Radars. Radars currently operate in the mmWave spectrum between 76 and 81 GHz and are used for applications like adaptive cruise control, cross traffic alerts, and lane change. Although they enable accurate detection and localization of the surrounding objects, they are relatively less suitable for object recognition and classification purposes. Cameras use visible light or infrared and have been used as an enabler of road signs recognition, enhanced blind spot detection and lane departure alert, but require large amounts of data to be processed (i.e., from 100 to 700 Mbps, according to the target precision of images). Light Detection and Ranging (LIDAR) sensors make use of laser beams to generate high-resolution 3D depth images for accurate detection, localization, and recognition of the surrounding objects. However, off-the-shelf LIDARs are quite costly, and their required data rate is comparable to that of automotive cameras [5].

---

[4]The Global Positioning System (GPS) provides an estimate of the current location of a vehicle in an Earth-coordinate frame. Standard GPS is accurate within 10 m, while differential GPS has improved accuracy, with errors limited to about 1 m. Besides spatial information, GPS also offers global time synchronization, with an accuracy of around ± 10 ns.

Traffic Conditions. Traffic conditions (and historical road traffic information) can be determined from GPS measurements and can be used to further improve the accuracy of the vehicle's path selection, while guaranteeing more efficient route planning.

Driving Commands. Driving signals (e.g., braking, accelerating, steering, turning signals) can help the neighboring cars to adjust their paths and speed (i.e., according to the current driving conditions) to reduce the probability of car accidents and traffic congestion. The sharing of such signals among neighboring vehicles can then help improve safety services.

Environmental Conditions. Safety and efficiency of the driving experience can be enhanced by alerting drivers about weather conditions, including heavy rain, snow, sleet, fog, smoke, dust, ice, and black ice [31]. Some recent cars are able to download weather information provided by national weather-alerting systems from surrounding infrastructures and adapt their driving parameters accordingly.

Location Attributes and Regional Information. Efficient traffic regulation in proximity of certain locations, such as schools or hospitals, is one of the promising use cases of V2X communications. Reducing the vehicle speed or even diverting traffic to alternative roads, possibly limited to the most critical hours, may significantly reduce the probability of accidents and the traffic congestion close to these sensitive areas. Similar types of signaling can also be used to indicate temporary events (e.g., concerts, city marathons, political demonstrations), which are usually crowded and therefore can create complex and dense traffic situations to be handled.

Vehicle Types. Trucks may contribute to congestion more than cars, as they occupy more road space, take more time to accelerate, decelerate and negotiate turns, and obscure vision [30].[5] Information on the type of vehicles can be exchanged among neighboring cars over V2X communications to (i) alert the surrounding cars about the approaching of special categories of vehickles like large trucks, tractors, or buses, and (ii) differentiate the exchanged data based on the type of the destination vehicles (i.e., certain prohibitions apply to only one category of vehicles).

Historical Data. Any available data referred to previously obtained knowledge can be turned into experience. As an automotive node is able to recognize a specific profile (e.g., a driver, a place, a road), it can access its saved historical data and exploit these statistics, e.g., to adapt its driving decisions accordingly. Such historical data may not be available from the vehicles currently on the road, but can be stored in infrastructure servers and downloaded when required.

In Table 2, we schematically provide a list of the above-described information sources that can be utilized to improve specific automotive services and/or communication protocols, with a focus on the limitations and issues of such collected data.

---

[5]The National Highway Traffic Safety Administration reported that, in 2015 in the US, motor vehicle deaths related to large truck crashes are 11% of the total, which is much higher that the percentage of large trucks among vehicles.

Table 2 Types of data for knowledge acquisition

Type of data	Utilization	Issues
GPS	• Geographic routing • Gas consumption rate • Speed prediction	• Limited accuracy
Radar	• Lane change assistance • Blind spot detection • Parking assistance	• Not suitable for object classification
Camera	• Road sign recognition • Congestion avoidance	• Huge data volume (100 ÷ 700 Mbps)
LIDAR	• Objects' depth map	• Data processing for high-resolution 3D images • Very costly
Traffic conditions	• Traffic estimation • Route planning • Gas consumption rate	• Privacy/security • Inaccuracy
Driving commands	• Impairment detection	• Data processing • Noisy/erratic signals
Environmental conditions	• Safety services	• Data obsolescence • Inaccuracy
Location attributes—regional information	• Efficient route planning	• Risk of stale information
Vehicle types	• Selective data broadcasting • Congestion avoidance	• Loose classification
Historical data	• Behavior profiling • Efficient route planning	• Huge database size • Limited accessibility

5.2 How to Measure the Utility of Data

Next-generation intelligent vehicles are required to intensively download/upload/exchange/distribute information to enable fundamental automotive applications and services. Therefore, investigating the actual "importance" of shared data to assess whether and which specific sensor information is worth transmitting (i.e., with the final goal of minimizing the network utilization and still deliver valuable information to the receivers) is an open research challenge. A fundamental role in this regard can be played by *machine learning*,[6] which offers tools to perform a variety of operations, including the following:

- **Learn Which Features Have Major Impact on Target Applications**. *Artificial Neural Networks* (ANNs) can be trained in an unsupervised manner to extract features from input vectors of different types of signals and provide a more compact

---

[6]The term Machine Learning (ML) generally refers to a wide set of data-driven algorithms that are generic in their definition, but can *learn* to perform specific tasks after proper *training*. If the training set is properly chosen, the ML algorithm should be able to generalize its behavior to previously unseen input data sequences, still providing a good estimate of the utility function [V8].

representation of the input data, which makes it possible to reduce the amount of data to be exchanged, thus saving transmission capacity and reducing the load. Generally, the reliability of the learning process increases with the number of relevant ANN entries in the input set [29].

- **Detect Correlation Among Signals**. By considering an input including several sources, a *Generative Deep Neural Network* (GDNN) [2] may reveal the presence of interdependencies among the readings of multiple sensors generated by vehicles in the same geographical area. The generative model can then be used to estimate the output samples from the input set. The accuracy of such predictions provides a way to measure the mutual information contained in different combinations of data.

- **Extract Information Features from General and Heterogeneous Signals**. Once a GDNN has been trained with measurements related to the quality of the radio link (e.g., the strength of the received signal power, the bitrate, the error probability, the outage probability) and the model of the input data has been learned, the generative property of the GDNN can be exploited to predict the evolution of the input vector, or part of it, in future time instants (e.g., to predict the channel quality in the next slot and proactively adapt all protocol layers accordingly). Endowing GDNNs with *reinforcement learning* features [33] can also develop generative models that link actions (e.g., settings of link parameters) and effects (e.g., the corresponding performance metrics), thus making it possible to automatically find optimization actions tailored to the specific operational scenario, according to a self-configuration-self-optimization paradigm.

The importance of data content can also be assessed based on the *cost* (in terms of network resource consumption) to collect the data and to exchange it among the nodes. Which measurements and data are easier to predict and/or more useful to combine or share is, however, an open and challenging question.

5.3 Knowledge Distribution

While assessing the importance of different types of data plays a significant role in the efficient minimization of the network resource consumption, network utilization can be further optimized by a synergistic exploitation of multiple radio interfaces (with totally different propagation characteristics and features). More specifically, *multi-connectivity* (MC) [9] enables each vehicular and/or infrastructure node to integrate wireless technologies, including 3G, 4G-LTE, Wi-Fi, DSRC, mmWave, VLC, to support a variety of V2X services and benefit from the strengths of each radio technology, with the final goal of efficiently and reliably exchanging different types of data contents. Some relevant hybrid networking solutions include the following:

- **Selective Transmissions**, in which data contents are transmitted through a single, dynamically selected radio interface. For instance, connected cars can maintain several signal paths to different infrastructures, operating at different frequencies, so that drops in one link can be overcome by switching data paths.

- **Parallel Transmissions**, in which data contents are duplicated and sent over different types of radios to add redundancy, making the message delivery more robust, but using more communication resources.
- **Hierarchical Transmissions**, in which a specific technology is used to provide a basic level of service, while different types of radios/paths are exploited to deliver supplemental information to improve the QoS of designated applications.

However, how to implement efficient multi-connectivity systems on next-generation connected cars is still an open issue. In particular, among the challenges that need to be addressed, the definition of an intelligent network selection mechanism, driven by a distributed or centralized/cloud-assisted decision process, must be engineered, to allow high-quality V2X applications to meet their requirements.

6 Emerging Protocols for Privacy Management and Secure Data Access Control in Vehicular Networks

Like any other computing system, vehicular communication networks can be plagued by vulnerabilities: connected nodes must thus be designed with security in mind, in order to limit the adversaries' ability to endanger vehicle operations, as well as driver and passenger safety. Investigating the main security/privacy issues related to vehicular communication systems and designing protocols and techniques for privacy management and secure data dissemination are therefore important research topics.

6.1 Security Concerns in Vehicular Networks

One of the most serious threats for security in next-generation vehicular networks originates from the tens of electronic control units (ECUs) that cars will incorporate. A solution may come from consolidation, integration, and virtualization of ECUs, with the final goal of reducing the total number of electronic components and increasing the number of functions and the complexity of the software.

However, the attack surface of future automotive systems extends beyond the car itself, touching most in-vehicle systems and an increasingly wide range of external networks. The authenticity and integrity of data transmitted across networks can be improved by providing secure storage systems for key exchange and encryption, to protect against unauthorized software or firmware updates [16]. Moreover, enhanced security mechanisms in which cars will connect to smart infrastructures (e.g., toll roads, gas stations) without disclosing personally identifiable information should be developed.

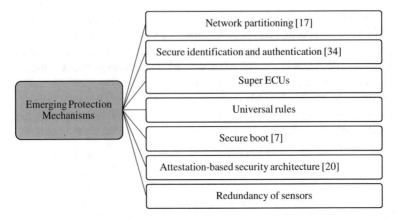

Fig. 5 Proposed protection mechanisms for vehicular networks

Lack of sufficient bus protection is another relevant security-related concern. In fact, the Controller Area Network (CAN) bus lacks the necessary protection to ensure robust data integrity [15]. Messages on the CAN-bus are not protected by any Message Authentication Code or digital signature and can be read by other nodes that can physically access the bus.

Finally, protection of data as it moves through the cloud and to data centers is another fundamental security feature that must be provided by transportation suppliers. Reliable automotive driving experience and connected communication capabilities can be supported by optimized data encryption and by guaranteeing embedded security features in the hardware of cars [16].

6.2   Emerging Protection Mechanisms for Vehicular Networks

Recently, countermeasures have been developed to face the increasingly threatened security in next-generation connected cars. As summarized in Fig. 5, a list of emerging protection mechanisms for vehicular networks includes the following.

1. *Network partitioning*: Security can be achieved by slicing the network, one partition being responsible for the safety-critical ECUs, while the other providing "comfort" functions [17].
2. *Secure identification and authentication*: Effective software protection can be guaranteed by securely implementing authorization functions in a trusted environment [34]. Cryptography solutions can also be implemented to allow each counterpart to verify the claimed credentials.

3. *Super ECUs*: With the increasing complexity of vehicular networks, one trend is to integrate several different applications on one (more powerful) ECU. However, low-cost devices are not able to run most of these complex operating systems.

4. *Universal rules*: Malicious attacks can also be theoretically prevented with good programming practices and by following the existing security recommendations protocols. However, ECUs usually come from different manufacturers, having potentially different protection specifications. The definition of universal defense mechanisms is therefore essential to enable secure transmissions.

5. *Secure boot*: This mechanism checks the digital signature of the software, prior to execution [7]. If an asymmetric algorithm is used, the public key has to be secured only against manipulation, but not against extraction. For both types, hardware support for the key storage is necessary.

6. *Attestation-based security architecture*: By comparing the result of specific hash functions with a list of authorized hashes, only successfully validated ECUs will be able to exchange symmetric keys for further encrypted communication [20].

7. *Redundancy of sensors*: The source of the sensor data is often not properly protected, and hence the signals might still be forged. One standard approach is to use redundant sensors and authentication checks in the ECUs. In the ideal case, there are two or more sensors measuring the same physical quantity (e.g., speed) in different ways, and a cross-check ensures the plausibility of the data.

Despite the increasing efforts of the automotive industry, there are still many security-related challenges to be considered in the near future. In general, an overall standardized approach to security, accepted by industry and legislation, is still missing and is therefore a challenging research topic.

7 Conclusions

In this chapter, we highlighted the challenges raised by next-generation automotive services, with reference to the design of the communication protocol stack.

In general, in order to compensate for the increased isotropic path loss experienced at higher frequencies (i.e., at mmWaves), next-generation automotive communication systems must provide mechanisms by which the vehicles and the infrastructure determine suitable directions of transmission to exchange sensory information. In this context, the design of enhanced communication protocols (i.e., at the MAC, network, and transport layers) is fundamental to meet the requirements of next-generation V2X services. In particular, the performance of intelligent vehicles in highly mobile mmWave scenarios strictly depends on the specific environment in which the vehicles are deployed, and must account for several automotive-specific features such as the vehicle's speed, the beam tracking periodicity, the node density, and the embedded antenna configuration.

Moreover, network resource minimization is another important issue for future intelligent vehicular systems. One possible way to achieve efficient communication is through synergistic orchestration among the multiple interfaces that are expected to be integrated in future intelligent vehicles, and by measuring the importance of data contents.

Security is another key concern for automotive networks. In this chapter, we analyzed the most serious security concerns and threats in next-generation connected cars and surveyed the most recent emerging protection mechanisms for secure data access control in vehicular networks.

Most of these research challenges, as well as many others, are still largely unexplored, so that additional investigation is needed toward the design of fully autonomous driving cars.

References

1. 5G-PPP. (2007). 5G Automotive Vision [White paper]. Retrieved October 4, 2017: https://5g-ppp.eu/white-papers/.
2. Bengio, Y., *et al.* (2013). Representation Learning: A Review and New Perspectives. *IEEE Transactions on Pattern Analysis and Machine Intelligence*, 35(8), 1798–1828.
3. Brandt-Pearce, M. (2014). The future of VLC. *Proceedings of the 1st ACM MobiCom workshop on Visible light communication systems (VLCS)*.
4. Cai, L., *et al.* (2009). Optimizing Geographic Routing for Millimeter-wave Wireless Networks with Directional Antenna. *Proceedings of the 6th International ICST Conference on Broadband Communications, Networks, and Systems*.
5. Choi, J., *et al.* (2016). Millimeter-Wave Vehicular Communication to Support Massive Automotive Sensing. *IEEE Communications Magazine, 54*(12), 160–167.
6. Fleming, B. (2008). New Technologies for Automotive Electronics [Automotive Electronics]. *IEEE Vehicular Technology Magazine, 3*(2), 10–12.
7. Freisleben, B., *et al.* (1990). Capabilities and Encryption: The Ultimate Defense Against Security Attacks? *Workshops in Computing Security and Persistence*.
8. Garcia, N., *et al.* (2016). Location-aided mm-wave channel estimation for vehicular communication. *2016 IEEE 17th International Workshop on Signal Processing Advances in Wireless Communications (SPAWC)*.
9. Giordani, M., *et al.* (2016). Multi-Connectivity in 5G mmWave cellular networks. 2016 *Ad Hoc Mediterranean Networking Workshop (Med-Hoc-Net)*. For an extended version see http://arxiv.org/abs/1610.04836.
10. Giordani, M., *et al.* (2018). "Vehicle-to-Network Communication: Millimeter Wave vs LTE", in the 17th IEEE Annual Mediterranean Ad Hoc Networking Workshop (Med-Hoc-Net).
11. Giordani, M., *et al.* (2018). "Feasibility of Integrating mmWave and IEEE 802.11p for V2V Communications", under submission for the IEEE Connected and Automated Vehicles Symposium (CAVS).
12. Giordani, M., *et al.* (2016). Initial Access in 5G mmWave Cellular Networks. *IEEE Communications Magazine, 54*(11), 40–47.
13. Giordani, M., *et al.* (2017). Millimeter wave communication in vehicular networks: Challenges and opportunities. *2017 6th International Conference on Modern Circuits and Systems Technologies (MOCAST)*.
14. Hartenstein, H., and Laberteaux, K. (2010). *VANET: vehicular applications and internetworking technologies*. Chichester: Wiley.
15. Hoppe, T., *et al.* (2011). Security threats to automotive CAN networks—Practical examples and selected short-term countermeasures. *Reliability Engineering & System Safety, 96*(1), 11–25.
16. Intel Security. (2017). Automotive Security Best Practices [Report].

17. Mahmud, S. M., and Alles, S. (2005). In-Vehicle Network Architecture for the Next-Generation Vehicles. *SAE Technical Paper Series*.
18. Ni, Y., *et al.* (2016). Delay Analysis and Message Delivery Strategy in Hybrid V2I/V2 V Networks. *2016 IEEE Global Communications Conference (GLOBECOM)*.
19. Ott, J., and Kutscher, D. (2004). Drive-thru internet: IEEE 802.11b for "automobile" users. *IEEE Conference on Computer and Communications (INFOCOM)*.
20. Oguma, H., *et al.* (2008). New Attestation Based Security Architecture for In-Vehicle Communication. *IEEE Global Telecommunications Conference (GLOBECOM)*.
21. Polese M., *et al.* (2017). Improved Handover Through Dual Connectivity in 5G mmWave Mobile Networks. *IEEE Journal on Selected Areas in Communications, 35*(9), 2069–2084.
22. Polese, M., *et al.* (2017). TCP and MP-TCP in 5G mmWave Networks. *IEEE Internet Computing, 21*(5), 12–19.
23. Paul, B., *et al.* (2011). VANET Routing Protocols: Pros and Cons. *International Journal of Computer Applications, 20*(3), 28–34.
24. Gonzalez-Prelcic, N., *et al.* (2016). Radar aided beam alignment in MmWave V2I communications supporting antenna diversity. *2016 Information Theory and Applications Workshop (ITA)*.
25. Qiao, J., *et al.* (2011). Enabling Multi-Hop Concurrent Transmissions in 60 GHz Wireless Personal Area Networks. *IEEE Transactions on Wireless Communications, 10*(11), 3824–3833.
26. Raza, U., *et al.* (2017). Low Power Wide Area Networks: An Overview. *IEEE Communications Surveys & Tutorials, 19*(2), 855–873.
27. Rappaport T., *et al.* (2013). Millimeter Wave Mobile Communications for 5G Cellular: It Will Work! *IEEE Access, 1*, 335–349.
28. Sharef, B. T., *et al.* (2014). Vehicular communication ad hoc routing protocols: A survey. *Journal of network and computer applications*, 40, 363–396.
29. Sepulcre, M., *et al.* (2013). Exploiting context information for estimating the performance of vehicular communications. *2013 IEEE Vehicular Networking Conference*.
30. The Organisation for Economic Co-operation and Development. (2010). The future for interurban passenger transport: Bringing citizens closer together [Report].
31. Tawab, S., *et al.* (2009). Real-time weather notification system using intelligent vehicles and smart sensors. *2009 IEEE 6th International Conference on Mobile Adhoc and Sensor Systems*.
32. Va, V., *et al.* (2016). Millimeter wave vehicular communications: A survey. *Foundations and Trends® in Networking, 10*(1), 1–113.
33. Watkins, C. J., and Dayan, P. (1992). Q-learning. *Machine learning, 8*(3–4), 279-292.
34. Wolf, M., *et al.* (2007). State of the Art: Embedding Security in Vehicles. *EURASIP Journal on Embedded Systems, 2007*(1), 074706.
35. Yamamoto, A., *et al.* (2008). Path-Loss Prediction Models for Intervehicle Communication at 60 GHz. *IEEE Transactions on Vehicular Technology, 57*(1), 65–78.
36. Zhang, M., *et al.* (2016). Transport layer performance in 5G mmWave cellular. *2016 IEEE Conference on Computer and Communications Workshops (INFOCOM WKSHPS)*.
37. Zhang, M., *et al.* (2016). The Bufferbloat Problem over Intermittent Multi-Gbps mmWave Links. *arXiv preprint* arXiv:1611.02117.

An Overview of 5G Technologies

Huu Quy Tran, Ca Van Phan and Quoc-Tuan Vien

Abstract Since the development of 4G cellular networks is considered to have ended in 2011, the attention of the research community is now focused on innovations in wireless communications technology with the introduction of the fifth-generation (5G) technology. One cycle for each generation of cellular development is generally thought to be about 10 years; so the 5G networks are promising to be deployed around 2020. This chapter will provide an overview and major research directions for the 5G that have been or are being deployed, presenting new challenges as well as recent research results related to the 5G technologies. Through this chapter, readers will have a full picture of the technologies being deployed toward the 5G networks and vendors of hardware devices with various prototypes of the 5G wireless communications systems.

1 Introduction

Intelligent devices are developing daily from personal and household equipment, such as smartphones, washing machines, fridges, air-conditioners, etc., to bulky items in factories. These devices are keeping changing to accommodate the upcoming fifth-generation (5G) of cellular networks. The 5G networks can therefore be regarded as an infrastructure to accelerate the process of social change and the industry.

The 5G networks are promising to meet the demands of various individual applications with a significant increase in size, content, and rate. It is also a platform for innovation to deal with millions of applications. The 5G networks, however, raise a number of issues that need to be tackled. For instance, how to guarantee that the

H. Quy Tran · C. Van Phan
Ho Chi Minh City University of Technology and Education, Ho Chi Minh City, Vietnam
e-mail: 1627002@student.hcmute.edu.vn; tranquyhuu@iuh.edu.vn

H. Quy Tran
Industrial University of Ho Chi Minh City, Ho Chi Minh City, Vietnam

Q.-T. Vien (✉)
Middlesex University, London, UK
e-mail: Q.Vien@mdx.ac.uk

© Springer Nature Singapore Pte Ltd. 2018
K. V. Arya et al. (eds.), *Emerging Wireless Communication and Network Technologies*,
https://doi.org/10.1007/978-981-13-0396-8_4

devices are interacting with each other with a latency of less than one millisecond? Although such concern is not a critical requirement in general telecommunication systems with only voice or data services, it is vital in some specific areas with particular services, such as healthcare, military, and disaster communications systems.

The applications of 5G cellular networks will spread across smart city infrastructure with high capacity storage, intelligent transportation, and smart communication systems. The 5G evolution is being fueled by a number of factors such as the explosion of mobile data traffic, the increasing demand for high data rates, and the growth in connected and searchable devices for low-cost, energy-saving, and environmentally friendly wireless communications.

This chapter is devoted to outlining various research directions for the 5G networks that have been or are being studied and examined with new design challenges as well as their relevant works will be provided.

2   Evolution of Mobile Technologies from 1G to 5G

Starting with the first-generation (1G) mobile communications systems launched by Nippon Telegraph and Telephone (NTT) for the first time in 1979, the 1G systems were then employed worldwide in the 1980s. In the 1G systems, analog wireless access with narrowband frequency division multiple access (FDMA) is employed with a channel spacing of around 25–30 kilohertz (kHz). Then, the second-generation (2G) systems, i.e., North American Interim Standards 54 and 136 (IS-54/136), European standards Global System for Mobile (GSM), and Japan standards Personal Digital Cellular (PDC), were deployed in the 1990s, all of which adopted time division multiple access (TDMA) with the channel spacing ranging from 25 to 200 kHz [1].

Along with FDMA and TDMA, a wireless access technique based on code division multiple access (CDMA) was developed and standardized first in the IS-95 by Qualcomm in 1995 with its initial version called cdmaOne. With the CDMA, the channel spacing is 1250 kHz, which is much wider compared to the traditional 2G systems [1]. The data transfer rate in the 2G systems is however only around 9.6 kilobits per second (kbps) and does not meet the requirements of multimedia communications with high resolution. This accordingly motivated the development of the third-generation (3G) systems. The 3G systems were expected to provide advanced services with much higher data rates of megabits per second (Mbps) [2–4]. The first 3G networks were introduced by NTT DoCoMo in Japan in 1998 and were commercially launched in October 2001 also by NTT DoCoMo. The 3G systems were then deployed worldwide; for instance, in South Korea in January 2002 by SK Telecom, in the United States in 2002 by Monet Mobile Networks, and in the United Kingdom in 2003 by Hutchison Telecom.

With higher data rate requirements, the cellular networks kept evolving from the 3G to the fourth-generation (4G) systems. As a candidate standard for the 4G systems, the first-release Long-Term Evolution (LTE) standard was commercially deployed in

Norway and Sweden in 2009. The 4G systems enable a very high data transmission rate of up to 1–1.5 Gigabits per second (Gbps) for low-mobility communication, such as pedestrians, stationary users, nomadic and local wireless access, and up to 100 Mbps for high-mobility communication, such as mobile access from trains and cars. The 4G technologies were regarded as the future standard of wireless devices, allowing users to download and transfer high-quality multimedia. There exist two standard core technology of the 4G networks, including Worldwide Interoperability for Microwave Access (WiMAX) and LTE using different frequency bands (see [5, 6] and references therein). The LTE has switched to LTE-Advanced (LTE-A) since the fall of 2009 with many various LTE services started to launch in South Korea, the United States, and the United Kingdom in 2012.

Beyond the current 4G systems with LTE-A standards, the fifth-generation (5G) wireless systems have been proposed to be the next telecommunications standards aiming at providing a higher capacity, a higher reliability, and a higher density of mobile broadband users [7–9]. In order to address the high traffic growth and increasing demand for high-bandwidth connectivity, the development of 5G networks becomes crucial to support a massive number of connected devices with real-time services and high-reliability communications in critical applications [10–13]. By providing wireless connectivity for a diversity of applications from wearable devices, smartphones, tablets, and laptops to utilities within smart homes, transportation, and industry, the 5G networks are promising to provide ubiquitous connectivity for any kind of devices, enabling and accelerating the development of Internet of Things (IoT).

Further than improving solely the maximum throughput, the 5G systems are expected to provide lower power consumption dealing with battery issues, concurrent data transfer paths, lower outage, and better coverage for cell-edge and high-mobility users. Moreover, the 5G systems are required to be more secure, better cognitive functionality via software-defined radio (SDR), and artificial intelligent (AI) capabilities. With the employment of the SDR, a lower infrastructure cost could be favorably achieved, which might help reduce the traffic fees while the users can experience high-quality multimedia beyond 4G speeds. The evolution of mobile technologies from 1G to 5G is illustrated in Fig. 1.

3 5G Trends, Targets, Requirements, and Challenges

Despite some uncertainties, the 5G wireless systems have drawn attention to publicity, generating a call for innovative designs and philosophies from researchers in academia to mobile operators and communications service providers in industry. Expected to be commercially deployed around 2020, the 5G systems are being learnt with various proposed technologies to create more effective and financially viable business models [14]. Some widely known use cases of the 5G systems can be listed as in Fig. 2, which consist of broadband experience everywhere anytime, smart

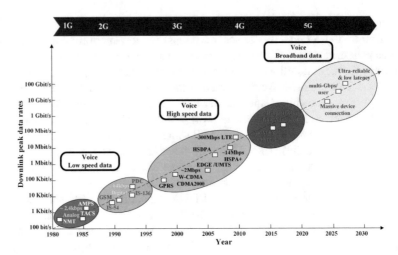

Fig. 1 Evolution of mobile technologies

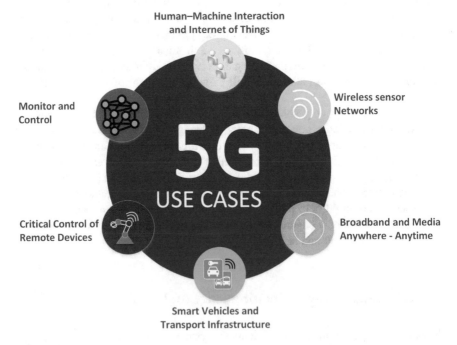

Fig. 2 5G use cases

vehicles transport and infrastructure, media everywhere, critical control of remote devices, and interaction between human and IoT.

Given the above use cases, it is totally optimistic that the 5G systems will offer significant benefits to all users and operators in the future mobile networks as long as all the challenges could be overcome. In the following, specific trends, targets, and requirements of the 5G networks will be briefly presented along with some of emerging challenges and major technical issues to be confronted.

3.1 5G Trends

As one of the key new services in the 5G networks, the IoT will influence the number of connected devices that will enter the marketplace. Additionally, the IoT will have a significant impact on 5G traffic patterns as well as their quality of service (QoS) requirements, all of which will also affect backhaul requirements and specifications. The aggregate data demand of the IoT should be very different from that of smartphone-oriented experiences. Capacity demands will grow and more base stations will have to be deployed to achieve the required QoS which are absolutely necessary for the IoT to be successful.

There are five key trends of the underlying 5G networks as the IoT expands [14], which can be listed as follows:

- **More capacity per device**: In order to meet the requirements of ultrahigh capacity per end device, as the first trend of the 5G networks, either more spectrums for improved spectrum efficiency or enhanced technologies are required.
- **More devices of different types**: The average number of devices per person is anticipated to rapidly increase in both quantity and diversity with a variety of device types for different services, such as smartphones, tablets, and wearable devices like smart watches and glasses. The 5G systems are therefore required to overhaul such exponential augmentation of devices.
- **Higher capacity for denser networks**: With the increased number of devices, the 5G systems aim at increasing the site capacity by up to 1000 times of the current networks. Although such task sounds feasible given numerous evolved technologies, several challenges need to be addressed. For instance, the current wireless backhaul links need to support a data rate of ten Gigabits per second (Gbps) or higher and are also required to cover denser networks.
- **Enhanced backhaul capability for critical applications**: Many new service types are developing in mission-critical networks for government, transportation, public safety, healthcare systems and military services. For these critical applications, the coverage, ultralow latency and strict security are dramatically the needs of wireless backhaul infrastructure in the 5G systems to lessen the risks of communication failure.
- **Diverse virtual and cloud-based services**: With cloud technology, it is promising that capital expenditure and operating expenses would be saved along with the openings of potential markets for a variety of virtual and cloud-based services. The 5G wireless systems will therefore give the operators and researchers the opportunity of reviewing and adapting the current technologies to the cloud platform.

3.2 5G Targets

The mobile data traffic is tremendously growing every year as per increasing users' demands over a number of applications and services with different smart devices. The present 4G networks, although have been shown to be satisfactory, may not be able to cope with such rapid growth in future. So, will the 5G networks be able to support a million connected devices per square kilometer with a download rate of up to 10 Gbps and a latency of less than one millisecond? The 5G systems are full of promise, incorporating a number of targets that require a lot of efforts. Some of the 5G targets are as follows:

- **Enhanced user experiences**: The 5G systems will not simply an enhanced 4G systems as an evolution, but they will aim at bringing new network and service capabilities given limited bandwidth and power resources. With novel design in the 5G networks, user experiences will be enriched guaranteeing that users can continuously access mobile broadband networks, especially in critical circumstances; for instance, in high-mobility trains, airplanes, dense areas, etc.
- **Platform for IoT**: The 5G systems will be driven toward providing a platform for IoT. A massive number of smart sensors will be connected to deliver various kinds of service in our daily life given an inevitable fact that they have severely limited power and short lifetime.
- **Improved mission-critical services**: The 5G systems will be destined for mission-critical services, such as public safety, healthcare, disaster, and emergency services, which require high-reliability communications with low latency and high coverage.
- **Unified network infrastructure**: The 5G systems will be tailored to meet the requirements of various network infrastructures in order to bring them together in a unified infrastructure. This integration will not only provide scope for optimizing all networking, computing, and storage resources but also enable dynamic usage of these resources along with convergence of services.
- **Incorporated market for operators**: The 5G systems will be directed to enable operators to collaborate over a digital or virtual market by taking advantage of cloud computing. Such market will make room for further development of the 5G networks.
- **Sustainable and scalable network**: The 5G systems will be particularly focused on energy consumption reduction and energy harvesting, targeting at compensating the radical increase of energy usage. With automation integration and hardware optimization, the operational cost will be expected to considerably reduce for sustainable and scalable network model.
- **Ecosystem for innovation**: The 5G systems will be means for involving vertical markets in different sectors and areas, such as energy, transportation, manufacturing, agriculture, health care, education, government, and so on. This will be an excellent opportunity to encourage startups and innovations in these diverse trading businesses.

3.3 5G Requirements

Aiming at enabling new services as well as enhancing current services in the next few years, there are a number of expected requirements for the 5G networks, which are more diverse than those for the 4G networks. Specifically, the 5G networks need to meet the following requirements:

- **User experiences** should be consistently and ubiquitously delivered in the 5G systems at a high data rate and low latency with optional mobility support for specific user demands of certain services.
- **Networks/systems** are required to support massive connected devices with high traffic density, high spectrum efficiency, and high coverage.
- **Devices/terminals** are desired to be smarter allowing operator control capabilities with programmability and configurability, supporting multiple frequency bands, increasing battery life, and improving resource and signaling efficiency.
- **Services** are indispensable to provide connectivity transparency with seamless, ubiquitous and high-reliability communications for mobile users, improve localization with additional three-dimensional space attributes, protect users' data from possible cybersecurity attacks, as well as ensuring the availability and resilience of mission-critical services.
- **Network deployment, operational, and management** are all needed to provide the new enhanced services in a low cost and low energy consumption for ensuring sustainability of the 5G and beyond networks, and also should facilitate the future upgrade and innovation assuring flexibility and scalability.

3.4 5G Challenges

Although the 5G systems are optimistic encouraging researchers in both academia and industry to overcome limitations of the current standards and theories, there are several challenges that need to be tackled in order to meet the requirements as stated in Sect. 3.3 and also to achieve the proposed targets in Sect. 3.2.

As one of the critical issues in the existing technologies, energy performance needs to be improved with appropriate resource allocation. The resources should be optimized to facilitate better utilization in a dynamic and adaptable manner. Additionally, given the scarcity of spectrum resource, an efficient spectrum usage is crucial in the 5G systems to support massive connected devices of different kinds.

In particular, a higher coverage with a higher density of mobile broadband users needs to be coped with in the 5G systems. Indeed, the 5G systems will need to manage a very dense heterogeneous network. As illustrated in Fig. 3, the radio resource management will become a paramount problem to handle the dense deployment of small cells in coordination with existing macrocells and billions of connected devices.

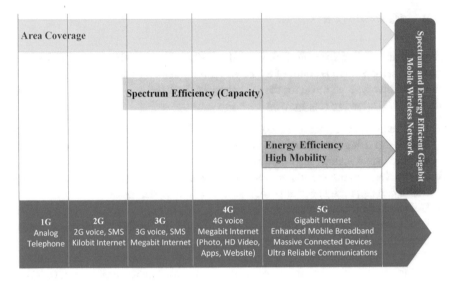

Fig. 3 Radio resource management in 5G

4   5G Enabling Technologies

Working toward 5G and beyond systems, a variety of enabling technologies have been being researched and developed, of which some are at still at early stage along with those well proposed in the literature as illustrated in Fig. 4. In this section, these technologies will be sequentially presented outlining their key concepts with relevant works.

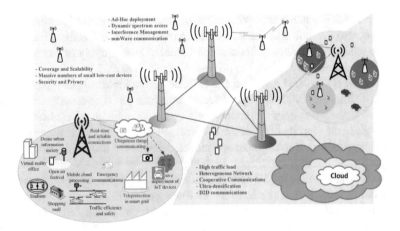

Fig. 4   5G enabling technologies

4.1 Massive MIMO

Massive multiple-input multiple-output (MIMO) (also known as large-scale MIMO or large-scale antenna systems) is an emerging technology in the next-generation mobile system, i.e., 5G and beyond, which has been upgraded from the conventional multiuser MIMO (MU-MIMO) technology. A typical massive MIMO architecture is illustrated in Fig. 5.

The massive MIMO has shown to be potential in dealing with the high sensitivity to blockages and distance-dependent propagation effects. In an effort to achieve all the gains of the MU-MIMO, the massive MIMO is promising to provide a larger scale in terms of energy and spectrum efficiency [15]. In particular, the massive MIMO exploits the spatial multiplexing gain to increase almost ten times of the capacity and 100 times of the energy efficiency compared with the MU-MIMO systems. In the massive MIMO, large arrays of antennas that contain a few hundred of antenna elements are deployed at base station (BS) to simultaneously serve several terminals using the same time–frequency resources.

In the MU-MIMO systems, to achieve both uplink and downlink spectral efficiencies, both the BS and the terminals must handle several complicated signal processing operations and have the channel state information (CSI) on the downlink which is accommodated by transmitting pilots in both directions [16]. With the increase in the number of antennas, the CSI process in the MU-MIMO systems is nevertheless unreasonable for the massive MIMO systems, especially in high-mobility conditions. In order to deal with such issue, the massive MIMO makes use of time division duplex (TDD) mode for pilot transmission based on an assumption that the uplink and downlink channels are reciprocal, and thus linear signal processing techniques can be employed to provide near-optimal performance with a low complexity [17].

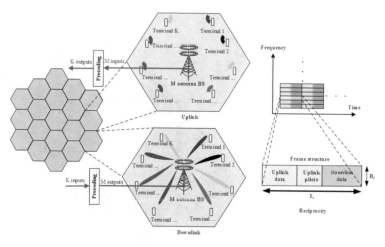

Fig. 5 Massive MIMO concept

4.2   mmWave Massive MIMO

Current mobile systems are all allocated in the microwave frequency bands of which most operate in the bands below 3 GHz and share the scarce spectrum resources of 600 MHz divided among operators. In contrast, mmWave frequency bands ranging from 3 to 300 GHz can offer multi-GHz of unlicensed bandwidth [18].

Some mmWave propagation measurements performed recently in both indoor and outdoor environments have similar general characteristics to the microwave propagation that reveals the great potential for small-cell communications [19]. Large arrays of antennas can eliminate the frequency dependence of path loss significantly compared with omnidirectional antennas. In addition, the narrow beams provided by adaptive arrays of antennas are able to reduce the impact of interference. In addition, the extremely short wavelength of the mmWave signals allows small antenna to direct them in narrow beams with enough gain to overcome propagation losses [20]. So, it is very likely to build a large number of antenna elements in a small area enough to fit into the mobile phones. This is the most important feature that helps to realize the massive MIMO at mmWave bands in realistic environments [21, 22].

mmWave massive MIMO has potential to provide ultra-large bandwidth and high spectrum efficiency that may significantly improve the overall system throughput in the future 5G cellular networks. An example of its deployment is illustrated in Fig. 6. However, due to the special propagation features and hardware requirements of the mmWave systems, there are several challenges when deploying the mmWave massive MIMO at the physical and upper layers. Particularly, the network architecture and protocols must be considered carefully in the network design to adapt signaling and resource allocation, as well as to cope with severe channel attenuation, directionality, and blockage [23].

Fig. 6 mmWave massive MIMO deployment

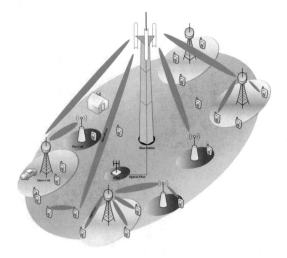

4.3   Cloud Radio Access Networks

In general cell site architecture deployed in the 3G systems, BS contains two separated sectors including remote radio head (RRH) or remote radio unit (RRU) and baseband unit (BBU) or DATA UNIT (DU). The RRH performs radio frequency (RF) processing, digital-to-analog conversion (DAC), analog-to-digital conversion (ADC), and power amplification and filtering, while the BBU provides baseband-processing functions. In contrast to the traditional radio access networks (RANs), cloud radio access network (C-RAN) has recently emerged as a novel architecture for the RAN in which the baseband processing is now centralized and based on cloud computing technology [24–29].

In C-RAN architecture, all baseband computational resources are processed and aggregated within a central pool, also known as a virtualized BBU Pool. The geographically distributed RRHs/antennas are connected to a cloud platform through an optical transmission network. This model allows reducing the number of BBUs while maintaining similar coverage and offering better services compared to the traditional RAN architecture [30–33]. In fact, several cell sites can effectively share the computation resources, and thus help save a lot of operation and management cost leading to a significantly reduced capital expenditure. In addition, the virtualization in cloud computing has also the potential to achieve load balancing and scalability. This means that it is able to allocate and utilize the resources more efficiently under busty traffic conditions thus reducing waste of computation resources and power consumption [34]. Moreover, the resource cloudification in the C-RAN allows network operators to provide the RAN as a cloud service [35].

A comparison of C-RAN architecture and the traditional BS architecture is illustrated in Fig. 7. In Fig. 7a, the antenna module is located a few meters from the BS and connected to the BS by using coaxial cables with high signal attenuation in the traditional BS. Another configuration is shown in Fig. 7b where the BS with RRH is separated into two parts including RRU and BBU, which are connected by fiber optic cable, while the coaxial cable is only used to connect the RRH and the antenna. Finally, the fully centralized C-RAN architecture, as shown in Fig. 7c, is characterized by a large number of the RRHs located at different antenna sites connected to a BBU pool cloud located in a centralized cloud server through an optical transmission network.

4.4   D2D Communications

Device-to-device (D2D) communications is one of the most important technologies in 5G systems. In D2D communications, mobile user equipment (UE) communicates with each other in short range without involving eNodeB or the core network on the licensed cellular network.

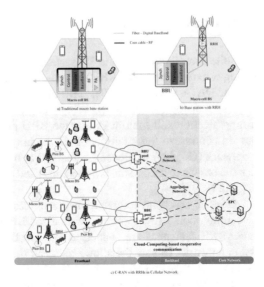

Fig. 7 Base station architecture evolution

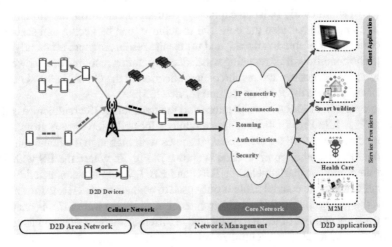

Fig. 8 Typical use cases of D2D communications in cellular networks

The D2D communications not only provides a flexible communication platform based on multiple radio access technologies embedded on mobile UEs but also promises to considerably improve energy efficiency, throughput, spectrum efficiency, and so on [36, 37]. In addition, it is feasible for integrating cooperative communication and cognitive radio along with performance optimization combining both ad hoc and centralized communications in the D2D communication. Figure 8 illustrates typical use cases of the D2D communications which can be found in smart building, healthcare, and public safety networks, e.g., [38, 39].

The D2D communication, however, faces several challenges including interference management, resource allocation, as well as delay-sensitive processing. In fact, in the 3GPP LTE architecture, there are many pairs of D2D UEs sharing cellular resources of eNodeB that may cause severe interferences including the interference from the eNodeB within the same cell, those from other co-channel D2D UEs in the same cell, and also those from the eNodeBs and co-channel D2D UEs within other cells [40, 41]. The resource allocation is also one of the most important concerns in the D2D communications given a limited number of subcarriers in the 3GPP LTE networks. Based on the quality of service requirement, the resource allocation technique will be selected consist of D2D mode or power control in cellular networks [42].

4.5 Ultradense Heterogeneous Networks

4.5.1 Heterogeneous Networks

In order to cope with the rapid growth of wireless traffic demands in 5G communications, the deployment of a large number of small cells (femtocell, picocell, and microcell) has been shown to be a feasible solution to achieve high capacity, leading to a heterogeneous network (HetNet). The HetNet is typically a multi-tier network architecture consisting of multiple types of infrastructure elements including macro-BSs, micro-BSs, pico-BSs, and femto-BSs with different transmission powers and coverage sizes.

In the HetNet, the powerful macro-BSs with high-power transmission are deployed in a planned way for covering large geographical areas whereas the small-cell BSs serving small coverage areas is used to complement the traditional macro-BSs. The range of a microcell or picocell is in the order of few hundred meters, whereas femtocells are used to provide indoor coverage within the range of few meters. A typical heterogeneous network is shown in Fig. 9 where the low-power BSs served for microcells or picocells which are deployed to cover a small area with heavy traffic such as a commercial center, airport, subway, and train station.

By deploying a variety of cells of different sizes, the HetNet architecture is highly probable for increasing the radio capacity, improving throughput, and serving several types of users with different QoS requirements in the next-generation cellular networks [43]. In addition, the deployment of low-power small-cell BSs in dense areas is one of the key solutions to enhance coverage and provide more capacity by covering smaller area than macro-BSs as well as improve the spectral efficiency of cellular networks. Moreover, the small cell integrated with macrocells provides a potential opportunity to decouple the control plane and user plane in which low-power small-cell BSs handle the control plane, while the overall control signaling to all users and cell-specific reference signals of small-cell BSs can be delivered to powerful high-power macro-BSs. Therefore, HetNets have advantages of serving hotspot customers with high data rates and busty traffic [44–48].

Fig. 9 Heterogeneous
cellular network

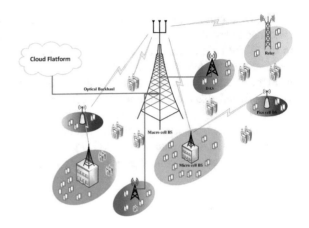

Along with a number of advantages, the HetNet is facing a critical issue when too
dense low-power small-cell BSs underlaid with macro-BSs reusing the same spectral resources could incur severe inter-tier interferences [45, 46]. Hence, advanced
signal processing techniques are vital to fully obtain the potential gains of Het-
Nets. Specifically, the advanced coordinated multipoint (CoMP) transmission and
reception techniques have been proposed to suppress both intra-tier and inter-tier
interference and improve the cell-edge user throughput [49, 50]. Another technique
for enhancing the performance of the HetNet is co-locating massive MIMO BS and
low-power small cell access in which the massive MIMO ensures outdoor mobile
coverage whereas the small cell access equipped with cognitive and cooperative
functionalities enables the HetNet to provide high capacity for indoors and outdoors
with low-mobility users [26].

4.5.2 Distributed Antenna Systems

Another approach in dealing with dense networks is an employment of distributed
antenna systems (DAS). The DAS has shown to be the high potential providing
more uniform coverage, especially in shadowed and indoor areas, as well as enhancing the transmit capability of BS by adding multiple remote antenna units (RAUs)
geographically distributed in a macrocell [51].

In the DAS, the spatially separated antenna units are connected to a BS or a central
unit (CU) by using a high-bandwidth low-latency dedicated link that can be coaxial
cable or optical fiber [52]. In this way, the BS can operate as a multiple-antenna
system, although the antennas are located in different geographic locations. The DAS
can accordingly improve indoor and outdoor coverage, reduce the outage probability,
and increase the capacity of cellular systems in a variety of configurations. A typical
DAS is illustrated in Fig. 10 where spatially separated antenna elements are connected
to a macro-BS via a dedicated fiber/microwave backhaul link. This configuration
indeed provides a better coverage since the terminals can connect to nearby antenna

Fig. 10 Distributed antennas systems

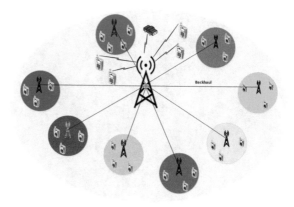

elements, and thus a higher capacity gain can be achieved by exploiting both macro- and micro-diversities.

Comparing with co-located-antenna systems, the DAS has been shown to achieve a much higher sum capacity due to higher water-filling gain and multiuser diversity gain. Moreover, DAS technique allows shortening considerably the radio transmission distance between the transmitter and receiver leading to support high data rate transmission and achieve significant improvement in power efficiency [53, 54]. Particularly, fully distributed antennas also result in higher sum rates than having multiple antennas at each RRU with the same number of antennas. Since all the RAUs in a macrocell are connected to a CU remotely, spatial diversity and spatial multiplexing can be exploited in the DAS in order to improve the system performance.

4.5.3 Ultradense Heterogeneous Networks

Demand for high-speed data traffic in the mobile and ubiquitous computing era has been growing explosively and exponentially in recent years. For instance, in apartments, enterprises, and hotspot environments where users with high traffic demand are densely distributed. Deploying ultradense heterogeneous small cells has been widely recognized as a promising technique to address such exponential traffic growth with enhanced coverage especially in indoor and hotspot environments [55, 56]. The networks with a large number of densely distributed heterogeneous small cells, also known as ultradense HetNets, are illustrated in Fig. 11.

In ultradense HetNet architecture, the low-power small-cell BSs are densely deployed within the coverage area, which is served by the high-power macrocell BS to enhance the spectrum efficiency and thus increase the network capacity. The ultradense HetNet has also been regarded as a network in which inter-site communications occur at very short distances with low interferences. In particular, the distances between the access nodes in the newly envisioned small cells range from a few meters for femtocells deployed in indoor up to 50 m for microcells or pico-

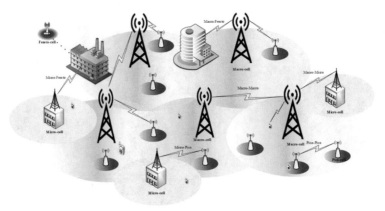

Fig. 11 Ultradense HetNets

cells with outdoor deployment [57]. Over the short distances between the users and the small-cell BSs, the received power of the desired signal at the user increases considerably, promising to provide a significantly enhanced network capacity.

Although the deployment of ultradense HetNets has been well identified as a feasible solution to manage the increasing traffic demands, the dense and random deployment of the small cells and their uncoordinated operation also bring several challenges in such multi-tier networks [58–61]. In this distributed network architecture, both backhaul and fronthaul traffics need to be relayed to the destination. Hence, an efficient multi-hop routing algorithm becomes crucial for such scenario. Since the coverage of the small cells in ultradense HetNet is less than that of the macrocell in the conventional cellular networks, the frequent handover in small cells causes a considerably increased redundant overhead and also reduces the user experiences. In addition, the mmWave antennas with beamforming technique equipped in the small-cell BS can provide strong directivity having the advantage of high-speed transmission but revealing the disadvantage in supporting the high-speed mobile users.

4.5.4 Security Issues

The security of mobile devices access in ultradense heterogeneous network is largely based on the specific features and architecture of the network system. We may find that the vulnerabilities of ultradense heterogeneous networks can occur in some cases, such as IP spoofing, interference management attack, handover management attack, unauthorized cell identification, RFID Tag, etc.

From there we can define security domains to access the network without fail, network accounts are not attacked, networks are stable, and quality of service meets the increasing needs of users. A typical security architecture of the user ultradense heterogeneous networks (UUDHN) is illustrated in Fig. 12 [62]. In this security

Fig. 12 User ultradense
heterogeneous network
security architecture

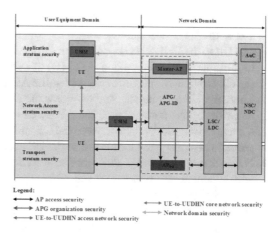

architecture, the UUDHN is a wireless heterogeneous network in which the access
point (AP) density is comparable to the user density. The UUDHN organizes an
access point group (APG) as the following coverage to serve each user seamlessly
without user's involvement, and there are many security feature groups.

- The AP access security: The APs are very familiar to the user and even may be
  deployed by the user. The security threats of AP deployment which the UUDHN
  facing is the same as home evolved node B (HeNB) in long-term evolution (LTE)
  network. The HeNB supports a device validation method with either certification
  based or universal subscriber identity module (USIM) based mutual authentication,
  which helps prevent the attacker from exploiting the HeNB as a springboard to
  access the LTE networks [63]. In order to access the UUDHN, the APs also need
  to simultaneously authenticate between them and the UUDHN networks. When
  the authentication is successful, the APs then can enter into the working status.
- The APG organization security: The overall security of the UUDHN will be threatened by a malicious AP which counterfeits the APG. To cope with the APG security threats, two security aspects should be considered, including: (i) A security
  refresher for new APs joining the APG or AP registration leaving the APG, and
  (ii) A secure communication (e.g., collaborative signaling and data exchange, etc.)
  between the APG members.
- User equipment (UE) to the UUDHN access network security: Because the APG
  is refreshing, there are so many threats have emerged. The members of the APGs
  will be changed and the AP wireless connections may be entered by attackers.
  For instance, when the data transferred between the UE and the APs (APG), the
  attackers can eavesdrop on or manipulate the signaling and user data. The user's
  mobility between the APs may be also found or discovered where a particular user
  is located. These can pose a huge threat to user's privacy. It is the effective measures to protect the user's private data with the keys based on specific encryption
  algorithms. For instance, it is relied on the APG interim key to derive the keys for
  ciphering of user plane (KUPint) and also protects the integrity and encryption of

the radio resource control (RRC) signals between the UE and AP/APG (KRRCenc and KRRCint) is an effective way.

- The UE to the UUDHN core network security: In the UUDHN, the AP may pertain to multiple APGs (in another APG-ID) at certain time at a local service center (LSC). Obviously, there are threats that the APG or the AP may be counterfeited. Then, the UE may be attacked to access the other the APG or the UUDHN core networks by the counterfeited the AP. So as to prevent these threats, the mutual authentication mechanism between the UE and the LSC (APG-ID defined therein) needs to be considered to make sure the UE access security.
- Network access security: The set of security features providing users and entities with secure access to services and which particularly protect against attacks on the (radio) access link.
- Network domain security: The set of security features about the APG organization security, including the APG initiating, APG-ID/master the AP selection, the APG refreshing, the APG handover, and the AP security itself. They protect against attacks from the counterfeited the APs/APG.
- The APG domain security: The set of security features enabling entities to securely exchange signaling data and user data (among access network, serving network, and within access network) and protect against attacks on the wireline network.
- User domain security: The set of security features securing access to mobile stations.
- Application domain security: The set of security features enabling applications in the user and in the provider domain to securely exchange messages [64].

5 Conclusions

After nearly four decades since the birth of the first-generation networks, mobile communications networks have been continuously evolved as an important infrastructure offering distinct types of services in our daily life and activities. This chapter has sketched a picture for the evolution of the mobile technologies from 1G to 5G that has attracted interests of a number of researchers and developers. The upcoming 5G systems have indeed drawn their attention with a variety of trends, targets, requirements, and challenges to be tackled. Dealing with these challenges, this chapter has outlined different enabling technologies, including massive MIMO, mmWave massive MIMO, C-RAN, D2D, HetNet, DAS, and ultradense HetNet. These technologies have been shown to be promising candidates for the 5G wireless networks, meeting the strict requirements of high spectrum efficiency and energy efficiency as well as enhancing user experiences and services. With a number of evolved techniques, we are completely hopeful and optimistic that the upcoming 5G systems are going to fulfill the mobile users' demands in ultrahigh data rates, very low latency, high mobility, high coverage, long-life batteries, high reliability, and extraordinarily enhanced services.

References

1. K. S. Gilhousen, I. M. Jacobs, R. Padovani, A. J. Viterbi, L. A. Weaver and C. E. Wheatley, "On the capacity of a cellular CDMA system," in *IEEE Transactions on Vehicular Technology*, vol. 40, no. 2, pp. 303–312, May 1991.
2. W. C. Y. Lee, "Overview of cellular CDMA," in *IEEE Transactions on Vehicular Technology*, vol. 40, no. 2, pp. 291–302, May 1991.
3. F. Adachi, "Wireless past and future evolving mobile communications systems", in *IEICE Transactions on Fundamentals*, vol. E84-A, pp. 55–60, January 2001.
4. H. Viswanathan and M. Weldon, "The past, present, and future of mobile communications," in *Bell Labs Technical Journal*, vol. 19, pp. 8–21, 2014.
5. A. Damnjanovic et al., "A survey on 3GPP heterogeneous networks," in *IEEE Wireless Communications*, vol. 18, no. 3, pp. 10–21, June 2011.
6. S. Parkvall et al., "LTE-Advanced - Evolving LTE towards IMT-Advanced," in *Proc. IEEE VTC 2008*, Calgary, BC, 2008, pp. 1–5.
7. A. Osseiran et al., "Scenarios for 5G mobile and wireless communications: the vision of the METIS project," in *IEEE Communications Magazine*, vol. 52, no. 5, pp. 26–35, May 2014.
8. J. G. Andrews et al., "What will 5G be?," in *IEEE Journal on Selected Areas in Communications*, vol. 32, no. 6, pp. 1065–1082, June 2014.
9. A. Gupta and R. K. Jha, "A survey of 5G network: Architecture and emerging technologies," in *IEEE Access*, vol. 3, pp. 1206–1232, 2015.
10. R. Trestian, P. Shah, H. X. Nguyen, Q.-T. Vien, O. Gemikonakli, and B. Barn, "Towards connecting people, locations and real-world events in a cellular network," in *Telematics and Informatics*, vol. 34, no. 1, pp. 244–271, February 2017.
11. R. Trestian and H. Venkataraman, 5G Radio Access Networks: centralized RAN, cloud-RAN and virtualization of small cells, eds. CRC Press, 2017.
12. F. Boccardi, R. W. Heath, A. Lozano, T. L. Marzetta and P. Popovski, "Five disruptive technology directions for 5G," in *IEEE Communications Magazine*, vol. 52, no. 2, pp. 74–80, February 2014.
13. C. X. Wang et al., "Cellular architecture and key technologies for 5G wireless communication networks," in *IEEE Communications Magazine*, vol. 52, no. 2, pp. 122–130, February 2014.
14. D. Cohen, "5G and the IoT: 5 trends and implications," in *Microwave Journal*, September 2016.
15. E. G. Larsson, O. Edfors, F. Tufvesson and T. L. Marzetta, "Massive MIMO for next generation wireless systems," in *IEEE Communications Magazine*, vol. 52, no. 2, pp. 186–195, February 2014.
16. H. Q. Ngo, E. G. Larsson and T. L. Marzetta, "Energy and spectral efficiency of very large multiuser MIMO systems," in *IEEE Transactions on Communications*, vol. 61, no. 4, pp. 1436–1449, April 2013.
17. L. Lu, G. Y. Li, A. L. Swindlehurst, A. Ashikhmin and R. Zhang, "An overview of massive MIMO: Benefits and challenges," in *IEEE Journal of Selected Topics in Signal Processing*, vol. 8, no. 5, pp. 742–758, October 2014.
18. F. Boccardi, R. W. Heath, A. Lozano, T. L. Marzetta and P. Popovski, "Five disruptive technology directions for 5G," in *IEEE Communications Magazine*, vol. 52, no. 2, pp. 74–80, February 2014.
19. M. Shafi *et al.*, "5G: A tutorial overview of standards, trials, challenges, deployment, and practice," in *IEEE Journal on Selected Areas in Communications*, vol. 35, no. 6, pp. 1201–1221, June 2017.
20. A. L. Swindlehurst, E. Ayanoglu, P. Heydari and F. Capolino, "Millimeter-wave massive MIMO: the next wireless revolution?," in *IEEE Communications Magazine*, vol. 52, no. 9, pp. 56–62, September 2014.
21. T. E. Bogale and L. B. Le, "Massive MIMO and mmWave for 5G wireless HetNet: Potential benefits and challenges," in *IEEE Vehicular Technology Magazine*, vol. 11, no. 1, pp. 64–75, March 2016.

22. S. Rangan, T. S. Rappaport and E. Erkip, "Millimeter-wave cellular wireless networks: Potentials and challenges," in *Proceedings of the IEEE*, vol. 102, no. 3, pp. 366–385, March 2014.
23. H. Shokri-Ghadikolaei, C. Fischione, G. Fodor, P. Popovski and M. Zorzi, "Millimeter wave cellular networks: A MAC layer perspective," in *IEEE Transactions on Communications*, vol. 63, no. 10, pp. 3437–3458, October 2015.
24. J. Wu, Z. Zhang, Y. Hong and Y. Wen, "Cloud radio access network (C-RAN): a primer," in *IEEE Network*, vol. 29, no. 1, pp. 35–41, January-February 2015.
25. Q.-T. Vien, N. Ogbonna, H. X. Nguyen, R. Trestian, and P. Shah, "Non-orthogonal multiple access for wireless downlink in cloud radio access networks," in *Proc. EW 2015*, Budapest, Hungary, May. 2015, pp. 434–439.
26. M. Agiwal, A. Roy and N. Saxena, "Next generation 5G wireless networks: A comprehensive survey," in *IEEE Communications Surveys & Tutorials*, vol. 18, no. 3, pp. 1617–1655, third quarter 2016.
27. M. Peng, Y. Li, Z. Zhao and C. Wang, "System architecture and key technologies for 5G heterogeneous cloud radio access networks," in *IEEE Networks*, vol. 29, no. 2, pp. 6–14, March-April 2015.
28. Q.-T. Vien, T. A. Le, B. Barn, and C. V. Phan, "Optimising energy efficiency of non-orthogonal multiple access for wireless backhaul in heterogeneous cloud radio access network," in *IET Communications*, vol. 10, no. 18, pp. 2516–2524, 2016.
29. H. Q. Tran, P. Q. Truong, C. V. Phan, and Q.-T. Vien, "On the energy efficiency of NOMA for wireless backhaul in multi-tier heterogeneous CRAN," in *Proc. SigTelCom 2017*, Da Nang, Vietnam, January 2017, pp. 229–234.
30. C. L. I, J. Huang, R. Duan, C. Cui, J. (. Jiang and L. Li, "Recent progress on C-RAN centralization and cloudification," in *IEEE Access*, vol. 2, pp. 1030–1039, 2014.
31. M. Peng, Y. Li, J. Jiang, J. Li and C. Wang, "Heterogeneous cloud radio access networks: a new perspective for enhancing spectral and energy efficiencies," in *IEEE Wireless Communications*, vol. 21, no. 6, pp. 126–135, December 2014.
32. A. Checko *et al.*, "Cloud RAN for mobile networks - A technology overview," in *IEEE Communications Surveys & Tutorials*, vol. 17, no. 1, pp. 405–426, first quarter 2015.
33. Q.-T. Vien, N. Ogbonna, H. X. Nguyen, R. Trestian, and P. Shah, Performance evaluation of NOMA under wireless downlink cloud radio access networks environments. In: 5G Radio Access Networks: Centralized RAN, Cloud-RAN and Virtualization of Small Cells, CRC Press, pp. 67–84, 2017.
34. P. Demestichas *et al.*, "5G on the horizon: Key challenges for the radio-access network," in *IEEE Vehicular Technology Magazine*, vol. 8, no. 3, pp. 47–53, September 2013.
35. P. Rost et al., "Cloud technologies for flexible 5G radio access networks," in *IEEE Communications Magazine*, vol. 52, no. 5, pp. 68–76, May 2014.
36. N. Panwar, S. Sharma, A. K. Singh, "A survey on 5G: The next generation of mobile communication," in *Physical Communication*, vol. 18, Part 2, pp. 64–84, March 2016.
37. R. Trestian, Q.-T. Vien, H. X. Nguyen, and O. Gemikonakli, "ECO-M: Energy-efficient cluster-oriented multimedia streaming in a LTE D2D environment," in *Proc. IEEE ICC 2015*, London, UK, June 2015, pp. 55–61.
38. K. Ali, H. X. Nguyen, P. Shah, Q.-T. Vien, and N. Bhuvanasundaram, "Architecture for public safety network using D2D communication," in *Proc. IEEE WCNC 2016 - Workshop on Communications in Extreme Conditions (ComExCon 2016)*, Doha, Qatar, April 2016, pp. 206–211.
39. K. Ali, H. X. Nguyen, P. Shah, Q.-T. Vien, and E. Ever, "D2D multi-hop relaying services towards disaster communication system," in *Proc. ICT 2017 - Workshop on 5G Networks for Public Safety and Disaster Management (IWNDP 2017)*, Limassol, Cyprus, May 2017, pp. 1–5.
40. L. Wei, R. Q. Hu, Y. Qian and G. Wu, "Enable device-to-device communications underlaying cellular networks: challenges and research aspects," in *IEEE Communications Magazine*, vol. 52, no. 6, pp. 90–96, June 2014.
41. J. Liu, N. Kato, J. Ma and N. Kadowaki, "Device-to-device communication in LTE-Advanced networks: A survey," in *IEEE Communications Surveys & Tutorials*, vol. 17, no. 4, pp. 1923–1940, fourth quarter 2015.

42. D. Feng, L. Lu, Y. Yuan-Wu, G. Y. Li, S. Li and G. Feng, "Device-to-device communications in cellular networks," in *IEEE Communications Magazine*, vol. 52, no. 4, pp. 49–55, April 2014.
43. A. Ghosh *et al.*, "Heterogeneous cellular networks: From theory to practice," in *IEEE Communications Magazine*, vol. 50, no. 6, pp. 54–64, June 2012.
44. R. Q. Hu and Y. Qian, "An energy efficient and spectrum efficient wireless heterogeneous network framework for 5G systems," in *IEEE Communications Magazine*, vol. 52, no. 5, pp. 94–101, May 2014.
45. Q.-T. Vien, T. Akinbote, H. X. Nguyen, R. Trestian, and O. Gemikonakli, "On the coverage and power allocation for downlink in heterogeneous wireless cellular networks," in *Proc. IEEE ICC 2015*, London, UK, June 2015, pp. 4641–4646.
46. Q.-T. Vien, T. A. Le, H. X. Nguyen, and M. Karamanoglu, "An energy-efficient resource allocation for optimal downlink coverage in heterogeneous wireless cellular networks," in *Proc. ISWCS 2015*, Brussels, Belgium, August 2015, pp. 156–160.
47. R. Trestian, Q.-T. Vien, P. Shah, and G. Mapp, "Exploring energy consumption issues for multimedia streaming in LTE HetNet small cells," in *Proc. IEEE LCN* 2015, Florida, USA, October 2015, pp. 498–501.
48. Q.-T. Vien, T. A. Le, C. V. Phan, and M. O. Agyeman, "An energy-efficient NOMA for small cells in heterogeneous CRAN under QoS constraints," in *Proc. EW 2017*, Dresden, Germany, May 2017, pp. 80–85.
49. J. G. Andrews, "Seven ways that HetNets are a cellular paradigm shift," in *IEEE Communications Magazine*, vol. 51, no. 3, pp. 136–144, March 2013.
50. N. Bhuvanasundaram, H. X. Nguyen, R. Trestian, and Q.-T. Vien, "Sum-rate analysis of cell edge users under cooperative NOMA," in *Proc. IFIP WMNC 2015*, Munich, Germany, October 2015, pp. 239–244.
51. R. W. Heath Jr, T. Wu, Y. H. Kwon and A. C. K. Soong, "Multiuser MIMO in distributed antenna systems with out-of-cell interference," in *IEEE Transactions on Signal Processing*, vol. 59, no. 10, pp. 4885–4899, October 2011.
52. Shidong Zhou, Ming Zhao, Xibin Xu, Jing Wang and Yan Yao, "Distributed wireless communication system: A new architecture for future public wireless access," in *IEEE Communications Magazine*, vol. 41, no. 3, pp. 108–113, March 2003.
53. L. Dai, "A comparative study on uplink sum capacity with co-located and distributed antennas," in *IEEE Journal on Selected Areas in Communications*, vol. 29, no. 6, pp. 1200–1213, June 2011.
54. H. Zhu, "Performance comparison between distributed antenna and microcellular systems," in *IEEE Journal on Selected Areas in Communications*, vol. 29, no. 6, pp. 1151–1163, June 2011.
55. X. Ge, S. Tu, G. Mao, C. X. Wang and T. Han, "5G ultra-dense cellular networks," in *IEEE Wireless Communications*, vol. 23, no. 1, pp. 72–79, February 2016.
56. S. Chen, T. Zhao, H. H. Chen, Z. Lu and W. Meng, "Performance analysis of downlink coordinated multipoint joint transmission in ultra-dense networks," in *IEEE Network*, vol. PP, no.99, pp. 12–20.
57. R. Baldemair *et al.*, "Ultra-dense networks in millimeter-wave frequencies," in *IEEE Communications Magazine*, vol. 53, no. 1, pp. 202–208, January 2015.
58. Y. S. Soh, T. Q. S. Quek, M. Kountouris and H. Shin, "Energy efficient heterogeneous cellular networks," in *IEEE Journal on Selected Areas in Communications*, vol. 31, no. 5, pp. 840–850, May 2013.
59. T. Zhang, J. Zhao, L. An and D. Liu, "Energy efficiency of base station deployment in ultra dense HetNets: A Stochastic Geometry Analysis," in *IEEE Wireless Communications Letters*, vol. 5, no. 2, pp. 184–187, April 2016,
60. S. F. Yunas, M. Valkama and J. Niemelä, "Spectral and energy efficiency of ultra-dense networks under different deployment strategies," in *IEEE Communications Magazine*, vol. 53, no. 1, pp. 90–100, January 2015.
61. D. Calabuig *et al.*, "Resource and mobility management in the network layer of 5G cellular ultra-dense networks," in *IEEE Communications Magazine*, vol. 55, no. 6, pp. 162–169, 2017.

62. Z. Chen, S. Chen, H. Xu, B. Hu, "Security architecture and scheme of user-centric ultra-dense network (UUDN)," in *Transactions on Emerging Telecommunications Technologies*, vol. 28, no. 9, pp. 2161–3915, February 2017.
63. 3GPP, "Technical Specification Group Radio Access Network; Small cell enhancements for E-UTRA and E-UTRAN – Physical layer aspects (Rel. 12)," in 3GPP TR 36.872, V12.1.0, September 2013.
64. S. Chen, F. Qin, B. Hu, X. Li, Z. Chen, "User-centric ultra-dense networks (UUDN) for 5G: Challenges, methodologies and directions," in *IEEE Wireless Communication Magazine*, vol. 23, no. 2, pp. 78–85, April 2016.

Design and Application for Reliable Cooperative Networks

Dinh-Thuan Do

Abstract The fast progress of mobile devices and modern wireless networks result in the explosive demand for wireless data transfer. Such wireless communications have to meet numerous challenges such as spectrum sharing, energy scarcity, and security to deal with the dramatic growth in wireless data which shift the focus of research directions to fifth-generation (5G) networks. To address these challenges in wireless design, researchers need to come up with energy and spectrum management solutions in 5G networks. The key technologies for 5G considering reliable cooperative networks such as full-duplex networks, energy-efficient communications so-called energy harvesting, and secure networks with physical layer considerations. In general, to guarantee quality-of-service (QoS) it is required high-speed data rate and reliable transmission in design of new paradigms in 5G networks. This chapter focuses on cooperative networks applied in several emerging wireless technology. In addition, these cooperative networks have been well-known models for improving the coverage of wireless systems. The key principle of cooperative network is that one or some relay nodes are installed to forward messages from the source node to its destinations in case of the direct transmission is inaccessible. As popular schemes, we introduce dual-hop scheme and multi-hop scheme with one or many relaying links from the source node to the destination node in this section. In each hop, the relaying node first process received signal of the signal from the previous hop and then used to relay the signal to the next hop. To examine the performance of relay networks, various relay selection protocols and key fundamental relaying schemes including amplify-and-forward (AF) and decode-and-forward (DF) are investigated. In recent applications of Internet of things (IoT) or wireless sensor networks (WSNs), the source and relay are usually energy-constrained nodes which result in limitation of operation time and the network performance. To overcome the challenge of replacing or recharging batteries in such wireless nodes, energy harvesting has been proposed architecture in relaying networks to prolonging the lifetime of these mobile nodes. Motivated by the recent benefits of self-interference elimination procedures of possible full-duplex (FD) transceivers. As an important model, this chapter further explores the reliable cooperative networks to enhance system performance. In such

D.-T. Do (✉)
School of Engineering, Eastern International University, Thu Dau Mot, Vietnam
e-mail: thuan.do@eiu.edu.vn

© Springer Nature Singapore Pte Ltd. 2018
K. V. Arya et al. (eds.), *Emerging Wireless Communication and Network Technologies*,
https://doi.org/10.1007/978-981-13-0396-8_5

networks, FD entities and energy harvesting-assisted relay node can be examined to confirm advantages of FD transmission architecture. Furthermore, with the aim to deal with high traffic volume and optimize spectral efficiency, two-way relaying scheme is widely considered. In particular, this chapter presents system model of relaying networks to improve the spectral efficiency of the network, and the outage performance for the network is analyzed. Finally, due to the broadcast nature of wireless transmission, secure communication is compulsorily required in emerging network design, especially in scenarios of increasing number of the unintended receivers which is very challenging mission. Through this secure model, the physical layer without cryptography is investigated to examine positive secrecy capacity. In particular, the secrecy performance of an energy harvesting relay system is studied, in which a legitimate source transfers data to a legitimate destination via the support of unreliable relays.

1 Relaying Protocol for Collaborative Transmission in Cooperative Networks

1.1 The Fundamental of Relay Protocols

Many cooperative wireless networks or emerging wireless networks have been introduced in the past years because of their great potentials for wireless transmission with specific benefits such as improving the throughput in Wi-Fi access point, improved quality of coverage area in cellular networks. In terms of applicability, such cooperative networks technology can be applied in future mobile broadband communications networks such as 3GPP LTE-Advanced (version 11), IEEE 802.16j, and IEEE 802.16m. The cooperative wireless networks can be mainly classified as:

- Amplify-and-Forward (AF), in which the intermediate (relay) node forwards the signals without decoding the signal.
- Decode-and-Forward (DF), which allows the signal to be decoded and re-encoded the information before forwarding the signal to the destination node.

Future cellular architectures will benefit from the deployment of small coverage networks (including micro-, pico-, and femto-cells) to facilitate the increasing demand for high-speed transmission [1, 2]. Thus, the deployment of small coverage network is based on serving the user using an intermediate base station acting as relay to adapt higher signal quality in small area coverage. It is possible to design relaying networks with multiple intermediate nodes to improve network performance. The benefits of collaborative diversity in a wireless network are the reduced efficiency of using spectrum when the source nodes and the relay nodes are transmitted in the orthogonal channel. However, the efficiency of the use of transmission channel resources can be improved by the relay selection solution. A fundamental model for solving this problem is the best relay node selection protocol. In this model, a single best relay is selected to send the signal to the destination. Typically, the principle of selecting the relay node is by selecting the node with best signal quality between the source and destination nodes (i.e., the best signal-to-noise ratio). However, in some applications such as ad hoc networks and sensor networks, monitoring of the connections of all transmission routes affects the delay time of information transmission in the networks. In order to limit such challenges, it has developed alternate partial node selection options, which require the partial channel state information (CSI) of only in the source-relay transmission link or relay-destination transmission link.

We first introduce signal transmission architecture using amplify-and-forwarding (AF). Each relay node in this method receives a version of the received signal that has been transmitted by the source node to it and then amplifies the received signal and forwards it to the destination node as described in Fig. 1, in which the destination node will combine the information sent by the multiple relay nodes, and it will make the detector to extract the transmitted signal. More specifically, we denote h_S, h_D are the channel coefficients transmitted between the source node and the relay (i.e., S-R link) and the relay node and the destination node (i.e., R-D link) following the flat Rayleigh flat channel model, d_1, d_2 are denoted as distance of S-R link and R-D link, respectively. Although the noise of the signal is also amplified by the cooperative mode (AF), the destination node receives only the version of the original signal being fading independently, and it can make better decisions about detecting the information. It is assumed that, in AF method, the destination node knows the channel coefficients due to assuming perfect channel estimation procedure between the source node and the relay node to perform optimal coding, so the channel information exchange mechanism must be incorporated into feedback signal transmission implementation. Another big challenge is that decode, and hence decode-and-forward (DF) protocol is considered as high complexity requirement in system implementation. However, the method of AF scheme is a simple method that we use it to analyze cooperative networks. In the fixed gain AF relay protocol, which is often referred to simply as the AF protocol, the relay node receives weak signals and transmits an amplified version of it to the destination node. The signal transmitted from the source is captured at the relay node as follows:

$$y_R = \frac{1}{\sqrt{d_1^m}} \sqrt{P_s} h_S x_S + n_R, \tag{1}$$

where x_S is the signal transmitted between the source and the intermediate node (relay node), m is path loss factor, the noise term denoted as $n_R \sim CN\left(0, \sigma^2\right)$, and noise is assumed as the white Gaussian noise at the relay node with zero mean and variance of σ^2. In this protocol, the relay node amplifies the signal from the source and forwards it to the destination to balance the effect of the source node and the relay node. The relay node is implemented by amplifying the received signal by a factor of inversely proportional to the transmitted power, denoted by G as below [1–4]

Fig. 1 System model of
cooperative network

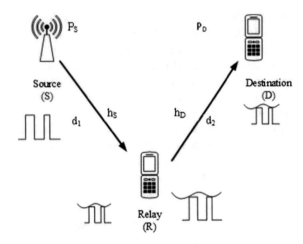

$$G = \frac{\sqrt{P_R}}{\sqrt{\frac{P_S|h_S|^2}{d_1^m} + \sigma^2}}, \tag{2}$$

in which we denote P_S, P_R are transmit power at the source, and the relay node.
The signal transmitted from such a relay node is sent to the destination node of
the form $G y_R$ and transmitted power at the relay node is P_R. In the second phase,
the relay node amplifies the received signal and forwards it to the destination with
transmitted power at the relay node. The signal received at the destination node in
(1) is represented by

$$y_D = \frac{\sqrt{P_R} h_D G}{\sqrt{d_2^m}} y_R + n_D, \tag{3}$$

in which $n_D \sim CN\left(0, \sigma^2\right)$ denoted as the white Gaussian noise at the destination
node with zero mean and variance σ^2. In this chapter, it is worth noting that we
assume noise at relay and destination node is the same value for simple analysis. In
this phase, the received signal is given by

$$y_D = \frac{\sqrt{P_R P_S} h_S h_D x_S}{\sqrt{P_S|h_S|^2 d_2^m + d_1^m d_2^m \sigma^2}} + n_D. \tag{4}$$

1.2 Calculation of End-to-End Signal-to-Noise Ratio (SNR) for Relaying Networks and System Performance Evaluation

It is assumed that we determine the signal part and noise part in the received signal, and then we compute the SNR in each hop as below

$$SNR_D = \frac{E\{|Signalpart|^2\}}{E\{|Noisepart|^2\}}, \tag{5}$$

where $E(.)$ denotes as expectation function.

It is assumed that it can be computed the SNR at the destination. As a result, we can determine the instantaneous transmission rate with the obtained SNR_D for the end-to-end SNR in such system as below

$$R_{AF} = \frac{1}{2}\log_2\left(1 + SNR_D\right). \tag{6}$$

To determine the transmission rate in the AF scheme, we must find the outage probability by computation of instantaneous rate below the fixed rate of the source information based on exponential channel distributions. The general outage probability equation is defined as follows:

$$P_{out} = Pr(R_{AF} < R_0) = Pr\left(SNR_D < SNR_0\right), \tag{7}$$

where $Pr(.)$ denotes probability function, SNR_0 is the threshold SNR.

In this case, SNR_0 can be computed by $SNR_0 = 2^{2R_0} - 1$.

1.3 Considerations on DF Scheme in Relaying Networks

Another relaying processing technique is the decode-and-forward (DF), where the relay node decodes the received signal, re-encodes it, and then transmits it to the receiver. This type of architecture is called fixed DF, or it is often referred to simply as DF. Note that the decoded signal at the forwarding node may be incorrect. It is worth noting that in case of signal is decoded incorrectly at the relay node that is forwarded to the destination node, then the decoding at the destination is meaningless. It can be confirmed that in this mechanism, the diversity obtained is only one, because the performance of the system is limited by the worst link from the source-relay link and relay-destination link.

Although DF relaying has advantages over AF relays in reducing the effects of additional interference at the relaying node, it requires the ability to forward detected free-error signals to the destination, such errors can result in reduced system performance. To determine the throughput in the DF paradigm, we must determine the ergodic capacity between the weakest link of source-relay pair and relay-destination pair. More specifically, ergodic capacity for transmitting and decoding related to the channel coefficients can be given by

$$C_{DF} = \min \left(E_{h_S} \left\{ \log_2 \left(1 + SNR_R \right) \right\}, E_{h_S, h_D} \left\{ \log_2 \left(1 + SNR_D \right) \right\} \right). \tag{8}$$

2 Simultaneous Wireless Information and Power Transfer-Aware Relay

2.1 Policies for Wireless Power Transfer

The deployment of huge amounts of sensor nodes and a large number of Internet of things applications will lead to huge information exchanges that consume a large amount of energy in order to maintain the system's performance. Wireless powered technique is a promising energy solution for dense and heterogeneous networks in the future with great impact in a wide variety of applications. It refers to the communications networks at which specific nodes provide energy to sustain their activities by collecting the received electromagnetic radiation. The basic block in implementing this technology is rectena, and this is a diode-based circuit that converts RF radio signals into DC. The first important architecture is wireless power transmission (WPT) in which a designated RF transmitter (which is a dedicated radio energy transmitter) transmits radio power to the equipment [5, 6]. Mobile users or sensors in such WPT-based network that consumes less power. In contrast to the energy collection of the surrounding environment, the WPT can be constantly and completely controlled, and therefore, it attracts applications with strict quality-of-service. The second network architecture is a wireless powered cooperative network (WPCN) where a dedicated RF transmitter disperses power on the downlink and WPT-based devices transmit information on the uplink. The third basic architecture is simultaneous wireless information and power transfer (SWIPT) information, and RF transmitters simultaneously transmit data at downlink devices. Because of the practical constraints, SWIPT cannot be made from the same signals without attenuation, and the actual operation divides the received signal into two parts: one being used for transmitting information and the rest is used for power transmission. The energy harvesting dividing factor can be specified as time domain with time switching (TS), and power domain with power splitting (PS) [1]. A special case of SWIPT with appropriate characteristics for SWIPT-based cooperative networks with energy/information relaying paradigm. In this network structure, a battery-free relay node extracts both information and energy from the source signal and then uses the energy collected to forward the source signal to its destination.

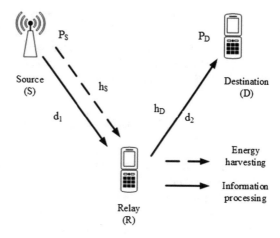

Fig. 2 Energy harvesting-aware relaying network

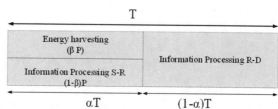

Fig. 3 The TPSR energy harvesting protocol

2.2 Energy Harvesting in Relaying Network

In the energy harvesting network, the relay node collects energy from all surrounding signals and supports communication between the source node and the destination node. In such model, we assume that a reference system model is depicted in Fig. 2, in which the energy collected from the source node is stored in the battery of the relay node and the relay node using the collected energy to forward the information. To evaluate the performance of the system, we will analyze the throughput achieved based on the proposed time power splitting relay (TPSR) protocol [3] as illustrated in Fig. 3.

According to the TPSR protocol, we will design two epochs: one called the energy storage (ES) stage and the information processing stage (IS). To perform wireless power transfer in the ES, the power of the received signal can be divided into two parts, one for the receiver collecting energy while the other for the receiver perform processing information in the TPSR protocol. We denote α, β are time and power allocation in the proposed TPSR protocol for energy harvesting function, respectively. Considering the receiver architecture that allows energy to be collected, the obtained signal at the input of the energy-gathering receiver is given by

$$y_R = \frac{1}{\sqrt{d_1^m}}\sqrt{\beta P_S}h_S x_S + \sqrt{\beta}n_R. \tag{9}$$

The energy harvesting block in the receiver at the relay node converts the signal into a DC current signal by a rectifier, which consists of a Schottky diode and a low pass filter (LBF). Then, this DC current signal will be used to charge the battery and use support for transmitting information to the destination node. We have the energy collected in the TPSR protocol given by

$$E_h^{TPSR} = \eta E\{i_{DC}\}\alpha\beta T = \eta\left(\frac{P_S|h_S|^2}{d_1^m}\right)\alpha\beta T, \tag{10}$$

where $0 < \eta \le 1$ is the energy harvesting efficiency fractions and it depends on the rectifier and its circuitry.

In the TPSR protocol, considering IS stage, the energy from the previous energy (ES) is fed to the IS. In this period, we assume that the relay node only receives support for transmitting information from the source node and then forwards it to the destination node. Thus, the received signal at the information receiver of the relay in the TPSR protocol is given by

$$y_R = \frac{1}{\sqrt{d_1^m}}\sqrt{(1-\beta)P_S}h_S x_S + \sqrt{(1-\beta)}n_R^A + n_R^C, \tag{11}$$

where $n_R^A \sim CN\left(0, \sigma^2\right)$ defined as white Gaussian noise (AWGN) at the relay node and n_R^C as the RF signal is converted from the passband to the baseband signal. With the AF relaying protocol, the relay node will amplify the information received by a coefficient G given by

$$G = \frac{1}{\sqrt{(1-\beta)\frac{P_S|h_S|^2}{d_1^m} + (1-\beta)\sigma^2 + \sigma^2}}. \tag{12}$$

After processing the received signal at the relay node, the relay node will amplify the received information from the source node and forward it to the destination node with transmitted power P_R, depending on the energy obtained at the ES stage. The signal obtained at the destination node y_D is given by

$$y_D = \frac{\sqrt{P_R}h_D G}{\sqrt{d_2^m}}y_R + n_D^A + n_D^C, \tag{13}$$

where $n_D^A \sim CN\left(0, \sigma^2\right)$ and $n_D^C \sim CN\left(0, \sigma^2\right)$ are denoted as Gaussian white noise (AWGN) at the destination node is obtained by the antenna and the RF signal is converted from the passband signal to the baseband, respectively. After some substituting steps, it can be obtained as

$$y_D = \frac{\sqrt{(1-\beta)P_R P_S h_S h_D x_S}}{\sqrt{(1-\beta)P_S|h_S|^2 d_2^m + d_1^m \sigma^2}}$$

$$+ \frac{\sqrt{P_R d_1^m h_D n_R}}{\sqrt{(1-\beta)P_S|h_S|^2 d_2^m + d_1^m d_2^m \sigma^2}} + n_D,$$

(14)

Here, we denote $n_{nR} \triangleq \sqrt{(1-\beta)}n_R^A + n_R^C$ and $n_D \triangleq n_D^A + n_D^C$ is the total of the AWGN noise and the converting noise in the relay node and destination node. Thus, the total variance of noise terms $\sigma_{nR}^2 \triangleq (1-\beta)\sigma^2 + \sigma^2$ at the relay node in the TPSR protocol scheme. On the other hand, the relay node transmits the amplified energy obtained during the ES energy-gathering period as a power source in the time portion of the information transfer to the destination $(1-\alpha)T$, the transmit power at the relay can be expressed by

$$P_R = \frac{E_h^{TPSR}}{(1-\alpha)T} = \eta \left(\frac{P_S|h_S|^2}{d_1^m} \right) \frac{\alpha\beta}{(1-\alpha)}.$$

(15)

By replacing P_R in (15) into the received signal at destination, the received signal at destination y_D can be formulated as

$$y_D = \underbrace{\frac{\sqrt{\eta \left(P_S|h_S|^2 d_2^m \right) \alpha\beta (1-\beta) P_S h_S h_D x_S}}{\sqrt{d_1^m d_2^m (1-\alpha)} \sqrt{(1-\beta)P_S|h_S|^2 d_2^m + d_1^m \sigma_{nR}^2}}}_{Signal part}$$

$$+ \underbrace{\frac{\sqrt{\eta \left(P_S|h_S|^2 d_2^m \right) \alpha\beta d_1^m h_D n_R}}{\sqrt{d_2^m (1-\alpha)} \sqrt{(1-\beta)P_S|h_S|^2 d_2^m + d_1^m \sigma_{nR}^2}} + n_D}_{Noise part}$$

(16)

We denote SNR_D as the end-to-end SNR at the destination node and it can be illustrated

$$SNR_D = \frac{\eta\alpha\beta(1-\beta)P_S|h_S|^2|h_D|^2}{\eta\alpha\beta d_1^m|h_D|^2\sigma_{nR}^2 + (1-\alpha)(1-\beta)d_1^m d_2^m \sigma_{nD}^2 + \frac{(1-\alpha)d_1^{2m}\sigma_{nR}^2\sigma_{nD}^2}{P_S|h_S|^2}}.$$

(17)

2.3   Investigation on Energy Harvesting System Performance

In this subsection, it can be shown that the outage probability of the system as P_{out}^{TPSR} given by

$$P_{out}^{TPSR} = Pr(SNR_D < SNR_0).\tag{18}$$

It is noted that SNR_0 can be computed by $SNR_0 = 2^{2R_0} - 1$ which corresponding the fixed rate R_0. In particular, in the case of high SNR, we can take the following approximation, because we can ignore the very small components, which means $\frac{(1-\alpha)d_1^{2m}\sigma_{nR}^2\sigma_{nD}^2}{P_S|h_S|^2} \approx 0$. Therefore, the approximate outage probability can be determined as follows:

$$P_{out}^{TPSR} \approx Pr\left(|h_D|^2 < \frac{(1-\alpha)(1-\beta)d_1^m d_2^m \sigma_{nD}^2 SNR_0}{\eta\alpha\beta(1-\beta)P_S|h_S|^2 - \eta\alpha\beta d_1^m \sigma_{nR}^2 SNR_0}\right).\tag{19}$$

It is noted that σ_{nD}^2 is the variance of the total noise terms at the destination node. From the formula (19), we will have the following result.

We first denote several parameters as $\omega = \eta\alpha\beta d_1^m \sigma_{nR}^2 SNR_0$, $\theta = (1-\alpha)$ $(1-\beta)d_1^m d_2^m \sigma_{nD}^2 SNR_0$, $\psi = \eta\alpha\beta(1-\beta)P_S$. The outage probability for TPSR protocol can be expressed as follows:

$$P_{out}^{TPSR} = 1 - \exp\left(-\frac{\omega}{\psi\lambda_{h_S}}\right)\sqrt{\frac{4\theta}{\psi\gamma_{h_S}\gamma_{h_D}}}K_1\left(\sqrt{\frac{4\theta}{\psi\gamma_{h_S}\gamma_{h_D}}}\right),\tag{20}$$

where $K_n(\cdot)$ denotes as a Bessel function with degree of n.

3   Full-Duplex Cooperative Networks

In this section, we present architecture of full-duplex (FD) communication for improving the spectrum efficiency, as in [7–9] the authors study FD communications to maximize the sum rate under controlling residual self-interference (SI). It can be confirmed that FD relay can bring advantages compared with half-duplex (HD) relay. But it can be degraded impacts of residual SI. The successful design of SI cancellation circuit leads to FD mode with higher spectrum efficiency usage and overall capacity of system. Combining wireless energy harvesting and cooperative network, we introduce new system model FD SWIPT (Fig. 5).

3.1 An Architecture of Full-Duplex Energy Harvesting Network

In this model, Fig. 4 shows a wireless dual-hop two-way full-duplex relaying network with DF protocol system is investigated. In such FD system model, we call A and B as two source nodes, while one intermediate relay node denoted by R. Especially, to adapt to FD scheme, each terminal is equipped with two antennas: the first one is used for signal transmission and the second one is used for signal reception, it can be operated in FD mode. It is assumed that A and B are fixed power resource (i.e., gird power) while the intermediate relay is required as free-battery device thanks to harvesting wireless energy, and hence such energy is used to forward received information toward the destination node. This system model also performs time switching-based relay (TS) protocol as studied in [9, 10] to operating wireless power transmission. This investigation presented this protocol for full-duplex bidirectional communication. Similar to the previous section, as illustrated in Fig. 3, T stands for block time of frame transmission, and $0 < \alpha < 1$ is time allocation for energy transfer (Fig. 5).

In the two-way circumstance, the wireless channels are assumed flat Rayleigh fading which including h_a, g_a are links corresponding for hop from source A to R and the second hop from R to A, while h_b, g_b are the two channel coefficients from

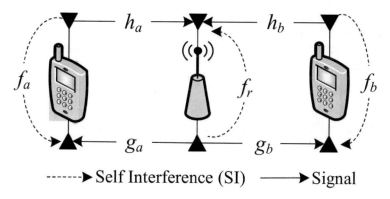

Fig. 4 System model of full-duplex energy harvesting network

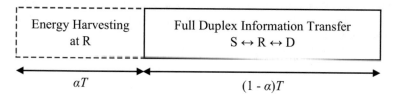

Fig. 5 Signal structure of energy harvesting protocol of FD SWIPT

source B to R and from R to B, respectively. We denote f_j, $j = a, b, r$ are the SI channel can be calculated at sources A, B, and relay R, respectively.

In FD architecture, the channel gains can be called by $|h_a|^2$, $|h_b|^2$, $|g_a|^2$, $|g_b|^2$, $|f_a|^2$, $|f_b|^2$, $|f_r|^2$, respectively. It is noted that such parameters are the exponentially distributed random variables (RVs) with variance $\lambda_{h_a} = d_1^{-m}$, $\lambda_{g_a} = d_1^{-m}$, $\lambda_{h_b} = d_2^{-m}$, $\lambda_{g_b} = d_2^{-m}$ and λ_{f_a}, λ_{f_b}, λ_{f_r}. In this model, for simplicity, it is assumed that d_1 as same as d_2, i.e., $d_1 = d_2 = d$. It is noted that $\lambda_{h_a} = \lambda_{h_b} = d^{-m} \triangleq \lambda_h$, $\lambda_{g_a} = \lambda_{g_b} = d^{-m} \triangleq \lambda_g$.

3.2 Full-Duplex System Analysis

In this section, we analyze the outage probability of concerned system in condition of Rayleigh fading channels. In energy harvesting phase, source A and source B transfer power signal to the intermediate node R thanks to collecting RF signals.

As a result, the received signal at R can be expressed as

$$y_{EH} = \sqrt{P_S} h_a x_a(t) + \sqrt{P_S} h_b x_b(t) + n_r(t), \tag{21}$$

where P_S is transmit power at source node, and x_a, x_b are the transmitted symbol from the source A and B, respectively. That with zero mean $E\{x_a(t)\} = 0$, $E\{x_b(t)\} = 0$ and unit power $E\{|x_a(t)|^2\} = 1$, $E\{|x_b(t)|^2\} = 1$, $n_r(t)$ is the additive white Gaussian noise (AWGN) at R with zero mean and variance σ^2.

Based on received RF signal in TS protocol, the harvested energy can be shown as [10, 11]

$$E_{EH} = \eta P_S \left(|h_a|^2 + |h_b|^2\right) \alpha T. \tag{22}$$

Following the principle of TS protocol, the R communicates in $(1 - \alpha)T$, and hence, the transmitted power from the R node, P_R is expressed by

$$P_R = \frac{E_{EH}}{(1 - \alpha)T} = \mu P_S \left(|h_a|^2 + |h_b|^2\right), \tag{23}$$

where $\mu \triangleq \eta \alpha / (1 - \alpha)$.

In the information processing phase, the node, namely, A and B forward signal to R. At the same time, the relay R transmit decoded signal in the previous epoch to A and B source node.

Similarly, the received signal at R is given by

$$y_r(t) = \sqrt{P_S} h_a x_a(t) + \sqrt{P_S} h_b x_b(t) + f_r x_r(t) + n_r(t), \tag{24}$$

where x_r is the transmitted symbol from relay with zero mean $E\{x_r(t)\} = 0$ and unit power $E\{|x_r(t)|^2\} = 1$; $n_r(t) \sim (0, \sigma^2)$ is the AWGN at relay.

In (24), DF scheme shows that relay node try to decode the source symbols, x_a and x_b, then, the forwarded symbol at relay is given below

$$x_r(t) = \overline{x_a}(t) + \overline{x_b}(t),$$ (25)

where $\overline{x_a}(t)$ and $\overline{x_b}(t)$ is decoded signal of the previous period at R with zero mean and unit energy, i.e., $E\{\overline{x_s}(t)\} = E\{\overline{x_d}(t)\} = 0$, $E\{|\overline{x_s}(t)|^2\} = 1$, $E\{|\overline{x_d}(t)|^2\} = 1$. In the two-way scenario, the received signal at two source nodes is expressed, respectively, as

$$y_a(t) = \sqrt{P_R} g_a \overline{x_a}(t) + \sqrt{P_R} g_a \overline{x_b}(t) + f_a \sqrt{P_S} x_a(t) + n_a(t),$$ (26)

$$y_b(t) = \sqrt{P_R} g_b \overline{x_b}(t) + \sqrt{P_R} g_b \overline{x_a}(t) + f_b \sqrt{P_S} x_b(t) + n_b(t),$$ (27)

where $n_a(t)$ and $n_b(t)$ are the AWGN at A and B, respectively, with zero mean and variance σ^2. We first compute instantaneous SINR at relay node, $\gamma_{a,b \to r}$, as below

$$\gamma_{a,b \to r} = \frac{\gamma_s \left(|h_a|^2 + |h_b|^2\right)}{\mu \gamma_s \left(|h_a|^2 + |h_b|^2\right) |f_r|^2 + 1} \approx \frac{1}{\mu |f_r|^2}.$$ (28)

Next, we derive the instantaneous SINRs of $A \to R \to B$ hop. The instantaneous SINRs from $A \to R$ hop and $R \to B$ hop, denoted $\gamma_{a \to r}$ and $\gamma_{r \to b}$, respectively, can be expressed as

$$\gamma_{a \to r} = \frac{\gamma_s |h_a|^2}{\mu \gamma_s \left(|h_a|^2 + |h_b|^2\right) |f_r|^2 + 1},$$ (29)

and

$$\gamma_{r \to b} = \frac{\mu \gamma_s \left(|h_a|^2 + |h_b|^2\right) |g_b|^2}{\gamma_s |f_b|^2 + 1},$$ (30)

where $\gamma_s = P_S / \sigma^2$. Similarly, the instantaneous SINR of $B \to R \to A$ hop is obtained, where the instantaneous SINRs of $B \to R$ hop and $R \to A$ hop, respectively, denoted $\gamma_{b \to r}$ and $\gamma_{r \to a}$, can be given as

$$\gamma_{b \to r} = \frac{\gamma_s |h_b|^2}{\mu \gamma_s \left(|h_a|^2 + |h_b|^2\right) |f_r|^2 + 1},$$ (31)

and

$$\gamma_{r \to a} = \frac{\mu \gamma_s \left(|h_a|^2 + |h_b|^2\right) |g_a|^2}{\gamma_s |f_a|^2 + 1}.$$ (32)

3.3 System Performance Analysis

In this subsection, we analyze the cumulative distribution function (CDF) of end-to-end SINR at each source node. Before further analyzing system performance, the CDF can be explored in the closed-form expression which is helpful for the later evaluations.

We first defined the function of complicated integral related to Bessel function as below

$$
\psi(\lambda, \kappa, \varepsilon) \overset{\Delta}{=} \int_0^\infty (\kappa x + \varepsilon) \times K_2 \left(2\sqrt{\kappa x + \varepsilon}\right) \exp(-\lambda x)\, dx
$$

$$
= \exp\left(\frac{\varepsilon\lambda}{\kappa}\right) \left\{ \frac{\kappa^2}{\lambda^3} \exp\left(-\frac{\kappa}{\lambda}\right) \Gamma\left(-2, \frac{\kappa}{\lambda}\right) \right. \tag{33}
$$

$$
\left. - \frac{1}{2} \sum_{m=0}^\infty \frac{(-\lambda)^m \varepsilon^{m+1}}{m! \kappa^{m+1}} G_{1,3}^{2,1}\left(\varepsilon \left| \begin{matrix} 2-m \\ 2, 0, 1-m \end{matrix} \right. \right) \right\}
$$

where $\Gamma(.,.)$ is the incomplete gamma function [12, Eq. (8.350.2)]. $G_{m,n}^{p,q}(x)$ is the Meijer G-function [12, Eq. (9.301)].

The tight lower bounded CDF of γ_i can be computed as [9]

$$
F_{\gamma_i}(t) = 1 - \frac{1}{\lambda_{f_i}} \left[1 - \mu\lambda_{f_r} t \left(1 - \exp\left(\frac{-1}{\mu\lambda_{f_r} t}\right) \right) \right] \times \psi\left(\frac{1}{\lambda_{f_i}}, \frac{t}{\mu\lambda_g\lambda_h}, \frac{t}{\mu\gamma_s\lambda_g\lambda_h} \right), \tag{34}
$$

where $\psi(a, b, c)$ as previous concerned expression.

4 Secure Communication Between Source and Destination via Untrusted Node

This section presents a system model in which two source nodes perform to send confidential signal. It is shown that RF energy harvesting can be powered for untrusted relay. Based on the principle of the cooperative network in expanding the information coverage, the relay in this concerned model is untrusted node as it might perform to decode the confidential information. However, to prevent the impact of the eavesdroppers on system performance, friendly jammer is designed. Under the total power constraint, the system need be calculated the optimal power splitting ratio to satisfy two roles of the energy harvesting and the information processing at the relay. We further examine the secure relaying performance in maximum energy harvesting protocol.

Fig. 6 System model for secure dual-hop relaying communication

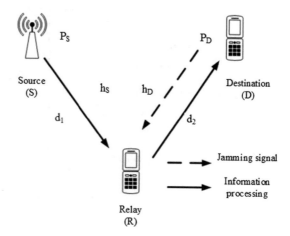

4.1 Secure Relaying Communication Model

This system model concerns an AF dual-hop cooperative network including of a source node (S) which transmit signal toward the destination node (D) via an untrusted EH relay (R), as illustrated in Fig. 6. In normal relaying system, the relay node performs information transmission cooperation, but in untrusted relaying network, S and D remain the information secret from R. To help S obtain confidential information from R, one require D sends a jamming signal to R. It is noted that in the same time S communicates with R. Different with previous FD model, each node in this model is equipped one antenna or deploying half-duplex mode. The direct link between S and D is not available. It is assumed that the direct link is blocked due to long-distance transmission or dense obstacles and hence relaying solution is selected. All communication links are assumed that independent and identical Rayleigh fading.

4.2 Maximum Energy Harvesting (MEH) Protocol

In the concerned MEH protocol [13], R collects energy from either the information signal of S or the jamming signal of D depends on which link can provide stronger energy signal. In particular, it is required that R compares the quality of both S-R link and R-D link to harvest energy. Therefore, MEH protocol obtains advantage from collecting the jamming signal from D (Fig. 7).

The energy harvested at R in this MEH scheme is expressed by

$$E = \eta\beta\max\left(P_S|h_S|^2, P_D|h_D|^2\right)(T/2). \tag{35}$$

We denote $P_S = P_k$, in which

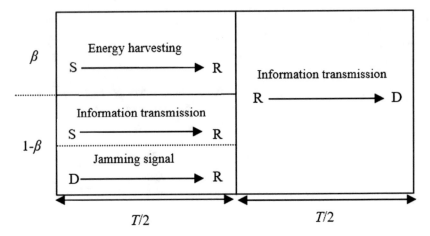

Fig. 7 Secure wirelessly powered system architecture with PS policy

$$k = \begin{cases} S, |h_S|^2 \geq |h_D|^2 \\ D, |h_S|^2 < |h_D|^2 \end{cases}. \tag{36}$$

Thus, the transmit power at the relay in the second phase is shown as

$$P_R = \eta \beta P_k \max\left(|h_S|^2, |h_D|^2\right). \tag{37}$$

In such model, the end-to-end SNR at destination is expressed by

$$SNR_D = \frac{\eta\beta\,(1-\beta)\,P_S P_k \max\left(|h_S|^2, |h_D|^2\right)|h_S|^2|h_D|^2}{\eta\beta\,(1-\beta)\,P_k|h_D|^2\sigma^2 \max\left(|h_S|^2, |h_D|^2\right) + \sigma^2\left[(1-\beta)\left(P_S|h_S|^2 + P_D|h_D|^2\right) + \sigma^2\right]}. \tag{38}$$

The strictly positive secrecy capacity (SPSC) for PS policy using MEH can be expressed as [13]

$$\Pr\left(R_{\text{sec}} > 0\right) = \begin{cases} \exp\left(-\frac{1}{\lambda_D}\frac{1}{\sqrt{\eta\beta\gamma}}\right) + \int_{\frac{1}{\sqrt{\eta\beta\gamma}}}^{\sqrt{\frac{2}{\eta\beta\gamma}}} \frac{1}{\lambda_D}\exp\left(-\frac{z}{(\eta\beta\gamma z^2 - 1)\lambda_S} - \frac{z}{\lambda_D}\right)dz - \\ \qquad -\int_{\frac{1}{\sqrt{\eta\beta\gamma}}}^{\sqrt{\frac{2}{\eta\beta\gamma}}} \frac{1}{\lambda_D}\exp\left(-\frac{\eta\beta\gamma z^3 - z}{\lambda_S} - \frac{z}{\lambda_D}\right)dz, \qquad\qquad\qquad \lambda_S \neq \lambda_D \\[4pt] \exp\left(-\frac{1}{\lambda}\frac{1}{\sqrt{\eta\beta\gamma}}\right) - \frac{1}{\lambda}\left(\frac{\Phi\left(\frac{1}{3}, \frac{\eta\beta\gamma}{\lambda}\left(\frac{2}{\eta\beta\gamma}\right)^{\frac{3}{2}}\right)}{3\left(\frac{\eta\beta\gamma}{\lambda}\right)^{\frac{1}{3}}} - \frac{\Phi\left(\frac{1}{3}, \frac{\eta\beta\gamma}{\lambda}\left(\frac{1}{\eta\beta\gamma}\right)^{\frac{3}{2}}\right)}{3\left(\frac{\eta\beta\gamma}{\lambda}\right)^{\frac{1}{3}}}\right), \qquad \lambda_S = \lambda_D \end{cases}. \tag{39}$$

where Φ is the incomplete gamma function.

5 MIMO Relaying Model and Challenges in Relaying Design

5.1 Wireless Powered Communication with Multi-antenna System

In this system model, hybrid access point (H-AP) with multiple antennas in scenario of multi-input multi-output (MISO) transmission mode is deployed to transmit signal to users with single antenna (U node) (as demonstrated in Fig. 8). The H-AP is required to energy transfer via N of transmit antennas. Nevertheless, relay can be operated under feed of wireless power. Such transmission conducts in three time slots including the uplink channel reserved for channel estimation, the downlink assigned for energy harvesting phase and the uplink wireless information processing phase. In principle, the H-AP uses a small part of the signal in the downlink to transmit wireless energy to the user by deploying energy beamforming. Interestingly, the harvested energy is reused to transmit data to the H-AP in the uplink while the downlink for energy transfer [14]. It is assumed that channel state information (CSI) available at H-AP, this model requires the maximum ratio transmission (MRT) with beamforming vector to be obtained optimal communication on condition as $\|\mathbf{w}_i\| = 1$

$$\mathbf{w}_i = \frac{\mathbf{h}_i}{\|\mathbf{h}_i\|}, \tag{40}$$

where $\|.\|$ stands for the Euclidean norm of a matrix. The received signal at U node in the downlink is thus expressed by

$$y_{U,i} = \frac{\|\mathbf{h}_i \mathbf{w}_i\|}{\sqrt{d^m}} \sqrt{P_{AP}} x_i + \mathbf{n}_{U,i}, \tag{41}$$

where x_i denotes as the symbol transmitted from the source at ith time index, $\mathbf{h}_i$ is denoted as the x_i channel vector for the downlink, $\mathbf{n}_{U,i}$ is the additive white Gaussian noise (AWGN) with zero mean and variance matrix of $\sigma_U^2 \mathbf{I}_N$, d stands for the distance between nodes and m is the path loss exponent (Fig. 8).

It worth noting that energy harvesting introduces the amount of harvested energy at the user node, and it can be examined

$$\mathcal{E}_{U,i} = \eta \alpha_i T \frac{P_{AP} \|\mathbf{h}_i\|^2}{d^m}, \tag{42}$$

where $0 \leq \eta \leq 1$ is energy conversion efficiency and such fraction is decided by the rectifier and the energy harvesting circuitry, wireless power transfer is occurred in time α_i.

Fig. 8 System model for wireless power transfer with multi-antenna scheme

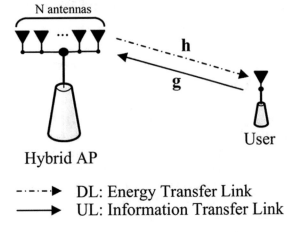

In uplink phase, the received signal can be calculated at AP node as

$$y_{AP,i} = \frac{\mathbf{g}_i}{\sqrt{d^m}}\sqrt{P_U}x_i + \mathbf{n}_{AP,i}, \tag{43}$$

where $\mathbf{g}_i$ denotes the channel vector for the uplink, $\mathbf{n}_{AP,i}$ is the AWGN at the AP node with variance matrix of $\sigma_{AP,i}^2\mathbf{I}_N$.

The system throughput performances in instantaneous are studied in this section.

In case of the maximum ratio combing (MRC) technique applied in the AP, it is noted that the received weight vector $\mathbf{w}_{R,i} = \mathbf{g}_i^T/\|\mathbf{g}_i\|$ (in which, $(.)^T$ is the transpose transformation of matrix).

Thus, the received signal at AP can be expressed as

$$y_{MRC,i} = \mathbf{w}_{R,i}y_{AP,i} = \sqrt{P_U d^{-m}}\|\mathbf{g}_i\|x_i + \mathbf{w}_{R,i}\mathbf{n}_{AP,i}. \tag{44}$$

Next, the achieved SNR in MRC scenario is determined by

$$\gamma_{MRC,i} = \frac{P_U\|\mathbf{g}_i\|^2}{d^m\sigma_{AP,i}^2}. \tag{45}$$

In case of selection combination (SC) scheme, the best instantaneous channel gain is selected to decode the emitted message. In this case, the received signal can be written as [14]

$$y_{SC,i} = \max\{y_{AP,i}\} = \sqrt{P_U d^{-m}}\max\{|\mathbf{g}_i|^2\}x_i + n_{AP,i}. \tag{46}$$

Thus, since AP node deploys the SC method, the SNR can be expressed by

$$\gamma_{SC,i} = \frac{P_U \max\left\{|\mathbf{g}_i|^2\right\}}{d^m \sigma_{AP,i}^2}. \tag{47}$$

For MRC scheme, the instantaneous throughput of MRC can be computed by [14]

$$C_{MRC} = \frac{\eta P_{AP}\|\mathbf{h}\|^2}{\eta P_{AP}\|\mathbf{h}\|^2 + P_U d^m}\log_2\left(1 + \frac{P_U\|\mathbf{g}\|^2}{d^m \sigma_{AP}^2}\right). \tag{48}$$

For SC scheme, the instantaneous throughput of SC is expressed as

$$C_{SC} = \frac{\eta P_{AP}\|\mathbf{h}\|^2}{\eta P_{AP}\|\mathbf{h}\|^2 + P_U d^m} \times \log_2\left(1 + \frac{P_U \max\left\{|\mathbf{g}_i|^2\right\}}{d^m \sigma_{AP}^2}\right). \tag{49}$$

5.2 Challenges in Design of Cooperative Networks

In real wireless cooperative communications, the design of relying mode can be met two main difficulties. The first one in MIMO scenario as in Sect. 5, perfect channel estimation cannot achieved, due to the limited length of training pilots, complexity of combining algorithm at the receivers and the time-varying nature of wireless channels. The channel estimation error naturally experiences serious system performance degradation resulted by the mismatch between the real channels and the filters. Therefore, a robust design is required to mitigate such performance loss, and hence, training pilot design is considered as essential part of the transceiver designs in MIMO cooperative model. The second one is that multi-hop relaying in 5G wireless networking. The quality of received signal can be low level at the last hop. Moreover, how can calculate the number of hop for keep system performance although multi-hop is foreseen to be a revolution in the way of connections with or without the direct intervention of the infrastructure. However, 5G networks will use multi hop relaying connection, i.e., besides the above traditional cellular topology, they will also apply the new paradigm of device-to-device (D2D) systems, in which users are allowed to communicate peer-to-peer through direct links that do not necessarily involve the control of the base station.

6 Conclusion

As a favorable method to lengthen the transmission range, the cooperative network is proposed to supply the higher performance of traditional wireless systems without required infrastructure design. This technique is applied in potential applications such as sensor networks, IoT, and cellular networks. Unfortunately, manually changing new batteries in various relay nodes in a traditional relaying network is not feasible, and energy harvesting is investigated as useful wireless charge model. Given a suggested outage threshold, the throughput performance, distance of each hop, and the number of relay are determined to show benefits of relaying networks to system designers in future 5G wireless networks.

References

1. Dinh-Thuan Do, "Power Switching Protocol for Two-way Relaying Network under Hardware Impairments," Radioengineering, Vol. 24 , No. 3, pp. 765–771, 2015.
2. Nguyen, H. S., Do, D. T., & Voznak, M., Two-way relaying networks in green communications for 5G: Optimal throughput and trade-off between relay distance on power splitting-based and time switching-based relaying SWIPT, AEU-International Journal of Electronics and Communications, Vol. 70, No. 12, pp. 1637–1644, 2016.
3. Dinh-Thuan Do, Time Power Switching based Relaying Protocol in Energy Harvesting Mobile Node: Optimal Throughput Analysis, Mobile Information Systems, Article ID 769286, 2015.
4. Dinh-Thuan Do, H-S Nguyen, "A Tractable Approach to Analyze the Energy-Aware Two-way Relaying Networks in Presence of Co-channel Interference", EURASIP Journal on Wireless Communications and Networking, 2016:271, 2016.
5. Liang, H., Zhong, C., Suraweera, H. A., Zheng, G., & Zhang, Z., "Optimization and Analysis of Wireless Powered Multi-antenna Cooperative Systems," IEEE Trans. on Wireless Communications, Vol. 16, No. 5, pp. 3267–3281, 2017.
6. N. T. Nguyen, Dinh-Thuan Do, P. T. Tran & M. Voznak, "Time Switching for Wireless Communications with Full-Duplex Relaying in Imperfect CSI Condition," KSII Transactions on Internet and Information Systems, vol. 10, no. 9, pp. 4223–4239, 2016.
7. K. T. Nguyen, Dinh-Thuan Do, X. X. Nguyen, N. T. Nguyen, D. H Ha , "Wireless information and power transfer for full duplex relaying networks: performance analysis," in Proc. of AETA: Recent Advances in Electrical Engineering and Related Sciences, pp. 53–62, 2015.
8. Xuan-Xinh Nguyen, Dinh-Thuan Do, "Maximum Harvested Energy Policy in Full-Duplex Relaying Networks with SWIPT," International Journal of Communication Systems (Wiley), 2017. https://doi.org/10.1002/dac.3359 (Online first).
9. X-X Nguyen, Dinh-Thuan Do, "Bidirectional Communication in Full Duplex Wireless-Powered Relaying Networks: Time-Switching Protocol and Performance Analysis," Wireless Personal Communications (Online first) 2017.
10. A. A. Nasir, X. Zhou, S. Durrani & R. Kennedy, Relaying Protocols for Wireless Energy Harvesting and Information Processing, IEEE Trans. on Wireless Commun., Vol. 12, No. 7, pp. 3622–3636, 2013.
11. Dinh-Thuan Do, "Energy-Aware Two-Way Relaying Networks under Imperfect Hardware: Optimal Throughput Design and Analysis", Telecommunication Systems (Springer), Vol. 62, No. 2, pp. 449–459, 2015.
12. I. S. Gradshteyn and I. M. Ryzhik, Table of Integrals, Series, and Products, 4th ed. Academic Press, Inc., 1980.
13. N. Q. Le, Dinh-Thuan Do, B. An, "Secure wireless powered relaying networks: energy harvesting policies and performance analysis," International Journal of Communication Systems (Wiley), 2017, https://doi.org/10.1002/dac.3369. (Online first)
14. X-X Nguyen, Dinh-Thuan Do, "An Adaptive-Harvest-Then-Transmit Protocol for Wireless Powered Communications: Multiple Antennas System and Performance Analysis," KSII Transactions on Internet and Information Systems, Vol. 11, No. 4, pp. 1889–1910, 2017.

Semantics for Delay-Tolerant Network (DTN)

Priyanka Rathee

Abstract Due to a huge increase in usage, the wireless networks face several kinds of link disruption based on the deployment and operating conditions. In such conditions, the DTNs are proved to be a reliable source of advancing the wireless traffic despite damaged nodes, hostile conditions and jamming. This chapter provides a glimpse of the delay-tolerant network. Initially, the chapter covers the introduction to DTN along with its applications and characteristics. The DTN architecture is given describing the system and the layered architecture. The main aim of DTN is to minimize the delay and maximize the delivery ratio. The routing and message dissemination are very important in DTN. Therefore, the types of routing and the methods of message dissemination are described. The security issues related to DTN are added in later section. The special category of VANET where the traffic is sparse and no direct end-to-end communication is available comes under vehicular delay-tolerant network. The chapter also consists of the vehicular environment in delay-tolerant networks and the protocols for vehicular delay-tolerant network (VDTN).

1 Introduction

Due to a huge increase in usage, the wireless networks face several kinds of link disruption based on the deployment and operating conditions. In such conditions, the DTNs are proved to be a reliable source of advancing the wireless traffic despite damaged nodes, hostile conditions and jamming. The delay-tolerant networking research group (DTNRG) was formed in 2002 in order to address the protocol design and architectural principles for the extreme environments. It was a part of internet research task force (IRTF). The DTNRG proposed the delay-tolerant network (DTN) architecture along with a communication protocol which is known as bundle protocol. In the case of traditional IP networks, the data can only be sent because it relies on end-to-end connectivity. On the other hand, the DTN keeps on transmitting data even

P. Rathee (✉)
Department of Computer Science, Shyama Prasad Mukherjee College,
University of Delhi, New Delhi, India
e-mail: rathee.priyanka124@gmail.com

© Springer Nature Singapore Pte Ltd. 2018
K. V. Arya et al. (eds.), *Emerging Wireless Communication and Network Technologies*,
https://doi.org/10.1007/978-981-13-0396-8_6

in the absence of an end-to-end connectivity and the identifiable path from source to destination. There are a number of networks where the TCP/IP model will not work directly. The DTN is the better substitute or modification for such network models. The DTN uses the irregular available links to communicate instantly. The examples of such networks include terrestrial mobile network, exotic media network, military applications and the sensors or actuator networks.

Terrestrial mobile network: The unexpected partitioning occurs in some of these kinds of networks because of the mobility of nodes or the change in signal strength. Some of the networks can be partitioned in a predictable and periodic manner.

Exotic media network: These types of networks consist of near-earth satellite communication, deep sea communication, very long distance radio communication, acoustic link in water and air, and free space optical communication. These types of systems suffer from predicted interruptions with high latencies, and outage because of the environmental conditions.

Military networks: The reason for network partitioning and disconnection in these networks may be the hostile environment that consists of node mobility, intentional signal jamming and the environmental factors. Because of the transmission of high priority voice data, other data may have to wait for long in the queue to transmit due to limited bandwidth. These networks are in need of special strong security mechanism.

Sensor or actuator networks: These networks have the features of limited memory, power and CPU capability. There exist thousands of nodes in the network. The communication within these networks needs to be scheduled in order to conserve the power. A group of nodes is addressed commonly again to ease the communication and to save battery life.

1.1 Significance of DTN

The delay-tolerant networks (DTNs) are completely different approaches as compared to typical connected wired or wireless networks. In DTNs, the end-to-end path availability is not required at any point of time to transfer data between the sender and the destination node pair. The DTNs came in the picture where the routes between a pair of nodes cannot be achieved, for example, sparse network scenario where end-to-end routes are not available, like military battlefields, there DTN provides the medium to place communication. It does not need any prior knowledge of the network in order to forward bundles from one node to another. It is based on store-carry-forward approach. At the time of natural hazards, the existing communication network fails and the contact to the affected areas is lost.

1.2   Features of DTN

The features of delay-tolerant networks are as follows:

1. Intermittent connection: The delay-tolerant network faces frequent disconnections because of the mobility in the network. The connection status and topology keeps on changing that is why there exists no guarantee to attain end-to-end communication path.
2. High delay: The direct end-to-end communication does not occur in delay-tolerant networks. The end-to-end delay can be calculated by summing up the total delay on the route caused by each hop. The delay includes the waiting, queuing and transmission time. The delay depends on the time period for which the connections are unreachable which results in a reduction of data rate and asymmetric characteristics in up down link.
3. Dynamic topology: The topology keeps on changing with the movement of the nodes from one location to another which results in the network partition and disconnections. The delay-tolerant networks deal with such partitioned and disconnected network in order to deliver the message successfully.
4. Heterogeneous interconnects: The DTN is an overlay network to transmit the asynchronous message. The bundle layer allows the delay-tolerant networks to run on heterogeneous networks.
5. Reliable transmission: The delay-tolerant networks use the store-carry-forward approach to forward the message toward the destination. It reduces the retransmissions and enhances the probability of successful delivery of a message to the destination.

1.3   Applications of DTN

The DTN is emerging as a vibrant paradigm for communication. The application of DTN includes packet switch networks, delay-tolerant event collection and social networks, etc. The nodes in DTN are mobile. Therefore, the network topology keeps on changing time to time with the movement of nodes. The major motivation behind DTN was the deep sea communication and interplanetary Internet. Apart from these two, the DTN has several applications like military application environment, acoustic networks, etc. A few of the applications of DTN are listed below:

1. **Deep sea network application**: The deep impact network (DINET) is an application of DTN in deep sea networking that is tested by NASA. It is a validation of an experiment for the interplanetary network. The NASA transmits the 200 space images having an approximate size of 14 MB to and from the space craft called as EPOXI—performs as a DTN router and situated at a long distance from the earth nearly 32 million kilometres away. Another deep sea application of DTN includes multipurpose end-to-end robotic operation network (METERON). It

focuses on simulating the specific scenarios like immersive remote control for the robot in orbit by the astronaut around the target object (like Moon and Mars).

2. **Wildlife tracking**: The DTN is used to monitor the wildlife. The 'zebranet' is very popular and effective communication system for delay-tolerant networks. It is used to keep track of the activities related to zebra. The system comprises of the nodes called 'tracking collars' to monitor zebras in grassland of Africa. The tracking collar comprises GPS, a wireless transceiver, flash memory and low power CPU. Another communication network like zebranet is 'SWIM'. It is used for underwater DTN communication that monitors the whales in the sea. This system comprises of the radio frequency device known as 'radio tag' and the SWIM station refers to the 'floating on water'. The radio tag consists of a wireless transceiver, flash memory and low power CPU.

3. **Village communication network**: The communication in remote villages is inconvenient and expensive because of non-availability of fixed infrastructure. At these places, the DTN can service a low-cost communication. The 'Darknet' is a DTN communication network widely used for villages. This network provides the asynchronous digital connectivity service using wireless technology. The Darknet was evolved by the researchers in MIT media lab and has been deployed successfully in the remote parts of India and Cambodia. It also comprises the Wi-Fi enabled kiosks, Internet access points and mobile access point vehicles.

4. **Health services**: Due to the bad medical facilities in some of the developing regions, the patient used to die suffering from the disease that may be cured if treated on time and in a proper manner. The solution of this problem is telemedicine, the expert doctor can guide the local doctors and the disease can be diagnosed by the expert remotely.

5. **Social-based mobile network**: It is an upcoming and emerging research area. The network offers social services like web-based social network service in the absence of the Internet. It also provides the social collaborations among nodes in order to offer the transmission data services. This network is called as a social-based network. The 'D-book' is an application for social-based mobile network that is used to create share and modify the profiles. These profiles comprise the basic user information, interests and contact information.

6. **Underwater or acoustic network application**: These networks are usually constructed by acoustically bound ocean sensors, the autonomous underwater vehicles and the surface stations that make available use of the links to the on shore control point. It is rapidly expanding network because of the benefits in disaster prevention, underwater robotics, under ocean tactical surveillance, harbour portal, monitoring of gas and oil pipelines, pollution monitor under sea, etc. In order to make these applications workable, there is a need of some kind of underwater communication network. The DTN is the well-suited communication network for the implementation of such networks.

7. **Smart phone application**: The DTN is also applicable for Android environments in smart phones in order to supply connectivity for the platforms which are deficient in network infrastructure. The execution for DTN protocol stack and the services for the android environment are called 'bytewalla' that permits the

use of android phones to transfer the data physically among the nodes in the network. The architecture of 'bytewalla' network comprises architecture of two networks, village network and city network, which can be inter-operated from remote locations.

2 DTN Architecture

The network architecture of DTN consists of the computing system called 'nodes'. The link between the nodes may go up and down. In case of link up, the source node can send the data to the destination node. This process in DTN is called 'contact'. There may exist more than one contact between the pair of nodes. The message will be buffered at an intermediate node if no further contact is available to reach the destination. The DTN does not follow continuous end-to-end connection. It follows store-and-forward method to send the data from the source node to the destination node.

Features of DTN architecture

The features of the DTN architecture are listed below:

1. The messages to be sent are the long and variable size in order to make the network able to select the good path and scheduling decision.
2. The use of naming and addressing scheme increases the interoperability.
3. It supports store-carry-forward approach in the network while transmitting the message.
4. It includes security methods to prevent unauthorized access to infrastructure.
5. It includes several features to improve the data delivery like delivery options, expressing data lifetime and coarse grained service class.

2.1 System Architecture

The DTN architecture presented in this section is the interoperability among the challenged networks. It is based on the message switching abstraction. The aggregation of messages is known as a bundle and the router used to forward the message is known as 'DTN gateway'.

Goals of DTN architecture: The goals of the architecture of the delay-tolerant network are as follows:

1. Sustain interoperability among the extremely heterogeneous type of network
2. Satisfactory performance in disconnected, high delay and erroneous network environment.

Components of DTN architecture: The components of delay-tolerant network architecture are given below:

1. A flexible naming approach along with the facility of late binding.
2. The API and message overlay abstraction. This includes the bundles and custody transfer.
3. Contact scheduling.
4. Routing.
5. Per hop authentication and reliability.

2.1.1 DTN API

Naming, Identifier, Endpoint, Late Binding and Regions
The whole network is divided into sections known as **regions**. Initially, the DTN architecture was designed to deal with the hierarchical identifiers to identify the end nodes and applications. The three-tuple **identifier** having the format (region, node, application) was used for identification. Therefore, the data was routed initially by the name of the region, followed by the name of the node and finally on the basis of the application. With the evolution advancement of DTN, the more flexibility was needed to incorporate various extreme, dynamic and heterogeneous environments. The nodes in such networks were mobile and a node was having multiple network interfaces. The **naming** system containing multiple naming spaces was in existence in IETF for DTN, called uniform resource identifier (URI). The URIs in delay-tolerant networks is known as **endpoint** identifiers (EIDs).

Another concept used in DTN is known as 'late binding'. Usually, the early binding is used to convert the machine name into the IP address at the source end using DNS when the data is transmitted. DNS incorporates both the early binding and late binding. A few DTN nodes are able to route data just using the names rather than address. When this data reaches near the destination, the name is then converted into the destination address. It is known as **late binding**. The late binding is permissible in DTN which makes DTN operate in high mobile networks.

2.1.2 Gateways

The router used to forward the message in DTN is known as 'DTN gateway'. It is also known as bundle forwarder. The DTN gateway inter-operates between two networks in order to make them compatible. The gateway of bundle layer focuses on virtual message forwarding rather than the packet switching. Figure 1 describes the gateway in delay-tolerant networks.

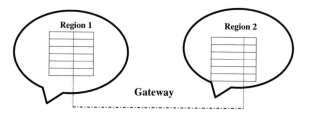

Fig. 1 DTN gateway

2.2 Layer Architecture

The TCP/IP is widely accepted and is applicable on a number of applications successfully. But in the unexpected network, the TCP/IP may not work well. There are other several ways to deal with such networks. One of them is 'link repair approach' and another approach may be to attach these networks only to the edge of the Internet with the help of a specific proxy agent. The layered architecture of the delay-tolerant network is given in Fig. 2.

Application layer: This layer describes the communication behaviour among processes in order to meet the requirements of the user. The example includes the FTP which helps in transmission of the bundle, NTP that is for time synchronization, application layer protocols for space communication, etc.

Bundle layer: It consists of bundle ALI, convergence layer protocols, authentication, routing, fragmentation and bundle encryption. The bundle layer protocols help in delivering the message using the concept of custody transfer.

Convergence layer protocols: The major purpose of convergence layer protocols is to furnish the abstraction over the protocol layers of underlying network and allow the maximum utilization of the links. The convergence layer can directly communicate to the link layer or can operate on top of the transmission protocols.

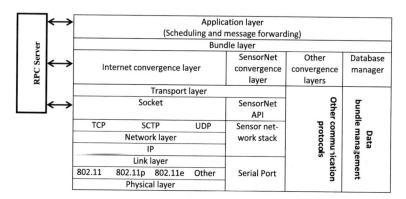

Fig. 2 Layered DTN architecture

The convergence protocol layers include Licklider transmission protocol and other convergence layer protocols.

Transmission layer: This layer aims to handle transmission and congestion control. The examples of transmission control protocol include TCP, UDP, etc.

Network layer: This layer helps in the finding, establishing and maintaining the route to transmit the messages.

Link layer: This layer has the responsibility of data framing, medium access control, frame checking and error control.

Physical layer: This layer aims to data modulation, send and receive data, channel selection and bit stream transmission over the physical media.

3 Transmission in Delay-Tolerant Network

This section will be covering the concept of contacts, bundle transfer, custody transfer and message switching approach including virtual message switching using store-carry-forward concept and the fragmentation. At the end of the section, the data delivery options are given.

3.1 Contacts

The contact is the time duration during which the capacity of the communication and network is very positive and high. The smart forwarding and routing decision can be performed if the contact is known before the transmission starts. The structure of one way contact is $(t_s, t_e, (S, D), C, D)$.

t_s start time
t_e end time
(S, D) Source and destination ordered pair
C Capacity
D Delay

3.1.1 Types of Contacts

Contacts are broadly categorized into following types.

1. **Persistent contact**: This type of contact is considered to be available every time. This contact is not needed to instantiate the connection. The cable modem connection or DSL can be taken as an example of persistent contact.
2. **On-demand contact**: The contact is instantiated when required. After instantiation, it also works as the persistent contact till the contact termination. The example of on-demand contact is the dial up connection.

3. **Intermittent scheduled contact**: It is a sort of agreement for a contact to be established for the particular duration at a particular time. The example is the link to earth orbiting satellite.
4. **Opportunistic contact**: Rather than scheduled, the opportunistic contacts present themselves unexpectedly. An example is the non-scheduled beaconing of flying aircraft for communication.
5. **Predicted contacts**: These are also not pre-scheduled but can be identified on the basis of the history of the contacts established previously.

3.2 Bundle Protocol

The bundle protocol (BP) is one of the major features of every network if the delay-tolerant networks. It is responsible for forming the store-and-forward overlay network. It contains a collection of well-defined protocol which helps in performing a store and forward communication. The main feature of BP is the custody transfer. The other characteristics of bundle protocol are:

(a) The ability to overcome intermittent if needed.
(b) The ability to deal with the long propagation delay in case it is required.
(c) The ability to avail positives of predicted, scheduled and opportunistic connection.

The bundle layer exists between the application layer and transport layer. The devices which help in the implementation of bundle layer are known as DTN nodes. The bundle layer is responsible for transferring the message hop by hop, but the acknowledgement is performed end to end. This layer also has several management and diagnostic features. The URI base flexible naming approach is the characteristic of bundle layer. This layer also includes basic security model, which is optional to be enabled with the aim to protect the unauthorized access of infrastructure. The DTN protocol stack shows the layers in Fig. 3.

The packet to be transferred is known as a bundle. It consists of data along with the signalling in order to transmit the data to transport layer that is called bundle convergence layer.

Licklider transmission protocol is a point-to-point protocol. It is designed to act as a convergence layer in support of the bundle protocol. It was specially designed for high latency scenario for deep sea communication as a convergence layer in support of bundle protocol.

The data in delay-tolerant networks is sent in the format of the variable length messages known as bundles. These messages are referred to as bundles because along with the data, there are other information for a destination like authentication data, protocol data, etc., which help for the completion of the transaction in one go. The bundle is made up of several blocks (Fig. 4). The first block is known as the primary block. It consists of version, source EIDs, destination EIDs, the length of

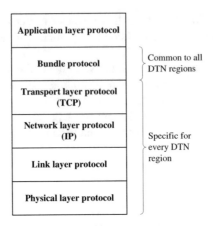

Fig. 3 DTN protocol stack

data and other fields needed to process the message. An example of other field is data dictionary which helps in decompressing the message.

Second is the fragmentation field, which is required in case when the complete bundle is not allowed to transmit. The complete bundle will be fragmented into smaller blocks and will be transmitted towards the destination.

The shortcomings of the bundle protocol include the absence of error detection and correction capabilities. It presumes that the errors will be handled by the TCP or UDP. The head of line blocking mechanism may be the problem which multiplexes the multiple bundles for the transport.

Application design principles of bundle layer
Following are the design principles for the bundle layer.

1. It must reduce the number of round trip exchanges.
2. The facility to resume the transmission in case of network failure.
3. The information of data life time and other related message delivery information should be given to the network.

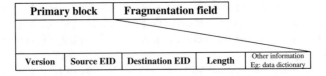

Fig. 4 Bundle block diagram

3.3 Custody Transfer

The custody transfer is defined as a responsibility of a bundle node to take the bundle up to the destination. The custody transfer is a component of DTN architecture which helps in enhancing end-to-end reliability of message delivery towards the destination. The node which is holding the custody of the message is known as custodians. If there is the single custodian for a message then it is called as sole custody. If there are multiple custodians for a message, it is called joint custody. Joint custody will come in the picture where the complete message is having multiple fragments and different fragments are in the custody of different nodes.

3.4 Flow and Congestion Control

The flow control means to assure that the rate of sending the messages will not be greater than the rate of receiving the messages at the destination. If the rate of sending the messages will be higher than the rate of receiving, the buffer will be over full after some time and the messages will be dropped.

The flow and congestion are controlled at the coarse timescales. In the case of very high delay, the pre-schedule or admission control mechanism is applied. For the small delays, the dynamic flow controls are possible. In the case of low delays, the region-specific transport can reinforce the own flow control. Logically, the flow control is performed hop by hop. Therefore, the hurdle is in conversion mechanism of bundle layer flow control to protocol specific flow control that depends on the message transport.

The congestion control in DTN is difficult as compared to another network because of two features. First, the DTN is the intermittent network. Until the next connection is found, the bundle cannot be forwarded. Second, the custody of the message expires in some extreme conditions till then the bundle will be stored by the node. The DTN applies first come first serve approach to assign the custody of the bundle.

Time synchronization
The purpose of time synchronization in DTN architecture is as follows.

1. Time computation of bundle expiration
2. Computation of expiration for application registration
3. Predicted or scheduled contact routing
4. Bundle or fragment identification

The identification and the expiration of a bundle are taken care by mentioning the creation timestamp along with the additional expiration field within every bundle.

3.5 Message Switching

The complete message keeps on switching between the nodes coming in the communication path from source to destination. The storage blocks associated with the nodes are able to store the large volume of data as compared to the other networks. These storage blocks are known as 'persistent source of storage'. This type of storage in DTN is needed for the following reasons.

1. Lack of communication link to another hop for long time duration.
2. To make the communication reliable and fast.
3. To reduce the retransmissions and errors.

The switching techniques permit to DTN node to know the size of the message while transmitting the whole message so that the storage and bandwidth requirement can be set beforehand. The process of message switching in DTN is given in Fig. 5.

Virtual message switching—store-carry-forward approach
In DTN applications, the data sent is of variable length known as application data units (ADUs). The order of sending messages is not stored. The ADUs are generally forwarded and transmitted in one unit as a whole. The ADUs are changed into the protocol data units (PDU) known as bundles. It is performed at the bundle layer. The DTN nodes are responsible for forwarding these PDUs. A bundle consists of two or more data units. Every block consists of the application data or information needed to send to the recipient.

During transmission, the bundle may be divided into more subparts known as 'fragments'. Each fragment itself is a bundle. The fragments may join or split anywhere in the network to form the new bundles as per the need of the network during transmission. The source and destination of a bundle are found with the help of endpoint identifiers. For a special case, the bundle format is having 'report to' ED which is generally used when there are multiple DTN nodes are associated with a single EID.

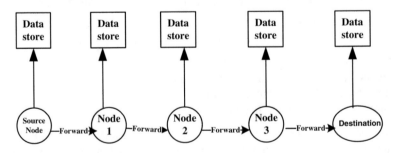

Fig. 5 Message switching in DTN

4   Routing in Delay-Tolerant Network

DTN belongs to the class of emerging networks which experience the long-term and frequent partitions. These networks face the absence of an end-to-end communication path. Therefore, routing plays a very important role in the performance of the network. In DTN, the types of routing generally used are unicast routing, multicast routing and anycast routing.

4.1   Unicast Routing

The unicast routing refers to the communication between the one dedicated sender and one specified receiver. The types of protocol following unicast routing are as follows:

Contact graph routing
This protocol takes benefit from the property of high contact frequency of nodes inside the community. If the time duration will be taken as an estimated value (EV), its benefit is that the message will not expire before reaching the destination. In this routing, the TTL is considered as an EV. Here, the routes which do not qualify enough TTL to reach up to the destination will not be considered. The CAR consists of inter-community and intra-community routing.

Inter-community routing: Every node spreads multiple copies of the message to the nodes belonging to different communities as earliest possible. The dissemination of the copies of the message is corresponding to the expected number of communities encountered to each pair of the encounter. For a single replica of messages which remains while propagation, the message will be sent to the node having the highest probability to meet at least one of the destination communities.

Intra-community routing: The copies of the messages are disseminated by the node to one of the destination communities to its encounter within the same destination community as per the proportion of the expected values (Tables 1, 2 and 3).

4.2   Multicast Routing

A number of DTN applications work on the basis of group forwarding. Therefore, multicasting is a very important and widely used concept in DTN. Because of the network partitions and disconnections, the multicast semantics in DTN need to be considerably different from conventional internet approach.

The multicast semantics in traditional networks like MANET and the internet are well defined in advance as the message will be sent to the dedicated number of the

Table 1 Epidemic routing protocols

The aim of this protocol: to achieve high delivery ratio	
A shortcoming of this protocol: the overhead increases dramatically with the increase in network size	
Vahdat et al.	Every message is copied or replicated and broadcasted in the network. The count of replicas of the messages increases very fast in the network which consumes a large amount of limited bandwidth and buffer space
Prophet	A node will duplicate and forward the message further to another node only if it has a higher probability of encountering the destination
Max prop	To optimize the limited buffer space, this scheme directs the nodes to schedule for the packets to be transmitted and the packets to be dropped. Therefore, the packets having a higher probability of reaching the destination will be replicated among nodes and the packets having a lesser probability to reach the destination, will be dropped when the buffer will be full
Delegation forwarding	The message will be forwarded to the encounter of the node only if the encounter is having a better metric. If n is the number of nodes in the network, delegation forwarding scheme can reduce the cost of the network to $O\sqrt{n}$ compare to $O(n)$
R3	The aim of R3 was to attain robust replication routing. It duplicates every packet along with a small number of paths in order to achieve more replication gain. The replication can also be stopped if it is noticed that the actual delay is very high than the estimated delay. At that time it will be switched to the single path forwarding

Table 2 Forwarding based routing protocols

The aim of this protocol: to overcome the high overhead in the network	
Shortcoming: it cannot attain high delivery ratio as the predictions cannot be correct all the times	
Jones at al.	The routing mechanism was designed for single replica routing. The base of the mechanism was minimum estimated expected delay. It was obtained on the basis of average meeting intervals among every pair of nodes within the network using Dijkstra's algorithm
Predict and relay (PER)	This scheme uses the semi Markov process. It depends on the prediction of future contacts Geo et al.: it is a forwarding based approach which exploits the contact patterns of the transient nodes. On the basis of these contact patterns, every node can predict accurately in order to take decision for data forwarding
Tour	In this mechanism, every node keeps a time sensitive forwarding set by taking the probabilistic contacts in the network. Here the messages will only be forwarded through the nodes in the forwarding sets in order to attain the expected optimal utilities

Table 3 Quota based routing protocols

	These protocols try to minimize overhead and maximize the delivery ratio. This is a tradeoff between the epidemic routing protocols and forwarding based routing protocol
Spray and wait	This mechanism works in two phases: spray phase and wait for phase. In spray phase, the copies of each message are disseminated. The node will enter into the wait phase if it is having only one copy of the message. It will wait till it meets the destination to transfer the message
Spray and focus	The spray phase does the same, i.e. disseminates the copies of the message. In focus phase, the node will enter when it is only having only one replica of the message. Here, the messages using single copy may be transferred to its encounter having higher utility in order to improve the performance Optimal probabilistic forwarding protocol: It was given by Liu and Wu. It is based on store and wait approach. The assumption for this protocol is that every node is aware of the mean inter-meeting times for every pair of nodes within the network. It is not a practical approach as the information in the network is completely distributed
Erasure coding based routing scheme	It was given by Wang et al. in this scheme the message is encoded at source node into m blocks. Each block will be duplicated for some n number of times up to the destination. Every message at the destination will be decoded only if all the m blocks will be reached successfully at the destination. Geo spray: This scheme is a combination of multiple replicas and single replica schemes. The routing decisions are taken on the basis of the geographic location of data
Expected encounter based routing protocol (EER)	The aim of the protocol is to build the replica distribution criteria among two encountering nodes which are based on the message's residual TTL. There are two phases of EER. These are multiple replica distribution and single replica forwarding. In multiple replica distribution phase, every node distributes the copies of the message to other nodes as earliest that may be achieved by disseminating the message copies on the basis of expected value 9 eV). The EV in multiple distribution phase is TTL. In the second phase, the minimum expected meeting delay (MEMD) is comparing to the destination in order to forward the message to the current encounter

group. But, this is not valid in DTNs. Therefore, it is not clear that how to describe the recipients to a multicast packet.

Multicast semantic model

As already discussed, because of the huge transfer delays, the membership of a group during transmission may change in DTN. Therefore, external temporal constraints must be defined in multicast routing for DTN. In this section, three multicast semantic models are given, which allows defining external temporal constraints.

1. **Temporal membership (TM) model**: To find the recipients of the multicast message, it is required to externally define the time during which the concerned

recipients are identified. The simple method is to specify the recipients of the message as a number of a group at the time for generating a message. In temporal membership model, the message incorporates the membership interval which defines the time duration for which the group members are specified. The recipients of the message are clearly defined. There is no time constraint to deliver the message. Therefore, the message may reach any time. The temporal delivery model permits the user to define the time-based features flexible for the recipient group.

2. **Temporal delivery (TD) model**: Unlike TM model, the temporal delivery model imposes a constraint on the delivery action of the message. Along with the membership interval, the message also consists of the delivery interval. It refers to the time period within which the message must be delivered to the receiver. It allows the user to have another control on message delivery.

3. **Current member delivery (CMD) model**: At the time of message delivery, the receivers in TM and TD may or may not be the member of the group. The additional field included by CMD model is the CMD flag along with the membership interval and delivery interval. In the case of CMD flag set, the message recipient must be the member of the group on the time of delivering the message. In case, CMD flag is not set, the CMD model will work like TD model.

Routing protocol for multicast

1. **Unicast-based routing (UBR)**: The source will be sending the replica of the message to all the intended receivers.

2. **Broadcast-based routing (BBR)**: It is also known as epidemic routing. In this approach, the messages will float within the network so that the messages can reach to all the intended receivers.

3. **Tree-based routing (TBR)**: The messages are sent along with the tree in DTN. The source is the root node and the receivers will be the leaf node. The messages are replicated only to the nodes having one or more outgoing paths.

4. **Group-based routing (GBR)**: In this scheme, the concept of forwarding group is used. Forwarding groups are the set of nodes having a path towards the destination. The message is flooded in the forwarding graphs in order to enhance the probability of message delivery.

4.3 Anycast Routing

Anycast permits the node to transmit the message to one member of the group. The key idea of anycast is that the sender wants to send the bundle to any one of the nodes providing the particular service without caring for any particular node. Anycast routing is applicable for applications like the discovery of resource in DTNs. The anycast in DTN requires the new semantic models as compared to the mobile ad hoc networks. The semantic models for anycast routing in DTN are as follows:

1. **Current membership model (CMM)**: The main objective is to decide the recipient of the message, the explicit identification of the time interval for which the recipient is determined. The message must be forwarded to the node that is the member of the group at the time of message arrival. The recipients of the message in DTN can be changed over the time. Using CM model, the message sent to request for the resource may be forwarded to any of the current index servers.

2. **Temporal interval membership (TIM)**: The additional information added in TIM is the temporal interval in order to indicate the time period during the selection of group members. Suppose G is the anycast group along with the temporal interval (t1, t2). The recipient should be the group member G at the time of interval.

3. **Temporal point membership model (TPMM)**: This model includes additional information as 'membership interval' along with the temporal interval. It refers to the time period within which the message must be delivered to the receiver. It allows the user to have another control on message delivery.

4.4 Buffer Management

Under store-carry-forward scheme, the incoming packets will be stored in buffer till the next hop towards the destination is met. The packets will be dropped if the next hop is down as per the buffer management strategies. It is necessary to drop the packet from buffer because the size of the buffer is limited and needs space for incoming packets. The transmitting node has to take the custody of the bundle till another node took custody or it is delivered to the destination successfully.

As we know that the buffer space is always limited in DTNs. Therefore, the management of utilizing buffer is very important. It plays a big role in the performance of the network. The buffer management strategies are:

1. **MaxProp**: In this scheme when the buffer is full, the node schedules to drop the packets having least probability to be delivered successfully.

2. **Adaptive optimal buffer management policies**: It was proposed by Li et al. In this scheme, the mobility model will be tuned on the basis of historical meeting information of the node.

4.5 Performance Metrics

There is a number of performance metrics for the routing in delay-tolerant networks. Some of them are listed below.

1. **The degree of centrality**: This metric belongs to the type of local variable. The degree of centrality refers to the number of edges connected to the node. This metric has the importance to identify the nodes and from the local viewpoint.

2. **Closeness to centrality**: This metric belongs to the type of global variable. It refers to the mean distance between the node and the other nodes that are reachable from the current node geodesically.
3. **Betweenness of centrality**: It is also a type of global variable. The betweenness means that the shortest path among all the nodes passes through the current node. It is used for measuring load for a particular node.
4. **Centrality bridging**: This metric lies under the category of a global variable. It tells about the network that up to what extent of correction, the edges or nodes are situated among the connected regions. This metric reports the information about the bridging of nodes and edges.
5. **Page rank**: it is also a global variable metric. It helps in measuring the importance of a node from the global perspective. This metric shows the important nodes especially hubs.
6. **The coefficient of clustering**: This metric belongs to the type of global variable. The coefficient of clustering refers to the measure of the degree of nodes within the network which wishes to cluster together. This metric is useful while forwarding information within a cluster.
7. **Similarity**: This is a local type of variable. The similarity metric illustrates the common features among the nodes. It is a very useful metric to be kept in consideration while transmitting the message.
8. **Selfishness**: This metric also belongs to the type of local variable. This metric refers to the measure of the degree of the cooperation and the interaction of a particular node with the rest of the nodes. It helps in identifying the willingness to forward the message.

5 Security in Delay-Tolerant Network

The specifications for the bundle security describe the confidentiality and the integrity mechanisms along with the other policy options. The LTP is a convergence layer protocol which adds some security in DTN. The threats while designing the process for DTN security mechanism are as follows.

1. **Non-DTN node threat**: The most common threat can come from the nodes which are not a direct part of DTN.
2. **Resource consumption**: Because of the lack of resources in DTN, the unfair use of the resources is a big issue. The DTN resources can be consumed unnecessarily by the entities in the following manner.
   (a) Unauthorized access by the entities
   (b) Uncertified control on DTN infrastructure
   (c) Transmission of bundles without permission
   (d) Unlicensed content manipulation for the bundle
3. **Denial of service (DOS) attack**: The DOS attack must be taken into account for DTN networks. The DOS attack can be scaled at any layer.

5.1   Security Issues

Key management: This is the major research open issue in DTN. The usage of the existing schemes is easy in DTN but the key distribution services and checking the security status online are not practical in high delay environments. There exist some identity-based cryptography schemes in order to solve the problems but they do not work up to the mark.

Traffic analysis: The protection on traffic analysis is generally not required in delay-tolerant networks. But in some specific case like hiding traffic, it is an important security issue that must be taken into consideration. The one open security issue in traffic analysis is that up to what limit it is a requirement for a generic scheme to protect the traffic analysis.

Routing protocol security: Definitely, as the routing protocols needed some modification for working in the DTN environment, the security for routing in DTN will also be applied differently. But it was the least focus on the security issue while writing the routing protocols for DTN. This again is an open research issue.

Multicast security: In multicasting, the message will be forwarded to all the recipients who are in the recipient group. In DTN, there is no mechanism which can identify the registration of a node as a multicast or anycast.

5.2   Fragments Authentication

DTN applicable schemes: The frequent network partitioned and disconnections are the features of DTN which conclude that the security services applicable for traditional security will not be applicable.

Confidentiality and integrity: The fragment can also be vulnerable from integrity and confidentiality point of view. The examples are

(a)   The source forging of bundle
(b)   Change in the desired destination
(c)   Modification of the control fields of the bundle.
(d)   Manipulation in payload and block fields
(e)   Manipulation of the data during transmission.

6   Delay-Tolerant Network and Vehicular Adhoc Netwwork

6.1   Overview of VANET Delay-Tolerant Networks

In a vehicular network, the DTN routing needs a vehicular model as it is closely linked with the mobile nodes and a population of the network. The vehicular networks can be divided into two subcategories, vehicular networks and vehicular delay-tolerant networks (vehicular network in sparse traffic). To make the applications feasible via vehicular networks, it is necessary to design the specialized networking protocols which may overcome the problems arose from the vehicular environments.

6.2   Characteristics of VDTN

The unique features of DTN and VANET will be combined to design the protocol for VDTN. A few of specific features of VDTN are discussed below.

1. **Vehicular applications**: Several safety applications in vehicular networks like the application of emergency brakes require hard delay constraints. For such applications, the DTN might not be proved optimal. For non-safety applications like exchanging entertainment information to other vehicles, parking lot payment, etc., where there is no hard delay constraints are required, the DTN principles can be applied.

2. **High mobility and frequent disconnections**: The mobility and high speed of vehicles in VANET results in a frequent change in topology and disconnection. For example, the two vehicles are moving in opposite direction at some speed. They will come in contact for a few seconds; the opportunistic window of communication will be for short duration. After that, they will not be in the range of each other resulting in disconnection. For low-density traffic, these two vehicles will remain in contact for a long duration, which is the case in VDTN. This concept of VDTN makes it more attractive and useful for such scenarios.

3. **Geographical awareness**: The geographical location of the vehicles can be identified and used for the algorithm of routing and message delivery. Even the trajectory of the vehicles can be determined. The examples are the route identification of the public transport like bus and train. There exist several message delivery paradigms in computer networking like unicast, multicast, broadcast and anycast. In vehicular networks, the special message delivery paradigm exists which is known as 'Geo-cast'. In geo-cast, the message will be sent to the group of vehicles belonging to common geographic location.

4. **Mobility prediction**: The vehicles are moving from one location to another location. But, the advantage of a vehicular network is that the vehicles will be moving on specified paths. On the basis of these paths, the future location and trajectory of the vehicle can be determined. Therefore, the mobility model must have the capability of determining the future location and path trajectory which will increase the performance of the network.

6.3 DTN Protocols for VANET Delay-Tolerant Networks

The VDTN is a special case of VANET for addressing the specific problems like network partitioning and frequent disconnections or where there is no guarantee of an end-to-end delivery of messages. The VANET also faces these problems because of high mobility in the network. The VDTN use store-carry-forward approach rather than an end-to-end message delivery. The conventional routing protocol designed for VANET will not work in VDTNs. The architecture of VDTN is based on three types of nodes (Fig. 6). These are terminal nodes called access points, the mobile node is known as vehicles and the relay nodes are referred as fixed device points at cross roads. The mobile nodes may transfer and receive data to and from other nodes. The VDTN architecture also separates the data and control planes to enhance the overall performance of the network, by transmitting long size message rather than the small sized packets.

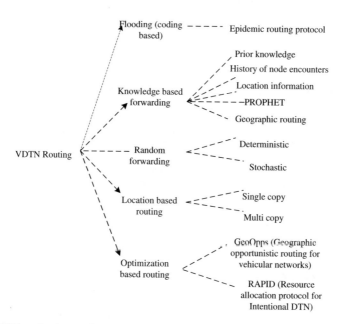

Fig. 6 VDTN routing protocols

7   Conclusions and Future Scope

The delay-tolerant-networks (DTNs) are completely different approaches as compared to typical connected wired or wireless networks. In DTNs, the end-to-end path availability is not required at any point of time to transfer data between the sender and the destination node pair. The DTNs came in the picture where the routes between a pair of nodes cannot be achieved, for example, sparse network scenario where end-to-end routes are not available, like military battlefields, there DTN provides the medium to place communication. It does not need any prior knowledge of the network in order to forward bundles from one node to another. It is based on store-carry-forward approach. The delay-tolerant network is very emerging and useful area. The important features and applications are already discussed in previous sections. Though, DTN has a number of applications, still it needs a lot of improvements. A few of the open research issues will be a discussion in the next section. There is a huge scope for improvement especially in the security of the DTN.

References

1. Tornell, S. M., Calafate, C. M. T., Cano, J. C., & Manzoni, P. (2015). DTN Protocols for Vehicular Networks: An Application Oriented Overview. IEEE Communications Surveys and Tutorials, 17(2), 868–887.
2. Caini, C., Cruickshank, H., Farrell, S., & Marchese, M. (2011). Delay-and disruption-tolerant networking (DTN): an alternative solution for future satellite networking applications. Proceedings of the IEEE, 99(11), 1980–1997.
3. Zhao, W., Ammar, M., & Zegura, E. (2005, August). Multicasting in delay tolerant networks: semantic models and routing algorithms. In Proceedings of the 2005 ACM SIGCOMM workshop on Delay-tolerant networking (pp. 268–275). ACM.
4. Soares, V. N., Rodrigues, J. J., & Farahmand, F. (2014). GeoSpray: A geographic routing protocol for vehicular delay-tolerant networks. Information Fusion, 15, 102–113.
5. Sobin, C. C., Raychoudhury, V., Marfia, G., & Singla, A. (2016). A survey of routing and data dissemination in delay tolerant networks. Journal of Network and Computer Applications, 67, 128–146.
6. Gong, Y., Xiong, Y., Zhang, Q., Zhang, Z., Wang, W., & Xu, Z. (2006, November). WSN12–3: Anycast routing in delay tolerant networks. In Global Telecommunications Conference, 2006. GLOBECOM'06. IEEE (pp. 1-5). IEEE.
7. Sollazzo, G., Musolesi, M., & Mascolo, C. (2007, June). TACO-DTN: a time-aware content-based dissemination system for delay tolerant networks. In Proceedings of the 1st international MobiSys workshop on Mobile opportunistic networking (pp. 83–90). ACM.
8. Benamar, N., Singh, K. D., Benamar, M., El Ouadghiri, D., & Bonnin, J. M. (2014). Routing protocols in vehicular delay tolerant networks: A comprehensive survey. Computer Communications, 48, 141–158.
9. Soares, V. N., Rodrigues, J. J., & Farahmand, F. (2014). GeoSpray: A geographic routing protocol for vehicular delay-tolerant networks. Information Fusion, 15, 102–113.
10. Paul, P. S., Ghosh, B. C., De, K., Saha, S., Nandi, S., Saha, S. & Chakraborty, S. (2016, January). On design and implementation of a scalable and reliable Sync system for delay tolerant challenged networks. In Communication Systems and Networks (COMSNETS), 2016 8th International Conference on (pp. 1–8). IEEE.

11. Wei, K., Liang, X., & Xu, K. (2014). A survey of social-aware routing protocols in delay tolerant networks: applications, taxonomy and design-related issues. IEEE Communications Surveys & Tutorials, 16(1), 556–578.

12. Ikeda, M., Honda, T., Ishikawa, S., & Barolli, L. (2014, November). Performance comparison of DTN routing protocols in the Vehicular-DTN environment. In Broadband and Wireless Computing, Communication and Applications (BWCCA), 2014 Ninth International Conference on (pp. 247–252). IEEE.

13. https://tools.ietf.org/html/rfc4838.

14. Chen, H., & Lou, W. (2016). Contact expectation based routing for delay tolerant networks. Ad Hoc Networks, 36, 244–257.

Part II
Wireless Technology
and Communications—Methodologies
& Implementations

Architectural Building Protocols for Li-Fi (Light Fidelity)

Mayank Swarnkar, Robin Singh Bhadoria and Karm Veer Arya

Abstract Light fidelity commonly known as Li-Fi is the technology emerged from visible light communication (VLC) that allows to transmit data through illumination, i.e., through light emitting diode (LED). Li-Fi varies in intensity faster than human eye to be followed, and therefore, Li-Fi is known for high data speed. With the increase in the number of wireless gadgets such as smartphones, tabs, smart wrist watches, etc., Li-Fi will surely be an incredible companion of Wi-Fi and an interesting subject for research. Therefore, in this chapter, we cover the evolution of Li-Fi, its architecture, requirement of Li-Fi, and its challenges. We also discuss the integration of Li-Fi protocols with existing protocols for integrated communication and conclude the chapter with the future scope of Li-Fi in the Internet.

Keywords Visible light communication · Light fidelity · Wireless communication · Light emitting diode · Modulation

1 Introduction

Wireless devices for communication are growing at a rapid pace because of its mobility and ease to handle. As per the reports of [1], the global mobile wireless penetration worldwide from 2008 to 2020 will increase from 66 to 99%. There are millions of devices using Wi-Fi to connect to Internet. One of the growing areas of wireless communication is now optical communication [2] in which communication medium is light. Visible light communication (VLC) [3] is an uprising and popular optical communication technology nowadays. Visible light has spectrum range from 400

M. Swarnkar (✉) · R. S. Bhadoria
Indian Institute of Technology Indore, Indore, India
e-mail: swarnkar.mayank@gmail.com

R. S. Bhadoria
e-mail: robin19@ieee.org

K. V. Arya
Dr. A.P.J. Abdul Kalam Technical University, Lucknow, India
e-mail: kvarya@gmail.com

© Springer Nature Singapore Pte Ltd. 2018
K. V. Arya et al. (eds.), *Emerging Wireless Communication and Network Technologies*,
https://doi.org/10.1007/978-981-13-0396-8_7

to 800 THz. One of the emerging VLC technologies is light fidelity or Li-Fi. It is becoming popular and can replace radio frequency communication technologies in the near future [4, 5]. We discuss about VLC and Li-Fi briefly in the following subsections.

1.1 Visible Light Communication

It is a communication technology which exploits light emitting diodes (LEDs) [6] emitted light for data transmission. It can be a supplement to radio frequency (RF) or cellular network communication. VLC uses visible light spectrum, and this spectrum is 10^4 times larger than the radio frequency spectrum. Therefore, it yields more bandwidth than radio frequency-based communication networks. This is one of the reasons VLC generates very high data transmission rates, making it competitive for radio networks. VLC Spectrum is shown in Fig. 1.

1.2 Importance of VLC

VLC is an emerging technology which has many advantages and following are the points which make it important to us:

- Radio frequency cannot regulate above 3 THz but can be regulated by VLC. Radio waves also suffer interference which VLC does not, and hence, regulation can be one easily in dense areas also.

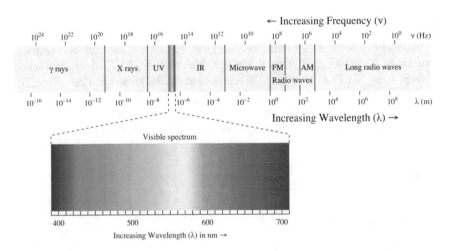

Fig. 1 Visible light spectrum

- VLC can be used almost everywhere because of its frequent use and easy and cheap availability. Radio waves are already used for general wireless communication. LEDs are used generally everywhere including houses and offices, which makes LEDs as ideal for pervasive data transmission. As per Zion Market Research, LED Lighting Market Share & Growth Will Increase $54.28 Billion by 2022 [7]. This makes VLC an easy and scalable way of wireless communication.
- Image sensors work as receivers in VLC. Image sensors are capable of detecting incoming data accurately as well as its direction from transmitter to receiver. This quality of Image sensors makes various new applications which are not possible in radio wave communication. It includes augmented reality, indoor navigation, accurate position measurement, and accurate control of robots or vehicles.

1.3 Modes of Communication in VLC

There are two modes of communications in VLC which are as follows:

- **Infrastructure-to-Device Communication**: It is a connection between a wired device and VLC device which intermediate protocols for connection. Usually, Internet backbone network (usually wired) is connected to VLC infrastructure in this mode of communication.
- **Device-to-Device Communication**: It is a connection between two VLC devices. It can be a LED to LED communication or a VLC sender to VLC receiver communication.

2 Light Fidelity

Light fidelity or Li-Fi is a type of VLC technology which uses common household LEDs for data transmission. Li-Fi is bidirectional, high speed, and broadcast network. Li-Fi works on the phenomenon of flickering of LED at a very high rate. Switching of states of LED is fast enough that it cannot be noticed by naked human eye. Since light waves of LED cannot penetrate walls which makes Li-Fi to work in shorter range, but this also works as an advantage because it is more secure from hacking compared to Wi-Fi. It is not necessary to have a direct line of sight for Li-Fi for signal transmission because light is reflected by the walls.

2.1 Need of Li-Fi

There are various reasons to promote use of Li-Fi. Some are practical examples explained below which shows the real need of Li-Fi in our daily life:

- **Speed**: Nowadays, high-speed Internet is the requirement of every person. High-quality video buffering, video calls, conference calls, etc., require high data speed. Li-Fi provides higher data transmission speed which is in Gigabits.
- **Cost**: LED are used in daily life by nearly all people who are economically capable of using Internet service. Currently, Internet is used by end users using Wi-Fi, and Ethernet CAT5 cables and dial-ups connections require costly hardware installations by users. Li-Fi removes all these hardware installation costs, which makes it overall cheaper than these types of connections.
- **Availability**: Internet connectivity can be provided to users at any place where white light or LEDs are available. Therefore, there is no point in neglecting the fact that Li-Fi can be provided anywhere, i.e., inside rooms, on roads using traffic lights, street lights, in shops, in hotels, at petrol pumps, etc.
- **Smart Device Add-ons**: Smart devices like smart television, smart rooms, smart cars, etc., use Internet services for intelligent performance. Instead of using radio waves for internet connectivity, the add-ons and simple Li-Fi receivers are integrated into the smart devices makes easy possible solutions.

2.2 Challenges for Li-Fi

Li-Fi definitely has many advantages but there are few challenges which need to be resolved for adopting this technology. Few are the following challenges needs to be taken care to install Li-Fi:

- **Line of Sight (LOS)**: LOS makes the signal stronger. Visible light signals can be reflected but do not get through opaque objects which is results in lack of coverage but provide a good security. Another disadvantage of line of sight is the prevention of the signal from dispersing among multiple rooms in any building. We also know that reflection absorbs energy and results in the limited communication rate without line of sight between the transceivers. In general, even if we use optical devices to spread signal, the power regulation cannot be strong enough to let reflected signals still preserve enough power for communication (reflection results in power loss). If light levels are low and receiver can collect photons, it can receive data at a lower data rate. Similar to radio wave technology that indirect signals have lower power and results in the reduction of data rate.
- **Multipath Distortion**: When the transmitter and receiver both are equipped with outspread beam, the facsimile of the identical signal from divergent paths arrive the destination with different delay, because paths differ in length from source to destination. This creates problem of multipath distortion which can cause inter symbol interference, and it results in critical degradation in the performance.
- **Simplex Communication**: VLC cannot have duplex connection, and LEDs can only be used for data transmission for either uplink or downlink. This needs to be isolated on the basis of code, time, wavelength, spatial isolation, or optical isolation. Due to high cost and bandwidth, visible light communication can be

implemented for downlink via Wi-Fi or IR may provide a good uplink where congestion is less probable, and it provides a high capacity uncongested downlink.

- **Power Wastage**: For VLC, LEDs needs to be on power all the time. During daytime, LEDs in domestic environments are usually switched OFF because of availability of natural light. Therefore in such environments, even if there is no requirement of LEDs, it needs to be turned ON which therefore consumes power which was not been used in Wi-Fi. Another issue is that if the LEDs are ON in domestic environment, the illumination will not be noticed because illumination level falls below ambient levels as the buildings are designed to have enough natural light in daytime. The power consumed is comparable with the radio transmissions but may have performance issues in daytime.

- **Transmitter Sources**: Solid state LED lighting is presently being sold supported its performance for illumination functions solely. Communications performance is not even a secondary thought; therefore, it is entirely impractical to expect the industry to side this into styles at this stage. In an exceedingly sensible sense, glorious results will be achieved with expensive and specifically designed LED devices. If higher devices area unit on the market for VLC then nice, otherwise to implement VLC, existing LED devices is to be thought-about which is able to sure enough have performance problems.

2.3   Comparison of Li-Fi and Wi-Fi

Li-Fi and Wi-Fi both are technologies which transmit data over wireless medium. Yet there are differences in both the technologies which are as follows:

- Li-Fi performs data transmission over visible light using LED bulbs whereas Wi-Fi performs data transmission over radio waves using wireless modems.
- Li-Fi has faster data transfer speed as compared to Wi-Fi. Data transfer with Li-Fi can be obtained over 1 Gbps whereas with Wi-Fi it is about 150 kbps. Data transfer speed of Wi-Fi is increased up to 2 Gbps using Wi-Gig or Giga-IR.
- Wi-Fi suffers interferences with other nearby wireless modems because of radio wave frequencies whereas there is no issue of interference in Li-Fi.
- Li-Fi uses IrDA [8] compliant devices whereas Wi-Fi uses WLAN 802.11 standards.
- Due to interference, Wi-Fi is not suitable for highly dense environment but Li-Fi suffers from no such issues.
- Wi-Fi has better coverage than Li-Fi. Wi-Fi ranges up to 32 m depending upon antenna and power whereas Li-Fi ranges no more than 10 m.

2.4   Architecture of Li-Fi

Architecture of Li-Fi is shown in Fig. 2. Li-Fi mainly comprises two components which are:

- **Transmission Source**: A bright white LED is the data transmission source of Li-Fi also known as LED luminaire. It consists of an LED source and LED Driver. The LED source may contain single or multiple LEDs. The LED source also includes a circuit which is used to control the current flowing through the LEDs which in turn controls the brightness known as LED driver. Whenever an LED is used for communication, a modified circuit is made in order to modulate the data through emitted light. There are generally two types of LEDs are used for generation of bright white light: Blue LED with phosphor and RGB light combinator to generate white light. An example of transmission source is a transmitter using ON-OFF keying modulation. Here, the data bits 0 and 1 can be transmitted by choosing two different levels of light intensity like 0 for OFF and 1 for ON.
- **Reception Element**: A silicon photodiode or image sensor are the reception element of Li-Fi also known as Li-Fi dongles. The photodiode is a semiconductor that converts light into current. It decodes the light illumination just in the reverse way of encoding by transmission source. An image sensor is also known as camera sensor and can be used as a reception to the transmitted visible light signals. An imaging sensor consists of many photodiode having certain topology in an integrated circuit. But the image sensor limitation is that it requires a number of

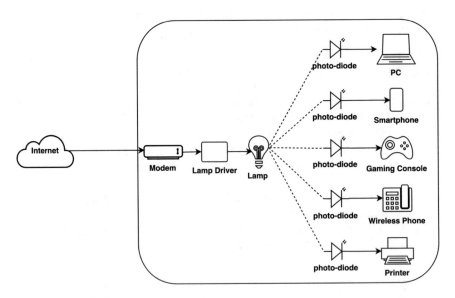

Fig. 2 Li-Fi architecture

photodiodes to obtain high-resolution photography. These photodiodes are directly connected to the end devices as the receptors generally in the form of Li-Fi dongles.

2.5 Difference Between Photodiode and Image Sensor

An image sensor is the combination of multiple photodiodes arranged in a manner in an integrated circuit. Photo-sensors receives one signal at a time whereas image sensor receives multiple signals simultaneously because image sensors contains multiple photodiodes. Figure 3 shows an example of image sensor using multiple photodiodes in which it collects multiple signals and then processes it. The way of receiving signals is totally dependent on the design of the image sensor.

2.6 Working of Li-Fi

Li-Fi has two main components such as transmitter and receiver. Usually, in Li-Fi, white light source is the transmitter which is generally LED bulbs. Signals generally transmitted in binary form also known as digital signal sent using digital signaling and digital modulation techniques. Li-Fi sends digital signals by coding luminous intensity to 0's and 1's. For example, LED at ON state is 1 and LED at off state is 0. The ON-OFF state change of LED bulb is so fast that the change is not noticeable by human eye. Another example is coding the signals by light color in visible light spectrum. A red light is 1 and blue light is 0. Here flickering in colors may be seen

Fig. 3 Working of image sensor using photo diode

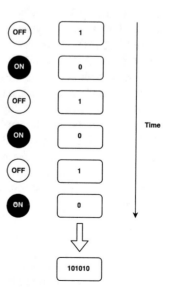

Signal   Encoding using LED   Encoding Description

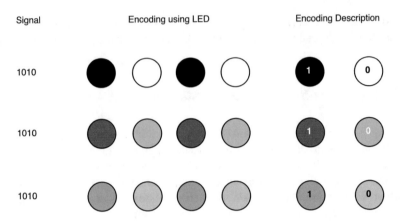

Fig. 4 LED encoding

by human eyes. One more example is using brightness of white light for mapping binary signals. The Li-Fi receiver is a photodiode or cluster of multiple photodiodes in some arrangement (image sensor) which decodes the LED in the same way it is encoded. In other words, it can be said that the decoders depend on the encoding scheme of encoders. These examples can be visualized in Fig. 4.

3 Li-Fi Standardization and Its Integrated Communication Protocols Solution

Li-Fi[1] follows VLC standardization which is IEEE 802.15.7. The IEEE 802.15.7 standard offers three physical (PHY) types for VLC. PHY I operation range: 11.67–266.6 kb/s, PHY II operation range: 1.25–96 Mb/s and PHY III operation range: 12–96 Mb/s. PHY I and PHY II are designed for a single source of light and support OOK modulation and VPP modulation. PHY III uses multiple optical sources with different frequencies and uses a fixed modulation format called color shift keying (CSK) modulation. Multiple modulation schemes at physical layer trade-offs between speed of data transmission and different dimming ranges of LED lights. Dimming or flickering of LEDs is important for efficient power utilization. It is also valuable because it can maintain connection even when the LED light dims because of natural light. Operating modes of PHY I, PHY II, and PHY III are shown in Tables 1, 2 and 3. The modulations used in PHY I, PHY II, and PHY III are as follows:

---

[1] We talk about physical layer communication of TCP/IP Suite and its integration with Li-Fi.

Table 1 PHY I operating modes

Modulation	Encoding method	Optical clock rate (kHz)	Data rate (kbps)
OOK	Manchester	200	11.67–100
VPPM	4B6B	400	35.56–266.6

Table 2 PHY II operating modes

Modulation	Encoding method	Optical clock rate (MHz)	Data rate (Mbps)
VPPM	4B6B	3.75–7.5	1.25–5
OOK	8B10B	15–120	6–96

Table 3 PHY III operating modes

Modulation	Optical clock rate	Data rate
CSK	12–24 MHz	12–96 Mbps

- OOK modulation with dimming,
- VPPM modulation with dimming, and
- CSK modulation with dimming.

3.1 OOK Modulation with Dimming

In OOK modulation [9], LEDs are flickered, i.e., turned on or off on the basis of binary values of data to be transmitted. It is the simplest modulation technique used in visible light communication. In this modulation, it is not a restriction to turn the light completely OFF; instead, the luminosity of the light can simply be reduced enough to distinguish between the ON and OFF levels of the LED light. OOK uses Manchester encoding [10] as digital signals encoding scheme.

3.2 VPPM Modulation with Dimming

In VPP modulation [11], bits are encoded on the basis of the change in the duty cycle per optical symbol. VPPM uses a variable term which represents the change in the duty cycle in response to the requested dimming level. VPPM exploit the position of the optical pulses to distinguish between the optical symbols. The logical symbols 0 and 1 are modulated pulse width and depend on the dimming duty cycle requirement of the LEDs.

3.3 CSK Modulation with Dimming

White LED lights are produced by using a combination of multiple colors by two different methods. White LEDs can be produced using blue LEDs and yellow phosphor. It has a little disadvantage that yellow phosphor slows down the switching response time of the white LEDs. Color shift keying modulation [12] is similar to FSK because both the modulation technique uses bit patterns encoded to color (wavelength) combinations. The spectrum of seven color bands is standardized as the IEEE 802.15.7 standard to define various colors for communication.

3.4 Hybrid Li-Fi: Integrating Li-Fi with Wi-Fi

There are few obstacles in Li-Fi connection uplink which can be taken care if Wi-Fi is used with Li-Fi. These obstacles are as follows:

- If we use LEDs for both uplink and downlink then both the links may interfere with each other.
- If transmitter is movable then focusing issue will arise between transmitter and receiver.
- LED as transmission from a device could also be uncomfortable to the users as a result of its high power consumption on finish device which can be running on battery as well as device can emit light every time.

A vital solution to overcome the above issues is to use radio transmission communication with visible light communication. One can use Wi-Fi for downlink and Li-Fi for uplink or vise versa. Another solution can be based on dynamic bandwidth usage, i.e., if a large file needs to be uploaded or downloaded then use Wi-Fi and Li-Fi for otherwise. There are few schemes proposed in literature for using Wi-Fi and Li-Fi together as integrated system [13–15]. However, the hybrid methods are in need of Wi-Fi should direct the uplink acknowledgments of communication frames which may form multiple small-packet traffic. As a result, it would reduce the throughput of shared Wi-Fi devices. In the same way, the latency and transmission rate are affected by the presences of Wi-Fi devices.

4 Conclusion and Future Scope

In this chapter, we discussed VLC and Li-Fi. It is true to say that Li-Fi is the future of wireless data communication and can work as a companion to Wi-Fi. Since Li-Fi transmits data using LEDs, therefore it can be used anywhere in the environment. Li-Fi is also a hope for a fast smart device communication and Internet of things. Li-Fi is a cheaper, safer, and green option for communication in the near future. Currently, there are issues need to be resolved for ideal working of Li-Fi in regular use in commercial manner and therefore an important topic of research for the researchers.

References

1. https://www.statista.com/statistics/232594/mobile-wireless-penetration-worldwide/
2. Gagliardi RM, Karp S. Optical communications. New York, Wiley-Interscience, 1976. pp. 1–445
3. Arnon S. Visible light communication. Cambridge University Press, 2015. pp. 1–206
4. Leba M, Riurean S, Lonica A. LiFi The Path to a New Way of Communication. IEEE, 12th Iberian Conference on Information Systems and Technologies (CISTI), 2017. pp. 1–6
5. https://www.smithsonianmag.com/innovation/li-fi-replace-wi-fi-180957320/
6. Sakai H, Kawamura T. Light-Emitting Diode. United States patent, US 4,698,730, 1987
7. https://globenewswire.com/news-release/2017/07/26/1062654/0/en/LED-Lighting-Market-Share-Growth-Will-Increase-54-28-Billion-by-2022-Zion-Market-Research.html
8. Williams SK, Millar I. The IrDA Platform. Hewlett-Packard Laboratories Technical Publications Department, 1995. pp. 1–7
9. Xie F, Furon T, Fontaine C. On-Off Keying Modulation and Tardos Fingerprinting. In Proceedings of the 10th ACM workshop on Multimedia and security, 2008. pp. 101–106
10. Fairhurst G. Manchester Encoding. Online Document. 2001
11. Stevens ML, Boroson DM, Caplan DO. A Novel Variable-Rate Pulse-Position Modulation System with Near Quantum Limited Performance. In IEEE 12th Annual Meeting on Lasers and Electro-Optics Society (LEOS), 1999. Vol. 1, pp. 301–302
12. Monteiro E, Hranilovic S. Design and Implementation of Color-Shift Keying for Visible Light Communications. Journal of Lightwave Technology. 2014. Vol. 15, pp. 2053–2060
13. Rahaim MB, Vegni AM, Little TD. A Hybrid Radio Frequency and Broadcast Visible Light Communication System. In IEEE GLOBECOM Workshops, 2011. pp. 792–796
14. Chowdhury H, Katz M. Cooperative Data Download on the Move in Indoor Hybrid (RadioOptical) WLANVLC Hotspot Coverage. Transactions on Emerging Telecommunications Technologies. 2014. Vol. 25, pp. 666–77
15. Huang Z, Ji Y. Design and Demonstration of Room Division Multiplexing-Based Hybrid VLC Network. Chinese Optics Letters. 2013. Vol. 11, pp 16–20

Infrastructure in Mobile Opportunistic Networks

Antriksh Goswami, Ruchir Gupta and Gopal Sharan Parashari

Abstract Opportunistic networks or OPNETs are challenged networks with sporadic communication opportunities and erratic link performance. OPNETs are simply considered complementary to traditional networks particularly at the time of disaster or at the locations without adequate network infrastructure. However, OPNETs are neither obsolete even after unprecedented infrastructural development in traditional networks nor ubiquitously deployed parallel to traditional networks. The need of the hour is to look at these networks with a newer vision considering current developed status of traditional networks. OPNETs can offer cellular network offloading, communication in challenged areas, proximity-based applications and censorship circumvention. This chapter will discuss the concept, protocol, architecture, cooperative framework, routing techniques, privacy issues and future directions of research in OPNETs.

1 Concept, Protocol and Architecture for Mobile Opportunistic Networks

Opportunistic networks are emerging networking paradigms where a source and a destination communicate on the fly depending on the availability of communication links. In these networks, connectivity is sporadic and the communication between mobile devices can be performed in the absence of direct route. This type of network

A. Goswami (✉)
Department of Computer Science and Engineering,
Indian Institute of Information Technology, Design and Manufacturing, Jabalpur, India
e-mail: goswamiantriksh@gmail.com

R. Gupta
Department of Computer Science and Engineering,
Indian Institute of Technology BHU, Varanasi, India
e-mail: ruchir.in@gmail.com

G. S. Parashari
Department of Economic Sciences, Indian Institute of Technology, Kanpur, India
e-mail: gopal.parashari@gmail.com

© Springer Nature Singapore Pte Ltd. 2018
K. V. Arya et al. (eds.), *Emerging Wireless Communication and Network Technologies*,
https://doi.org/10.1007/978-981-13-0396-8_8

does not require the previous knowledge of the network topology. The connection and the disconnection of the devices in the network results in the randomness of the network [1]. This networking paradigm heavily benefits from the heterogeneous networking and extant as well as future communication technologies. Hence, given the recent advances in wireless networking technologies and unprecedented proliferation of mobile devices, opportunistic network seems very much promising for a variety of future mobile applications.

The terms delay-tolerant networks (DTN) and opportunistic networks are often used interchangeably. A number of research challenges have been introduced by mobile opportunistic networks in the field of communication, computing and networking. Messages transferred from a source reaches to a destination device by using various routing protocol [2]. In the next section, we will have a look at all these protocols.

1.1   Routing Protocols

The traditional routing techniques, which are employed in fixed networks, are not applicable in opportunistic routing. In literature, a significant amount of routing techniques has been proposed for opportunistic networks [3]. Normally, it is assumed that in a network, a pre-existing end-to-end path exists between the nodes. However, some mobile opportunistic networks may not adhere to this assumption. Mobile sensor nodes do not always remain active due to limited battery power. Other networks like vehicular network, pocket switched networks and battlefield networks also experience similar disconnections due to mobility, less battery power and node failure. One solution for successful message delivery in such networks is that the source postpones the delivery until the destination comes within the communication range of the source. As destination comes under the range, message is directly delivered to destination. Another solution is to forward the message to all the nodes, which are in communication range. The receiving node carries the message and also forwards the message to the nodes which are in their communication range. Advantage and disadvantage are also bestowed with these solutions. The first method, albeit uses very low resources but has low delivery rate at the same time. The second method has a very high delivery rate but it exploits large amount of resources. The routing protocols can be described by using two types of probabilities, viz. transfer probability and replication probability. Transfer probability is the probability of transferring the message to another node if the nodes meet and replication probability is the probability of retaining the copy of message by the sender after transmitting it. In the first solution, for successful message delivery, the transfer probability for the destination node is one and zero for others, whereas the replication probability is always Zero. The second solution uses transfer probability as one for all nodes, and the node replicates the packet each time it meets another node.

1.1.1 Direct Delivery Protocol

Under this protocol, the delivery of message happens only when the destination comes within the communication range of the source.

1.1.2 Epidemic Routing Protocol

In this protocol, the source sends copy of the messages to each node within its range. Nodes after receiving the message also send the copy of the message to each node within their range. This protocol, though very simple, may consume significant resources like battery power of each node, communication link and storage capacity of each node which is used to keep the copy of message.

In this protocol, the message is expired after some amount of time or after some hop count. A node may receive the same message from multiple nodes, and therefore, it wastes the resources of the network. For this, each node first sends the index message which contains the IDs of the messages which it has already received. Nodes only after receiving this index message transfers the messages that are not yet received by other nodes.

1.1.3 Random Routing

This routing method has the transfer probability to each contact between 0 and 1. In this method, replication probability also lies between 0 and 1, and the message is transferred every time the transfer probability is greater than a threshold value.

1.1.4 PRoPHET Protocol

This is probabilistic routing protocol using history of past encounters and transitivity, which is used to estimate each node's delivery probability for each other node. The delivery probability of node i when it meets node j is updated by

$$p'_{ij} = (1 - p_{ij})p_0 + p_{ij}, \tag{1}$$

where p_0, a design parameter for a given network, is an initial probability. When node i is not in contact with j for some time, the delivery probability decreases by

$$p'_{ij} = \gamma^k p_{ij}. \tag{2}$$

Here, k is the total amount of time units since last update, and γ is the ageing factor ($\gamma < 1$). In PRoPHET protocol, nodes exchange their index messages as well as their delivery probabilities. Each node computes the transitive delivery probability of all the nodes whose delivery probabilities they received from other nodes. For example, when node i receives delivery probabilities for node j, node i may compute the transitive delivery probability through j to z with

$$p'_{iz} = (1 - p_{iz})p_{ij}p_{jz}\lambda + p_{iz} \tag{3}$$

with λ being a design parameter for the impact of transitivity.

1.1.5 Link-State Protocol

This approach assigns the weights to each path which connects the source with destination. The weights may be the median inter-contact duration or exponentially aged inter-contact duration. The formula for exponentially aged inter-contact duration between node i and j is

$$q'_{ij} = \alpha q_{ij} + (1 - \alpha)(t - t_{ij}), \tag{4}$$

where α is the ageing factor, t is the current time and t_{ij} is the time of last contact. Whenever nodes come into the communication range, they share their link-state weights and messages are forwarded to the neighbour which has route with lowest link-state weight to the destination.

1.1.6 Binary Spray and Wait [4]

In spray and wait routing technique, there are two phases spray phase and wait phase. In *spray phase*, source node initially spreads and forwards L copies of every message, where L is any arbitrary number. In *wait phase*, if the destination is not found in the spraying phase, each of the L nodes that is carrying a copy of the message forwards the message only to its destination. As this mechanism does not tell how to spread initial L copies of the message, a new protocol, viz. *binary spray and wait* is proposed. In this protocol, the source initially transfers the L copies of the message. After this any node A that has $n(>1)$ copies of message, and encounters another node B with no copies, hands over the $n/2$ message copies to B and keeps $n/2$ for itself. When it is left with only one copy in this process, it switches to direct transmission.

2 Cooperative Framework for Building Opportunistic Networks

Ad hoc mobile networks use mobile data for taking decisions on various types of works like building sensing maps, taking the atmospheric decisions, taking environmental decisions. This data is usually transferred by the Internet connections.

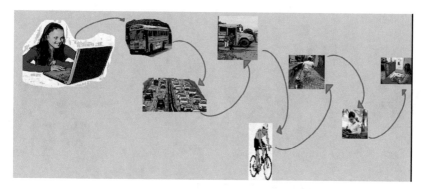

Fig. 1 Opportunistic routing with cooperation

However, an Internet connection is not always available due to poor connectivity or expensive network access (e.g. in disaster or war like scenarios). Due to this, there should be a solution which can function in all critical scenarios [5]. The network of mobile nodes without fixed infrastructure can help in such type of information-intensive applications as shown in Fig. 1 as example. Existing data forwarding techniques are not sufficient for these types of applications as spatial–temporal correlation among the sensed data has not been explored [6]. For implementing forwarding scheme which can manage these sensing data with low delay and energy consumption, a cooperative sensing and data forwarding framework is necessary. Cooperative data scheme can eliminate sampling redundancy and hence save energy. Two proposed cooperative data forwarding schemes are epidemic routing with data fusion and spray—wait data with fusion [4]. These techniques take into the consideration that there is no central infrastructure or communication management. Nodes when forwards the data also integrates the data on the basis of their proximity and temporal information. This is because when users require the sensory data, they are only interested in the aggregated results of the sensory data. For example, only the average value of the relevant parameter may be of interest. This scheme removes the redundancy in the data and hence also save the energy for forwarding the redundant data. For example, the data of close proximity can be of high correlation, which can be integrated to remove redundancy. In this scheme, the forwarding is based on correlated data packets. Which means, independent data packet forwarding is not assumed, and this makes it difficult to analyze theoretically. An ordinary differential equation model is proposed and used for analysis. In addition to the above-discussed application of opportunistic network, opportunistic computing [7] may be another application. The pervasive mobile devices can come together and collectively solve some problem. This is only possible with cooperative devices in the opportunistic network. Although wireless LAN, cellular network and other wireless technologies are available, these are not efficient for opportunistic computing [7]. By applying the opportunistic computing using opportunistic networking, the vision is to increase the computing power and enrich the mobile devices with better processing abilities.

Fig. 2 Opportunistic computing in healthcare scenario

This computing can be used in the pervasive healthcare applications, military communication systems, intelligent transportation system and crisis management systems [7]. In the intelligent healthcare system, for example, in Fig. 1, a patient who wants treatment from the healthcare centre has his own sensor which is sensing different parameters of the body. The sensed data of the patient is collected in the mobile device with patient. This data can also be enriched by combining with the other environmental sensor information and other devices information which are surrounding the patient. This data may be processed with OC (opportunistic computing) or without OC (left part of Fig. 2). A human will have his own personal device with other equipment required for sensing different body parts for building personal body area network. Required signals sensed from the personal body area network will be sent to the healthcare centre. In this process, the sensed signals can be enriched by taking the sensed data of the neighbouring users of the patient and sensed data from the environmental sensors. The additional data informs the healthcare centre about the environmental conditions of surroundings of the patient. This helps the computing facility at healthcare centre in producing more accurate results. Healthcare centre will perform computing on the received signals and results are sent back to the user device for triggering some actions. In this way, the cooperation of the other neighbouring nodes involves but it is not efficient. Opportunistic computing solution simplifies this as in the right frame of Fig. 2. In this, the patient device using opportunistic computing will automatically exploit all the signal from his body sensors and from the other surrounding sensors. The signals from the devices around him will be given to particular service components available on one of the devices. These components are composed and executed in distributed fashion. In this way, more effective solution can be produced as the devices around the patient can identify the services available on the devices around the monitored person which can be combined the information with the patient's own sensors. Moreover, the overhead to the network traffic can be reduced using OC.

The social network of mobile devices can also be used for computing facilities [7]. The resource of a mobile device can also be used and shared with other devices when come into contact opportunistically. In this way, a richer functionality of the different resources available in the network can be composed. Other applications of cooperative mobile opportunistic network include executing the task remotely and exchanging the information.

3   Recent Advances in Routing Methodologies

In the last decade, routing protocols of mobile opportunistic networks attracted considerable attention. This has resulted in many advances in routing methodologies. A wide range of methodologies has been designed which covers from modification to wired network routing algorithms to new routing algorithms. These protocols mostly use the multi-hop routing [8] which enables them to forward the message to a long range in comparison with a limited single node communication range, by exploiting neighbour nodes as relay nodes. Multi-hop forwarding techniques also provide multiple paths between the source and destination. Even if there is direct path available between two nodes, it may be possible that it is already congested or link quality of path is poor. Multi-hop forwarding techniques strengthen communication path between source and destination [8]. But it does not exploit all the benefits of wireless communication, viz. broadcasting. This is due to the fact that in multi-hop protocol design, a node discards all the packets at data link layer that is not destined to this node. In this way, a node may remain unexploited which is closer to the destination than the source. With the use of forward error control (FEC) and automatic repeat request (ARQ), multi-hop techniques oppose the time-variant impairment of wireless propagation [9].

For example, as in Fig. 3, if in a scenario there are three nodes 'A', 'B' and 'C'. Node 'A' finds a packet for node 'B'. As compare to node 'A', another node 'C' is closer to node 'B'. If node 'A' uses the multi-hop transfer mechanism for transferring the packet and it chooses node 'B' as next hop. In this, if node 'A' is not reachable to node 'B' then routing cannot be successful although there is a path available to node 'B' via 'C'. Whereas if node 'A' chooses the node 'C' as next hop, which may be most reliable link, then it does not take advantage of wireless propagation in which 'B' directly receives packet. The point in this example is that in both the

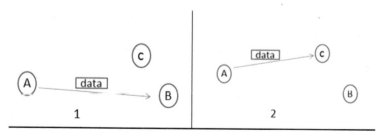

Fig. 3 Multi-hop routing

decision of choosing the next hop, node 'A' is not exploiting all the benefits of wireless communication. In contrast to this, opportunistic routing only requires the node to directly transfer the packet without selecting the next hop. In this way, most of the possible advantages of the wireless network can be exploited.

Node mobility and wireless network instability provide the ad hoc networks with spatial diversity and distributed nature. Due to this property of the ad hoc networks, a neighbour that was selected as a next hop previously for data forwarding may move outside, and a more favourable node may now become a new and unexplored neighbour. By using opportunities of wireless propagation in this network, even if more favourable hop moves outside the network, some other less favourable node may propagate the packet to the destination or may also take advantage from a more favourable node that just came into the visibility of the sender and still unexplored. Such opportunities increase the probability of successful transmission of packet but on the cost of routing overhead.

3.1 Challenges in Opportunistic Routing

Achieving the maximum routing progress with minimum number of copies of the same packet and minimum coordination overhead is the major challenge of opportunistic routing [3]. In order to utilize the possible benefits of opportunistic routing and to avoid the above-mentioned problems, an effective protocol requires to implement the following tasks [3]:

1. Candidate selection,
2. Forwarder election,
3. Forwarding responsibility transfer and
4. Duplicate transmission avoidance.

Candidate selection selects the possible next hop. The next hope may be any neighbouring node whose distance to the destination is lesser than the forwarder. Indeed, it is not a good practice to forward the message to the nodes which are farther away from the destination than the actual forwarder, as in that case, message instead of traversing towards the destination moves away from it. This is to be noted that the more precise is the forwarder election, the smaller is the amount of coordination overhead. The *forwarder election* provides a mechanism to choose the node that is nearest to the destination, among all the nodes selected from the candidate selection phase and who have successfully received the packet. Particularly, this process allows the node to select the next hop at the receiver side. Obviously, better this process is, more is the increase in throughput.

The forwarder node and candidate nodes are jointly allowed by *forwarding responsibility transfer* to become aware of the winner of the election. Here, the unique feature that helps in differentiating opportunistic routing from flooding is responsibility transfer. Though in both, flooding and opportunistic routing, several nodes receive the same packet but opportunistic routing allows only single node per time interval to be in charge of packet forwarding, which is totally different from flooding. If implemented efficiently, it causes less duplicate transmissions, which in turn reduces overhead generated by the duplicate transmission avoidance mechanism. However, if implemented incorrectly, several packet transmissions may occur in spite of only one innovative packet transfer(by winning forwarder). This will cause throughput loss to a network implying need of an effective mechanism to stop useless transmissions. This mechanism is known as *Duplicate transmission avoidance* mechanism.

3.1.1 Game-Based Data-Forward Decision Mechanism for Opportunistic Networks [10]

The routing techniques discussed till now, take decision at the sender level, i.e. whether to forward the message or not. In this technique, the receiving node is also involved in taking the routing decision that whether it will accept the forwarded message or not. Aiming at the problem of forward decision of the routing pattern, By selecting four important context parameters and combining them with *Kalman filter*, this mechanism provides a utility function. By using this utility function, it proposes the two-player game between sender node and receiver node. With this mechanism, the receiver node also involves into the decision process and optimizes the performance and balances its load. This routing mechanism exhibits the high delivery ratio and full consumption of forwarding capacity.

3.1.2 Game Theory-Based Routing Scheme (GTR) [11]

Routing scheme that we have studied so far are mainly of two categories, viz. multicopy-based routing and single copy-based routing. Multicopy-based routing schemes, though robust in delivery of data to the destination, consume a large amount of resources of the network. Delivery ratio of such schemes are high and the transmission time is very low. The disadvantage of these types of mechanisms are only that these consume a large amount of resources. Opposite to these mechanisms, single copy-based routing mechanism consume less amount of resources but there delivery ratio is low and transmission time is very high. Hence, a mechanism which provide the trade-off between these two types of mechanism will be efficient. *GTR* is such type of mechanism that takes this trade-off into account. For this, the residual resources (energy, bandwidth and buffer space) of a node are defined after time t. The delivery probability for each node, similar to PRoPHET, is also calculated in this mechanism. After this, a multiplayer bargaining game is defined. These games are non-zero-sum games in which participating players try to achieve win-win situation. Sender node computes the payoff of all other nodes within its communication range by utilizing the delivery probability and residual resources. After calculating this payoff, the sender node takes the routing decision based on this payoff. By using this mechanism, a node calculates which message can be delivered to which node, *when to make duplicate copies of a message and how many duplicate message copies should be delivered in the networks.* GTR performs better than some routing techniques in terms of packet delivery ratio, average hop and average delay.

3.1.3 Social Graph-Based Routing

Some routing techniques use the social information of the node like moving pattern and knowledge about its interaction with other nodes [12]. By using this information, the routing can become simpler. However, these techniques raise the privacy and security issue of the nodes. These issues will be discussed separately in the later section.

4 Mobile Peer-to-Peer Content Dissemination Model in Load Balancing

Nodes in opportunistic networks communicate with the help of error-prone and unstable links. Each node transfers the messages to their connected node by using these channels and some routing techniques as discussed in previous section. As all these channels' capacity is limited, therefore the load balancing to these links is required [13]. Thus, the load balancing mechanism becomes crucial to the network along with the routing techniques. Load balancing and reliability cannot be achieved together as most of the load balancing techniques are not robust for high link-failure rate [14]. There are two types of load balancing techniques available in literature, viz. local load balancing and global load balancing. In local load balancing technique, for informing the energy change, the 'hello' message is sent to each neighbour by each node. By using the interval of this 'hello' message, the control overhead and the timeliness of the energy change can be determined. By using local load balancing technique, a data packet may enter into 'energy bottleneck' region where the energy of nodes is less whereas the energy of the node outside the region may be high. Due to this reason, global load balancing is used so that the higher energy path can be achieved.

Mobile opportunistic network can itself be used for load balancing in cellular network [15]. In this, the cellular network traffic produced from mobile devices is delivered over the complementary network made up of opportunistic network. Although this traffic is originally produced for transmission over cellular connection, it is actually transferred over opportunistic network. This is called mobile data offloading. The offloading process exploits the automatically build social network in mobile ad hoc networking. The devices which are in the network can be seen as member of the social network. This whole network can be called as mobile social network (MoSoNet). In the mobile social network, the devices which are in the vicinity of any device is called the neighbours of the later. Any device first communicates

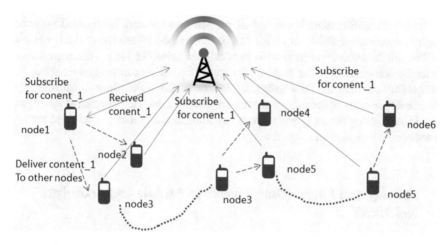

Fig. 4 Mobile data offloading

within its own social network to make up its neighbours connection. Then, neighbours communicate with their neighbours and so on. In this way, the information, that any device wants to transmit, flows. A number of schemes are proposed in the literature for mobile data offloading which uses MoSoNets. In one of the schemes [15] of mobile data offloading, first k target users of opportunistic network are selected for transferring the cellular data to these users. For example, in Fig. 4, the value of k is 1. The objective of this selection is to minimize the traffic over the cellular network or can be translated as to maximize the expected number of receiver of the information over opportunistic network. As the dissemination of information in mobile ad hoc network is analogous to susceptible-infected-recover (SIR) epidemic spreading model in social network, this maximization problem can be seen as maximizing the number of infected nodes. The target set is identified by monitoring the pattern of human mobility. The history of mobility is used for information delivery in the future. For example, for a given time period, based on the user mobility pattern, target set is identified and then in future during the same time period this target set is used.

The data which can be offload over mobile social network includes weather forecast, movie trailers, multimedia newspapers, etc. which is generated by the content service providers. Multimedia newspapers include written text, photos, audio, video clips and some interactive games. To take the leverage of the delay-tolerant nature of non-real-time information and application, service providers selects only a few users (target user) for transferring the information. This minimizes the mobile network's traffic. Hence, in the process of mobile data offloading, at first, the bootstrapping occurs in which the target set (node 1 in Fig. 3) is selected for the mobile data offloading. These users then further propagate the information to the subscribed users in their social network, when their mobiles are within the transmission range of each other. In this way, the load on the mobile cellular network can be minimized.

The content dissemination in mobile peer-to-peer network is capable to support various emerging applications. This can be understood by analysing the behaviour of content dissemination in mobile peer-to-peer networks [16]. This requires the investigation about the time required for spreading the data in the given region, the probability to reach the data within a given time and to the given destination, the probability bounds of the minimum time required for reaching the data at the region that is sufficiently far away and the rate at which such a probability tends to zero as the distance increases to infinity.

5 Privacy and Forwarding Models in Mobile Opportunistic Networks

Social network information is used for effective routing in social network routing protocols. However, this may cause privacy breach and security threats for users, and this in turn discourages the participation [17]. Moreover, the confidentiality of the payload carrying by data packets is also required. If confidentiality is not ensured, message contents may get disclosed which enables any untrusted or unintended user to read the message. This threat may be alleviated using encryption.

Another threat, communication patterns can be determined by tracing the messages as they progress through the opportunistic network. This does not require the attacker to read the message content. Only by intruding on multiple encounters and detecting the same message is transmitted can provide the communication pattern. Location of any node can also be determined by exchanging the message since the message can only be exchanged when both nodes are in physical proximity.

Humans carry mobile devices and their mobility generates communication opportunities. Whenever user devices are not in direct communication range and if one wants to transfer data, then path is created opportunistically using other devices and data is transferred. This allows the users to communicate even if their direct connection is intermittent. In this way, the critical real-time application can also be handled as this type of applications continuously required connectivity. But this lead to another privacy threat, i.e. identification of location of the source devices. Therefore, there should be some location privacy-aware algorithms that may prevent this threat. For protecting the users' location and making them certain about superior privacy, a k-anonymity protocol (LPAF) [18] is proposed which is fully distributed meaning without any central server and synergetic. Location-based services (LBS) application requires the location information of the source for producing the desired results [18]. The lightweight multi-hop Markov-based stochastic model may be utilized for location prediction and to direct the queries towards the LBS location and to reduce the required resource in terms of retransmission overhead. This protocol ensures the location privacy for the LBS as it does not require user identity. The main idea in this protocol is to take the leverage of the social network formed by the mobile devices. It choose the path which hides the real sender of the data during the location obfuscation. Location obfuscation is a process to conceal the location information of the user from the location-based services. Zakhary et al. [18] propose a lightweight Markov model to implement the privacy-preserving protocol using a stochastic model for location prediction. The implementation is based on two different possibilities which depends on whether the nodes can locate them self (GPS coordinates) or not [18]. If the node can locate them self, then path prediction can be maintained locally using the Markov model proximity. If the node cannot locate them self, then recorded IDs of the sighting of fixed access points or GSM communication cell IDs can be used as the location identifier. The local predictions of nodes is being exchanged by the nodes when they meet. This exchange help them to obtain more accurate estimate of prediction [18]. In the location obfuscation, a node with some queries to obfuscate, explore best neighbours to forward these queries to increase the privacy by keeping the query within the social group.

6 Concept for Pocket Switched Networks, Amorphous and Semantic Opportunistic Networks

Devices, which can be kept in the pocket and remain always on, can collectively create a network communication paradigm which is based on opportunistic contacts between mobile devices and human mobility [19]. This network communication paradigm is called Pocket switch network. These networks use the contacts which is opportunistically created between mobile devices and the mobility of humans to spread the data without the depending on any infrastructure. It makes use of both human mobility and local connectivity in order to transfer information between different users' devices. It is also aimed at enabling network services for mobile users even when they are not in reach of direct network connectivity like MANETs, mobile social networking. devices, in pocket switched networks, include mobile phones and other hand handle devices.

6.1 Amorphous Opportunistic Networks

In some situations, a decentralized network is naturally formed due to the population of people with same interest. This type of networks is called amorphous opportunistic networks [20]. As these networks consist of a group of people who are unknown to each other, therefore, it is also insecure for the group members. Semantic opportunistic networks and Freenet are the most frequently used amorphous opportunistic networks where hosts create appropriate group of same issues to reveal conditional similarities [20]. This network has encouraged the research in many related fields. Involvement of social networking with these types of networks enables it to be a solution for content search network.

7 Future Research Trends in Mobile Opportunistic Networks

The advancement in the electronics and networking technology has made availability and connectivity of various handheld movable devices everywhere. These advancements have given rise to new computing possibilities in the form of Internet of Things (IoT) [21]. Internet of Things allows the development of various applications, of which only a few are currently available. As these applications will be equipped with objects of growing intelligence therefore these applications would definitely improve the quality of our lives. For example, application in the healthcare domains, in the transportation and in the logistics domains will require permanent connectivity of the mobile devices to the Internet so that they can communicate with each other and reveal the information sensed from the surroundings. MANET makes it simple to avail such type of requirement to these devices, and therefore, it is suitable to this type of applications. The handover of the mobile objects while they are moving between different subnetworks should be seamless in MANET so that they can support to sensible applications. The future research in this area can focus on reducing the handover time, so that it can become seamless, and controlling the network congestion. The solutions may be in the form of reactive ad hoc routing protocols and proactive agent advertisement approach. Along with this direction of future work, there are some issues that must also be investigated:

- Route holding time should be reduced on reactive protocols so that the declaration time of broken route can be reduced.
- As the handover occurs the management of IP addresses of mobile object should be done to handle the address collision discovery so that seamless handover becomes possible.
- While in handover the old registration should be maintained so that if actual one is lost, it can work as backup registration. Hence, further work on handover is necessary to assess the utilization of multiple agent registrations.
- To reduce the handover time, multiple routes can be maintained to the gateway. With this, agent will not lose the information even if the route is broken. The management of these multiple routes still requires some feasible solution.
- MANET is preferred type of network in 4G wireless systems. As these are far from structured networks, the performance during handovers must be investigated.
- Efficient localization of mobile devices is crucial for monitoring applications; therefore, some new and robust techniques should be explored.
- For opportunistic networks, all the protocols which consist of medium access layer, transport and routing should be reinvestigated. This is required to deal with the energy and bandwidth constraints. Additionally, it would also address the issue of intermittent connectivity.

8 Conclusion

Opportunistic networks are an emerging and rapidly growing way of communication. This chapter discusses issues, advancements and applications of opportunistic networks. Routing in opportunistic networking is has attracted considerable attention. Privacy and security is the main concern in using these networks. The advancement in the wireless technology communication making it widespread throughout the world. Due to the evolution of mobile social networks, opportunistic network is becoming more prominent and easy to use. This network is most suitable for different types of application domain, viz. e-commerce and emergency services. New proposals in the routing protocol, load balancing, content dissemination, privacy and security are making it stable and useful to different areas of communication. Due to the ease of manage infrastructure, this is fascinating the world. This networking style is most suitable for the information and application which are non-real-time and delay tolerant in nature. We can say that the opportunistic network will be the next generation of networking technology.

References

1. Denko, Mieso K., ed. Mobile Opportunistic Networks: Architectures, Protocols and Applications. CRC Press, 2016.
2. Loo, Jonathan, Jaime Lloret Mauri, and Jess Hamilton Ortiz, eds. Mobile ad hoc networks: current status and future trends. CRC Press, 2016.
3. Song, Libo, and David F. Kotz. "Routing in Mobile Opportunistic Networks." Mobile Opportunistic Networks: Architectures, Protocols and Applications (2016): 1.
4. Xue, Jingfeng, et al. "Spray and wait routing based on average delivery probability in delay tolerant network." Networks Security, Wireless Communications and Trusted Computing, 2009. NSWCTC'09. International Conference on. Vol. 2. IEEE, 2009.
5. Pelusi, Luciana, Andrea Passarella, and Marco Conti. "Opportunistic networking: data forwarding in disconnected mobile ad hoc networks." IEEE communications Magazine 44, no. 11 (2006).
6. Zhao, Dong, et al. "COUPON: A cooperative framework for building sensing maps in mobile opportunistic networks." IEEE transactions on parallel and distributed systems 26.2 (2015): 392–402.
7. Conti, Marco, Silvia Giordano, Martin May, and Andrea Passarella. "From opportunistic networks to opportunistic computing." IEEE Communications Magazine 48, no. 9 (2010).
8. Biswas, Sanjit, and Robert Morris. "Opportunistic routing in multi-hop wireless networks." ACM SIGCOMM Computer Communication Review 34.1 (2004): 69–74.
9. Mei, Haibo, Peng Jiang, and John Bigham. "Augment delay tolerant networking routing to extend wireless network coverage." Wireless Communications and Signal Processing (WCSP), 2011 International Conference on. IEEE, 2011.
10. Zhang, Cheng, Qing-sheng Zhu, and Zi-yu Chen. "Game-based data-forward decision mechanism for opportunistic networks." Journal of computers 5.2 (2010): 298–305.
11. Li, Li, Yang Qin, and Xiaoxiong Zhong. "A novel routing scheme for resource-constraint opportunistic networks: A cooperative multiplayer bargaining game approach." IEEE Transactions on Vehicular Technology 65.8 (2016): 6547–6561.
12. Mtibaa, Abderrahmen, et al. "Peoplerank: Social opportunistic forwarding." Infocom, 2010 Proceedings IEEE. IEEE, 2010.

13. Cetinkaya, Coskun, and Edward W. Knightly. "Opportunistic traffic scheduling over multiple network paths." INFOCOM 2004. Twenty-third Annual Joint Conference of the IEEE Computer and Communications Societies. Vol. 3. IEEE, 2004.
14. Ganjali, Yashar, and Abtin Keshavarzian. "Load balancing in ad hoc networks: single-path routing vs. multi-path routing." INFOCOM 2004. Twenty-third Annual Joint Conference of the IEEE Computer and Communications Societies. Vol. 2. IEEE, 2004.
15. Han, Bo, Pan Hui, VS Anil Kumar, Madhav V. Marathe, Jianhua Shao, and Aravind Srinivasan. "Mobile data offloading through opportunistic communications and social participation." IEEE Transactions on Mobile Computing 11, no. 5 (2012): 821–834.
16. Wang, Shengling, Xia Wang, Xiuzhen Cheng, Jianhui Huang, Rongfang Bie, and Feng Zhao. "Fundamental Analysis on Data Dissemination in Mobile Opportunistic Networks With Lvy Mobility." IEEE Transactions on Vehicular Technology 66, no. 5 (2017): 4173–4187.
17. Parris, Iain, and Tristan Henderson. "The impact of location privacy on opportunistic networks." World of Wireless, Mobile and Multimedia Networks (WoWMoM), 2011 IEEE International Symposium on a. IEEE, 2011.
18. Zakhary, Sameh, Milena Radenkovic, and Abderrahim Benslimane. "Efficient location privacy-aware forwarding in opportunistic mobile networks." IEEE Transactions on Vehicular Technology 63, no. 2 (2014): 893–906.
19. Hui, Pan, et al. "Pocket switched networks and human mobility in conference environments." Proceedings of the 2005 ACM SIGCOMM workshop on Delay-tolerant networking. ACM, 2005.
20. Rao, K. Srihari, Mrs Lokeswari, and J. Srikanth. "Security in unstructured opportunistic networks: Vague communication and counter intelligence functionality." Int. J. Comput. Sci. Telecommun 3.1 (2012): 54–61.
21. Leal, Bernardo, and Luigi Atzori. "Connecting Moving Smart objects to the internet: Potentialities and issues When Using Mobile Ad Hoc network technologies." Mobile Ad Hoc Networks: Current Status and Future Trends (2016): 313.

Generic Design and Advances in Wearable Sensor Technology

Siddig Gomha and Khalid M. Ibrahim

Abstract Rapid development in telecommunication, microelectronics, sensors, and material science creates new opportunities for the interaction between human body and wearable sensors. The wearable devices fixed on the human body can sense, analyze, and transmit data through a wireless to a terminal device for more processing; these collected data will provide a health care of human being. Wearable sensor technology has many applications in medical and health care, monitoring elderly and chronically sick persons, monitoring player in such sports, education and teaching assistance, tracking and monitoring human/animals, and security. However, to design and implementing wearable devices, a lot of obstacles will have to be overcome to inexpensively develop the active electrical devices on elastic substrate taking advantage of macroscale fabrication techniques. The wearable devices system is consist of sensors (e.g., temperature and pressure sensors, gyroscope), low-power embedded system, and wireless transceiver system (e.g., Wi-Fi, Bluetooth, ZigBee); all these devices are integrated into one system to transmit and receive data. This chapter will shade the light for these obstacles toward realizing a robust and reliable integrated wearable system.

Keywords Wearable sensors · Body area network · Healthcare monitoring
Power harvesting for wearable devices system · Smart glass · Augmented reality

S. Gomha (✉)
Pan African University Institute for Basic Science Technology and Innovation (PAUSTI),
Nairobi, Kenya
e-mail: siddig.gomha@gmail.com

S. Gomha · K. M. Ibrahim
Faculty of Engineering, University of Medical Sciences and Technology (UMST),
Khartoum, Sudan
e-mail: khmoibrahim@gmail.com

© Springer Nature Singapore Pte Ltd. 2018
K. V. Arya et al. (eds.), *Emerging Wireless Communication and Network Technologies*,
https://doi.org/10.1007/978-981-13-0396-8_9

1   Introduction of Wearable Sensor-Based Systems—Design and Architecture

Wearable sensors have received considerable interest over the past decade due to their tremendous promise for monitoring the wearer's health, fitness, and his surroundings [1]. Wearable sensor networks are considered as main part of a multistage system of Wireless Sensor Network (WSN) as illustrated in Fig. 1. The architecture of this network includes in its first stage the sensor nodes attached to the human body, which are integrated into the wireless Body Area Network (BAN). Each node contains embedded system to sense, sample, and process one or more physiological signals. The second stage (e.g., mobile tire) works as sink node to aggregate all transmitted data from the sensor nodes; in the last stage, the collected data is transmitted through GSM or Internet to the central remote server [2, 3].

Each node contains microcontroller, transmitter, and energy harvesting and storage system, in designing wearable electronic devices considering the different characteristics, such as low-power electronics, energy harvesting, energy storage, wireless transmission, microsensors, and system integration [4].

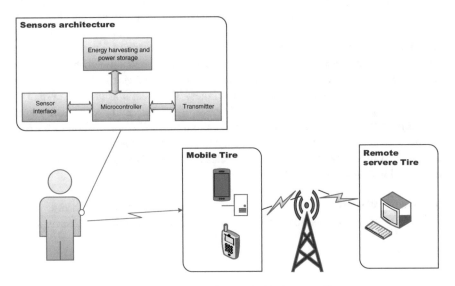

Fig. 1   The whole system architecture and main parts inside the wearable sensor

2 Challenges and Issue in Wearable Computing and Sensor Implications

The most important applications of wearable devices in our daily life are monitoring health care, tracking person's movements, and a wearable sensor system can provide a documented record for healthcare condition, as well as gives a feedback in real time; thus, the human being's life will be improved. Furthermore, integration of wearable systems with breakthroughs technologies [e.g., Internet of Things (IoT), Augmented Reality (AR), and Artificial Intelligence (AI)] comes with a pledge of carrying out an exciting new standard of people convenience.

Wearable devices, particularly used in combination with smart textiles, are a powerful candidate in getting to be the unique interface between people along with the digital community, substituting or just extending smartphone as well as other handheld linked items. Wearable devices are able to integrate these kinds of advantages of information economic system in lots of fields including, medical care, production, education and learning, energy, and safety [5].

The forecasting of wearables market in the next 5–10 years from now would be to increase into a multibillion-euro Internet marketing business, as reported by Gartner, product sales of wearables are going to grow from 275 million items in 2016 to 477 million items in 2020; this corresponds to a $61.7 billion money coming in by 2020, as well as stated by HIS forecasts, unit shipments are going to achieve 320 million by 2020, which represents approximately more than $40 billion in earnings [5].

Wearable devices industry faces many challenge and obstacles to design reliable, low cost, and high-performance wearable system. The key challenges toward the successful realization of effective wearable sensor systems are related to materials, power consumption, analytical procedure, communication, data acquisition, processing, and security. The criteria of wearable devices include small size, rigidity, flexibility, lightweight, and stability. The present wearable power sources are unable to meet all the requirements for wearable electronics owing to their low energy densities and slow recharging. There are several energy harvesting challenge issues, including unreliable power supply and limited stability. There are also major challenges relating to handling and securing the big data generated by wearable sensors [1].

The most significant issues are the improvement of intelligent signal processing, information technical analysis as well as interpretation, communication specifications interoperability, electronic modules performance, and system integration [6].

Regarding the wireless network of the wearable sensors, there are several challenges: the physical layer suffers from some challenges when it comes to implementing wireless body area networks, e.g., bandwidth limitations, receiver complexity, and power consumption that is higher in dynamic conditions, and many researches have been carried out to tackle these challenges [2].

3 Wireless Communication Standards Used in Wearable Technology

The main component of a Wireless Sensor Networks (WSNs) is the node, which is a small module contains small processing unit to process the signals received from the sensor; furthermore, it has a transceiver unit to transmit/receive data from the other nodes and the main central node, and in case of the Body Sensor Networks (BANs), nodes are distributed over the body area. In most cases, all the nodes are similar in the network. The information collected by the nodes flow in hierarchy form toward the sink node, or access point, and central node is usually more powerful and provides access to the WSN information.

In the overall context of the BANs technology, data needs to be transmitted over two transmission links: the first link is short-range communication between the nodes and central node, and in this tire, the gathered signals are transmitted to the central node; and the second transmission link is long-range communication for sending the collected data from the central node to a remote medical station (e.g., cell phone, computer) as shown in Fig. 2.

3.1 Short-Range Wireless Communication Standards for Wearable Sensors

Transmission of data in terms of short-range between nodes can be performed through physical wires or wireless links. Bluetooth IEEE 802.15.1 and Zigbee 802.15.4 are short-range wireless communication technology used to connect nodes in BAN networks.

ZigBee is a short-range wireless technology designed as remote control and wireless sensor applications, which is appropriate for operation in many radio environments. ZigBee technology is built on IEEE 802.15.4 standard; it works on frequency

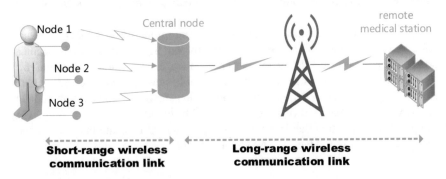

Fig. 2 The (short/long) wireless link between human body and medical station

of 2.4 GHz and 868 MHz/900 MHz bands; it has many features such as very low-power consumption and very low complexity; and it is not required to have wide bandwidth, but it is necessary to consume low power to save the battery lifetime for the low data rates nodes that have built-in battery. The IEEE 802.15.4 standard relates to the physical and media access controller layers, at the same time the ZigBee specifications cover network security, and application layers. The IEEE 802.15.4 standard for ZigBee has many features; can facilitate star and peer-to-peer topologies; uses 64-bit and 16-bit short addressing, providing up to 65,000 nodes per network; and supports up to 128 bytes packet size [7].

Bluetooth is another wireless technology for short-range application to connect (portable/fixed) devices; it has numerous features such as consume low power, inexpensive, and working in the free spectrum band 2.4 GHz; it uses more than 79 channels in the ISM (Industrial, Scientific, and Medical) radio band to reduce interference and fading; and Bluetooth technology has the ability to transmit/receive data for the distance up to 100 m, with data rate up to 3 Mbps in the enhanced mode. The supported topology network by Bluetooth is star network with one master and seven slaves; the Bluetooth consumes low energy which is suitable for the low energy systems. Furthermore, Bluetooth 3.0 utilizes the Wi-Fi (physical and MAC) layers for higher data throughput.

Infrared Data Association (IrDA) is a wireless technology for short-range BAN communication, is a cheap cost communication protocol for short-range data exchange over infrared light IrDA includes many technologies, infrared, the Medical Implant Communication Service (MICS), and Ultra Wideband (UWB). The advantage of IrDA is a little power consumption, and it has data rate up to 16 Mb/s, the main disadvantage of IrDA is the need of Line of Sight (LOS) communication, therefore making it unreasonable for Body Sensor Network applications. However, MICS is an ultralow power, free mobile service for transmitting low rate of data in medical devices. IrDA uses the frequency band from 402 to 405 MHz, and the radiated power is limited to 25 μW [8].

The main disadvantage for Zigbee and Bluetooth wireless technology are not able to support high data rate multimedia applications, in 2003 the IEEE 802.15.3 released high data rate standard (e.g., 11–55 Mbps) with a focus on MAC and physical layer specifications (IEEE 2009). The WiMedia alliance consists of more than 350 groups of company and organizations which have approved Multiband-Orthogonal Frequency Division Multiplex (MB-OFDM) based Ultra Wide Band technology (UWB) as the empowering technology for high rate Personal Area Networks (PANs) (WiMedia 2010). WiMedia technology based on wireless USB has the ability to transmit and receive data for the distance range 3–10 m with data rate speed from 110 to 480 Mbps, and frequency operation is in the range of 3.1–10.6 GHz. Recently, the WiMedia Alliance released an improved UWB PHY standard can handle high-speed data rate up to 1,024 Mbps [9]. Table 1 summarizes the most common short-range wireless communication standard.

Table 1 Summary of common short-range wireless communication standard

Tech. standard	Freq.	Range (m)	Data rate	Current drain	Cost/node
Zigbee (802.15.4)	868 MHz 915 MHz 2.4 GHz	10–75	20 kbps 40 kbps 250 kbps	Very low (20–50 μA)	Very low
WiMedia (802.15.3)	3.1–10.6 GHz	~10	480 Mbps	Up to 400 mA	Medium
Bluetooth (802.15.1)	204 GHz	10–100	1–3 Mbps	1–60 mA	Low
IrDA	IR	200	54 Mbps	–	Low

3.2 Long-Range Wireless Communication Standards for Wearable Sensors

Regarding second transmission link long-range communication between the sensor nodes and a remote station, there are many wireless systems existing that can fulfill that goal. Such technologies are included in the wireless network generations (e.g., 2G, 3G, 4G); these network generations can offer extensive network coverage. Furthermore, the advances of the future fifth-generation (5G) of mobile communication systems are on the way, and it was anticipated that would ensure worldwide consistent access to the web at much higher data rates, and thus to meet the demand for more efficiently, we need to gather medical information from wearable system in a real time and send the data to the remote location for more processing. Table 2 shows summary of wireless network generations [10].

Table 2 Summary of common long-range wireless communication standard

	Freq.	Data rate	Standard	Service
2G	800, 900, 1800, 1900 MHz	9.6–14.4 kbps	GSM, and (GPRS, EDG in 2.5G)	Digital voice, SMS
3G	800, 900, 1800, 1900 MHz	2 MBs	UMTS, 3GPP, CDMA	Broadband data, multimedia
4G	800 MHz, 2.6 GHz	Up to 1 GB	LTE standard, WiMAX Release 2	Complete IP-based, High stream multimedia

4 Design Antenna for Wearable Devices System

Design and implement antennas for a wearable device is another challenging issue because body tissues reduce the functionality of the antenna by absorbing the power radiated from the antenna and changing the input impedance of the antenna [11]. Several antennas have been developed in different shapes and techniques: textile antennas (e.g., shirts, suits) for GSM/PCS/WLAN communications [12], broadband textile-based UHF RFID tag antenna [13], stretchable and flexible textile antenna which can be implantable or wearable [14], and tunable antenna which was designed to work in ISM 433 MHz band that is customized for wearable wireless sensor applications; it is operating well across several body locations [15], and also in [11], a slot antenna has been implemented using 3D printing technology and CIP techniques. The slot was shaped by painting silver ink on the inner side of a 3D printed plastic box, the proposed antenna has 39% bandwidth at 10 dB and center frequency of 2.5 GHz, and the total radiation efficiency (simulated) of the antenna is 55% on body and 90% in free space.

5 Inferences of Body Area Network System Architectures

Body Sensor Network (BSN) architecture is shown in Fig. 3, and the system is constituted by three layers (tires) that can be identified as sensor tire, mobile tire, and remote server tire. The following sections show more detail about each tire.

5.1 Sensors Tire

A collection of sensors are fixed on the human body for collecting biomedical signals and sending them to the central receiver called sink node, in this tire a star network is commonly used due to the short distance ranges [16].

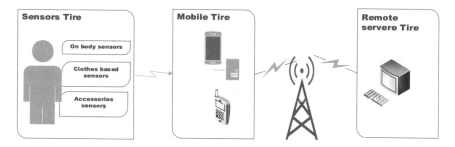

Fig. 3 Body Sensor Network (BSN) system tiers

There are many types of sensors attached to the human body, electrochemical sensors as wearable accessories (e.g., belt, armband), electrochemical sensors combined into dresses of human (e.g., shirt, gloves), and electrochemical sensors that are applied to the body (e.g., skin patch) [17].

The transmitted data from the sensors is gathered and sent to the central node of BAN, which can be a custom-designed microcontroller, PDA, smartphone, or pocket PC [8].

5.2 Mobile Tire

The mobile tier may consist of either Android smartphone or computer known as central hubs. The mobile phone works as a gateway node which aggregates the information flow from the sensor nodes and forwards it to the access point. The IEEE 802.11/Wi-Fi/GPRS standards are used as communication medium with the access point. The mobile tire responsible for collecting and storing the detected information, and manipulates the power usage; at a discrete time interval, the data sensed from the patients is sent to the remote server tier through GPRS networks to provide the healthcare monitoring services [18].

Furthermore, access points (AP) can be used for connecting the WBAN to various other networks like cellular network or the Internet, ad hoc-based architecture in which multiple APs will be sending information, and APs are in a mesh construction by which deployment is fast and flexible; through this method, coverage area also get improved from 2 to 100 m [19].

Due to the limited capacity of the battery storage, the gathered information are regularly sent to separated external devices (e.g., cloud computing) with more optimization power, and smartphones have become commonly used in this tire because they are powerful and provide all the technologies desired for several applications. Additionally, smartphones have highly secure encrypted transmission systems [20].

5.3 Remote Server Tire

Since both sensors and mobile tires have limited resource, collected information are commonly sent through GSM or Internet to the central remote server (cloud computer) for extensive data processing and long-term storage. Furthermore, cloud computing system has the ability to provide storage and analysis of the big data gathered from the wearable-based systems; they ease connection and share the resources in a universal manner, providing online on-demand services [20].

Remote server is a tire 3 which creates a communication path through tier 2. The design of this tier is application specific, for example, through Internet we are connected to the Medical Servers (MS) of some specific hospital where all our monitored is getting stored regularly [19].

6 Wearable Sensor Based on Advanced Materials

In the last few years, numerous new materials have been developed; the goal of this section is to briefly explore the current state of advanced materials that have been used as wearable sensors. The smart materials (e.g., nanomaterials) significantly improve the feasibility and efficiency of applying wearable device sensors in the medical fields and academic research; accordingly, modern wearable devices are becoming lighter, cheaper, smaller, and more reliable. The chart in Fig. 4 shows the main classification of the common wearable sensors.

6.1 Wearable Temperature Sensors

Temperature sensing devices can be classified into two types, resistometric or resistance change thermally (e.g., the operation concept depends on resistance changes of the sensing elements), and the second type is depending on measuring the thermal sensitivity for the Field Effect Transistors (FETs) (e.g., the operation concept of this type depends on sensing transfer characteristics for the FET because it is effecting by temperature). The FET-based temperature devices have some advantage than thermal resistance-based devices; it has a high level of sensitivity, and simply integrated into analogue IC (Integrated Circuits) having the ability of signal amplification [21].

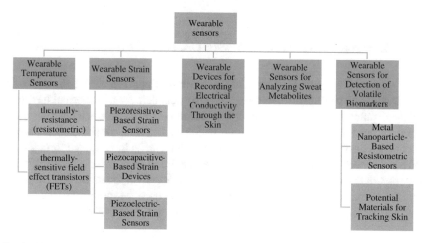

Fig. 4 Classification of the common wearable sensors

6.2 Wearable Strain Sensors

The mechanism of strain is explained as measuring the amount of changing in shape or distorting through the application of pressure.

Strain gauge is a well-known sensor used for measuring the level of strain, and this conventional type of strain sensor can be considered as a resistor or capacitor on a substrate which is take advantage of piezoelectric effect (e.g., piezoresistive-based strain sensor is consist of sensing thin film attached with a flexible substrate, any change in the electrical conductivity lead to changes in the bulk resistivity of the material as a function of applied force/load) [22].

Wearable strain sensors have valuable applications in the medical field, such as monitoring of heartbeat and respiration rate.

There are some variables which limit the utilization of wearable strain devices (e.g., shape, flexibility, size, and resistivity), also the surface on which it should be utilized. Usually, the sensor should be lightweight, reliable, and stretchable to match the mechanical properties of human skin. Therefore, there is a demand for advanced technology to empower the design and implement the high-performance and low-cost strain sensors [21, 22].

The comparison between capacitive and resistive type of strain sensors shows that the resistive-based sensors have better stretchability and higher sensitivity, but with some disadvantage (e.g., nonlinearity behaviors and hysteresis phenomenon). While, the capacitive-based strain sensor has advantages in linearity, stretchability, and hysteresis performance, but this type has one disadvantage, is very low sensitivity [23].

6.3 Wearable Devices for Sensing Change Through the Skin

High sensitive wearable sensors enable us to sense a small electrical change on the skin, it could be useful for monitoring the electrical activity of abnormal heartbeat, and these outcomes can help the doctor to decide whether there is a need for medication, or need Implantable Cardioverter Defibrillator (ICD), or some other surgical treatment. Furthermore, the human body condition can be investigated through skin humidity as well as the amount of sweat, which are indirectly indicating the health condition of human [21]. Glucose is the main component of the sweat that comes out from the human body, and sweat rate sensor can be used for monitoring the amount of sweat comes out after any physical activity for the human body [24].

6.4 *Nanomaterials-Based Wearable Sensors*

Graphene and Carbon Nanotube (CNT) are a promising material for the development of flexible and wearable tactile sensing system, functioning as active sensing materials of tactile sensors, as well as this nanomaterial have good properties nanostrips semiconductor nanowires, ZnO nanowires, metal nanowires, nanoparticles, and thin films [25]. Fiber-based sensors have valuable features, light, durable, and flexible, and therefore ideal for wearable devices, like ECG and EMG. Fiber-based sensor is made from CNTs, or metal alloys (e.g., stainless steel, copper, silver); also some of them are made from dielectric textiles by surface coating and lamination [26].

7 Power Harvesting for Wearable Devices System

Power harvesting and saving energy are very important issues in wearable sensor devices, because the energy is limited for supporting the whole sensor system and its operation. At first, each part of the system will be designed as a very low-power consumption. In the meantime, many types of batteries and energy harvesting devices are often selected for smart sensors. Specifically, rechargeable batteries and the storage capacitor are of great importance for the development of future smart sensors [25].

Several techniques have been used for harvesting and saving power, such as self-powered integrated systems, MEMS transducers, and human-powered energy harvesting, and also vibration energy can be converted into electrical power through electromagnetic, electrostatic transducers, and piezoelectric materials applied in resonators and oscillators [27].

7.1 *Self-power Devices*

Reducing power consumption is required in sensors design to extend battery life, sensors based on piezoelectric as well as triboelectric phenomena have the capability of self-power without the need for an external power supply, which is particularly suitable for the mobile wearable application. Several researches have been carried out for synthesis materials based on self-powered technology: one interesting example of self-powered tactile sensor is based on a piezoelectric potential change in the ZnO nanowire under small external force [28], and another example is a self-powered temperature sensor based on the thermoelectric effect [29].

7.2 Energy Harvesting from Heat of the Human Body

Human is classified as warm-blooded animals, or homeotherms, during metabolism human blood can produce heat energy; this energy is dissipated into these forms, IR infrared radiation, dissipated into the surrounding as heat, and dissipated into the form of water vapor. Moreover, only a small fraction of that heat flow can be converted into electricity by using a Thermoelectric Generator (TEG), thus can be used in a wearable device [30].

7.3 Rechargeable Micro-battery and Super-Capacitors for Energy Harvesting

The harvested energy should be well stored to guarantee constant power source without continuing harvesting; therefore, a compact rechargeable battery must be integrated into the system. For this purpose, several micro-batteries have been developed, e.g., micro-batteries based on micro-fluidic MEMS packaging Li-ion micro-batteries (active area $6 \times 8 \, mm^2$, 0.2–$0.4 \, mAh$) with interdigitated electrodes and glass housing have been fabricated and successfully tested [31], and also, a customizable thickness is in the range between 0.3 and 1 mm resulting in an area capacity between ca. 1 and $5 \, mAh/cm^2$ thin micro-battery is in development at the Fraunhofer IZM and TU Berlin, it is cost-effective and reliable technology that allows extreme miniaturization of batteries into silicon and glass chips and electronic packages [32].

7.4 Energy Harvesting from Electromagnetic Radio Frequency Spectrum

Radio Frequency Energy Harvesting (RFEH) is a process of converting Radio Frequency (RF) into power energy, GSM mobile system has frequency work (900/1800 MHz), and digital television works in frequency of 700 MHz; all these frequency spectra have a good source for radio frequency energy harvesting, and the energy harvested is enough for wearable sensors applications. Super-capacitors can be used as a tank for storing the harvested energy. Furthermore, the supercapacitor tanks have advantage than conventional battery in terms of accurate estimation of the remaining energy, and also the conventional battery has limited number of recharge/discharge cycles [33].

8 Use Case for Optics for Smart Glasses, Augmented Reality, and Virtual Reality Headsets

Augmented Reality (AR) is a technique that enables us to combine the digital information world with the real world; AR is a technology that has three key requirements: combines real and virtual content, interactive in real time, and registered in 3D [34].

AR displays can also be categorized based on where the display is placed; there are three approaches, positioning AR display between the user's eye and the real-world scene, wearable devices, and Head-Mounted Displays (HMD).

AR is guided by some pioneer technology companies (e.g., Microsoft, Google, and Facebook), smart glasses are wearable devices that combine both real and virtual information in the consumer's view field, and they will become the next evolution in the media technology [35].

8.1 Augmented Reality Smart Glasses

AR smart glasses are worn like regular glasses to combine virtual and physical information for user's view, there are three pioneer companies have started commercializing smart glasses, Google Glass (Project Aura), Microsoft Hololens, and Everysight.

Furthermore, smart glasses have been studied by several researchers from the viewpoint of advanced applications, as example application of smart glasses in medical field as international collaborations during surgeries, and also smart glasses can be used to improve the efficiency of logistical procedures in the shipment and storage of things in warehouses [34, 36]. Google Glass can be used in art gallery for assisting visitors to obtain augmented information whereas looking at pictures [35].

8.2 Using Smart Glass in Surgical Education

Smart glasses have many features that can be used for remote monitoring to improve medical training; the glass attached to a laptop/computer/mobile phone through a wireless (Wi-Fi) provides users the access to the World Wide Web; in addition, the glass comes with video recording (e.g., 720p high-definition (HD) camera) and audio (e.g., microphone), providing the user to manage with the aid of voice instructions [37]. Also, with some modification, the smart glasses can be used to recognize the veins in a patient's human body, and the glasses work with short-wave infrared radiation to look into the interior composition of the veins under skin, hence making it possible for fast and simple position and accessibility to the perfect vein for injection [38].

8.3   Wearable Contact Lenses

Nowadays, contact lenses used usually for visibility modification or for improve the appearance of eyes, whatever the purpose of wearing contact lenses, it comes up with smooth reach with our tear fluids, which means that provide a perfect wearable solution for optical diagnostics, at the same time independently, the wearable contact lenses sensor can be used to measure intraocular pressure, and glucose amount in tear liquid, according to distinct electrical reactions. However, electronics on soft contact lenses have a big challenge because the system requires flexibility, stretchability, reliability, and optical transparency for unobstructed vision [39].

9   Use Case in Health Monitoring and Prognosis

Wearable sensor devices have many applications in our daily life, health care and medical (46%), safety and security (27%), sport and fitness (21%), and general purposes (6%). In this particular part, we pay attention to the applications of wearable sensors in the healthcare and medical field [40]. Health monitoring systems based on wearable sensors have many advantages than conventional clinical diagnostic; the system can detect the health situations from the beginning, due to the fact that the medical diagnosis is usually done after the presence of primary health symptoms; moreover, historical medical information for the patient tend to be documented in the system. Two types of health monitoring system, preventive health monitoring systems, and responsive systems, the first one can give real-time feedback information to the consumer to optimize characteristics which may end up in resisting health situations later on, they support good behaviors and reduce the possibility of significant illness by instantaneously foretelling harmful recreation as well as notice the person regarding all of them. The second type is responsive physical condition supervising; systems can detect the health conditions at a preliminary stage by means of monitoring and examining multiple biomedical signals (e.g., EEG, ECG heart rate, blood glucose, and blood sugar) for long period of time, for example, the CodeBlue and MobiHealth are two projects performed in the past few years, but recently, there is a power efficient long-term consistent confidential overall health checking system proposed by [41]. The system is based on several biomedical sensors (e.g., heart rate, ECG, EEG, blood pressure levels, fresh air saturation, human body temperature, blood glucose, and accelerometer) [42].

Human health care can be carried out through various observations of the human body, wound fluid, sweat, and body odor. There are many researches that have been done in a health monitoring based on sweat analysis, and various sensing as well as imaging methods for injury treatment [40].

10 Conclusion

This chapter presents the wearable devices technology, and the application of wearable sensors has recently attained preferred development for regular monitoring of various physiological data associated with a human. Sensors linked to the human body are able to particularly identify the parameters, for which they are designed for.

Many challenges for body sensors network have been explored through this chapter, the network architecture for wearable system as well as wireless standards for short-range and long-range communication. The smart materials (e.g., nanomaterial, carbon nanotube) play a key role in body sensors manufacturing. Furthermore, power harvesting and compact battery storage are very important issues for supplying and saving power for the wearable system. Finally, Augmented Reality (AR) and smart glasses have been presented as cutting-edge wearable applications in surgical education, as well as applications of wearable sensors in health monitoring and prognosis.

References

1. A. J. Bandodkar, I. Jeerapan, and J. Wang, "Wearable Chemical Sensors: Present Challenges and Future Prospects," *ACS Sensors*, vol. 1, no. 5, pp. 464–482, May 2016.
2. H. Elayan, R. Shubair, and A. Kiourti, "Wireless sensors for medical applications: Current status and future challenges," *Antennas Propag.*, 2017.
3. A. Mosenia and S. Sur-Kolay, "Wearable Medical Sensor-based System Design: A Survey," *IEEE Trans.*, 2017.
4. M. Hubl *et al.*, "Embedding of wearable electronics into smart sensor insole," in *2016 IEEE 18th Electronics Packaging Technology Conference (EPTC)*, 2016, pp. 597–601.
5. G. Appelboom *et al.*, "Smart wearable body sensors for patient self-assessment and monitoring," *Arch. Public Heal.*, vol. 72, no. 1, p. 28, Dec. 2014.
6. M. Chan, D. Estève, J.-Y. Fourniols, C. Escriba, and E. Campo, "Smart wearable systems: Current status and future challenges," *Artif. Intell. Med.*, vol. 56, no. 3, pp. 137–156, Nov. 2012.
7. F. Brunetti *et al.*, "Communication Networks for Wearable Robots," in *Wearable Robots*, Chichester, UK: John Wiley & Sons, Ltd, pp. 201–234.
8. A. Pantelopoulos and N. G. Bourbakis, "A Survey on Wearable Sensor-Based Systems for Health Monitoring and Prognosis," *IEEE Trans. Syst. Man, Cybern. Part C (Applications Rev.*, vol. 40, no. 1, pp. 1–12, Jan. 2010.
9. E. P. Scilingo, A. Lanatà, and A. Tognetti, "Sensors for Wearable Systems," in *Wearable Monitoring Systems*, Boston, MA: Springer US, 2011, pp. 3–25.
10. M. Adnan and S. Hilles, "An Evolution to Next Generation Heterogeneous Cellular Networks," *Int. J.*, 2017.
11. M. Vatankhah Varnoosfaderani, D. V. Thiel, and J. W. Lu, "A Wideband Slot Antenna in a Box for Wearable Sensor Nodes," *IEEE Antennas Wirel. Propag. Lett.*, vol. 14, pp. 1494–1497, 2015.
12. Z. Wang, L. Z. Lee, D. Psychoudakis, and J. L. Volakis, "Embroidered Multiband Body-Worn Antenna for GSM/PCS/WLAN Communications," *IEEE Trans. Antennas Propag.*, vol. 62, no. 6, pp. 3321–3329, Jun. 2014.

13. S. Shao, A. Kiourti, R. J. Burkholder, and J. L. Volakis, "Broadband Textile-Based Passive UHF RFID Tag Antenna for Elastic Material," *IEEE Antennas Wirel. Propag. Lett.*, vol. 14, pp. 1385–1388, 2015.
14. A. Kiourti and J. L. Volakis, "Stretchable and Flexible E-Fiber Wire Antennas Embedded in Polymer," *IEEE Antennas Wirel. Propag. Lett.*, vol. 13, pp. 1381–1384, 2014.
15. J. L. Buckley, K. G. McCarthy, D. Gaetano, L. Loizou, B. O'Flynn, and C. O'Mathuna, "Design of a compact, fully-autonomous 433 MHz tunable antenna for wearable wireless sensor applications," *IET Microwaves, Antennas Propag.*, vol. 11, no. 4, pp. 548–556, Mar. 2017.
16. N. Fahier and W.-C. Fang, "An advanced plug-and-play network architecture for wireless body area network using HBC, Zigbee and NFC," in *2014 IEEE International Conference on Consumer Electronics - Taiwan*, 2014, pp. 165–166.
17. M. D. Steinberg, P. Kassal, and I. M. Steinberg, "System Architectures in Wearable Electrochemical Sensors," *Electroanalysis*, vol. 28, no. 6, pp. 1149–1169, Jun. 2016.
18. G. Huzooree, K. K. Khedo, and N. Joonas, "Wireless Body Area Network System Architecture for Real-Time Diabetes Monitoring," Springer International Publishing, 2017, pp. 262–271.
19. M. Saini and G. Pandove, "A Review Article on Issues and Requirements of Wireless Body Area Network (WBAN) with Fuzzy Logic," *Int. J. Adv. Res. Comput. Sci.*, vol. 8, no. 3, 2017.
20. A. Mosenia, S. Sur-Kolay, A. Raghunathan, and N. K. Jha, "Wearable Medical Sensor-Based System Design: A Survey," *IEEE Trans. Multi-Scale Comput. Syst.*, vol. 3, no. 2, pp. 124–138, Apr. 2017.
21. H. Jin, Y. S. Abu-Raya, and H. Haick, "Advanced Materials for Health Monitoring with Skin-Based Wearable Devices," *Adv. Healthc. Mater.*, vol. 6, no. 11, p. 1700024, Jun. 2017.
22. M. Farooq and E. Sazonov, "Strain Sensors in Wearable Devices," Springer, Cham, 2015, pp. 221–239.
23. M. Amjadi, K.-U. Kyung, I. Park, and M. Sitti, "Stretchable, Skin-Mountable, and Wearable Strain Sensors and Their Potential Applications: A Review," *Adv. Funct. Mater.*, vol. 26, no. 11, pp. 1678–1698, Mar. 2016.
24. S. C. Mukhopadhyay, Ed., *Wearable Electronics Sensors*, vol. 15. Cham: Springer International Publishing, 2015.
25. T. Yang, D. Xie, Z. Li, and H. Zhu, "Recent advances in wearable tactile sensors: Materials, sensing mechanisms, and device performance," *Mater. Sci. Eng. R Reports*, vol. 115, pp. 1–37, May 2017.
26. W. Zeng, L. Shu, Q. Li, S. Chen, F. Wang, and X.-M. Tao, "Fiber-Based Wearable Electronics: A Review of Materials, Fabrication, Devices, and Applications," *Adv. Mater.*, vol. 26, no. 31, pp. 5310–5336, Aug. 2014.
27. A. Lay-Ekuakille and S. C. Mukhopadhyay, Eds., *Wearable and Autonomous Biomedical Devices and Systems for Smart Environment*, vol. 75. Berlin, Heidelberg: Springer Berlin Heidelberg, 2010.
28. Z. L. Wang, "From nanogenerators to piezotronics—A decade-long study of ZnO nanostructures," *MRS Bull.*, vol. 37, no. 9, pp. 814–827, Sep. 2012.
29. Y. Yang, Z.-H. Lin, T. Hou, F. Zhang, and Z. L. Wang, "Nanowire-composite based flexible thermoelectric nanogenerators and self-powered temperature sensors," *Nano Res.*, vol. 5, no. 12, pp. 888–895, Dec. 2012.
30. A. Bonfiglio and D. De Rossi, Eds., *Wearable Monitoring Systems*. Boston, MA: Springer US, 2011.
31. R. Hahn, M. Ferch, K. Hoeppner, M. Queisser, K. Marquardt, and G. A. Elia, "Development of micro batteries based on micro fluidic MEMS packaging," in *2017 Symposium on Design, Test, Integration and Packaging of MEMS/MOEMS (DTIP)*, 2017, pp. 1–5.
32. M. Robert Hahn, Katrin Höppner, Marion Molnar, Marc Ferch and K. M. Lücking, Moritz Hubl, Giuseppe Elia, "Micro Battery Prototype Line And Micro Fluidic Electrolyte Handling," in *IDTechEx Energy Harvesting and Storage Europe 2016*, 2015.
33. F. J. Velez, N. Barroca, L. M. Borges, R. Chávez-Santiago, and I. Balasingham, "Radiofrequency energy harvesting for wearable sensors," *Healthc. Technol. Lett.*, vol. 2, no. 1, pp. 22–27, Feb. 2015.

34. M. Billinghurst, A. Clark, and G. Lee, "A Survey of Augmented Reality," *Found. Trends® Human–Computer Interact.*, vol. 8, no. 2–3, pp. 73–272, 2015.
35. M. C. Leue, T. Jung, and D. tom Dieck, "Google Glass Augmented Reality: Generic Learning Outcomes for Art Galleries," in *Information and Communication Technologies in Tourism 2015*, Cham: Springer International Publishing, 2015, pp. 463–476.
36. P. A. Rauschnabel and Y. K. Ro, "Augmented reality smart glasses: an investigation of technology acceptance drivers," *Int. J. Technol. Mark.*, vol. 11, no. 2, p. 123, 2016.
37. O. Moshtaghi, K. S. Kelley, W. B. Armstrong, Y. Ghavami, J. Gu, and H. R. Djalilian, "Using google glass to solve communication and surgical education challenges in the operating room," *Laryngoscope*, vol. 125, no. 10, pp. 2295–2297, Oct. 2015.
38. J. Ruminski, M. Smiatacz, A. Bujnowski, A. Andrushevich, M. Biallas, and R. Kistler, "Interactions with recognized patients using smart glasses," in *2015 8th International Conference on Human System Interaction (HSI)*, 2015, pp. 187–194.
39. J. Kim, M. Kim, M. Lee, K. Kim, S. Ji, and Y. Kim, "Wearable smart sensor systems integrated on soft contact lenses for wireless ocular diagnostics," *Nature*, 2017.
40. M. D. Steinberg, P. Kassal, and I. M. Steinberg, "System Architectures in Wearable Electrochemical Sensors," *Electroanalysis*, vol. 28, no. 6, pp. 1149–1169, Jun. 2016.
41. A. M. Nia, M. Mozaffari-Kermani, S. Sur-Kolay, A. Raghunathan, and N. K. Jha, "Energy-Efficient Long-term Continuous Personal Health Monitoring," *IEEE Trans. Multi-Scale Comput. Syst.*, vol. 1, no. 2, pp. 85–98, Apr. 2015.
42. A. Mosenia, S. Sur-Kolay, A. Raghunathan, and N. K. Jha, "Wearable Medical Sensor-Based System Design: A Survey," *IEEE Trans. Multi-Scale Comput. Syst.*, vol. 3, no. 2, pp. 124–138, Apr. 2017.

Realizing the Wireless Technology in Internet of Things (IoT)

Dimitrios G. Kogias, Emmanouel T. Michailidis, Gurkan Tuna
and Vehbi Cagri Gungor

Abstract The evolution of the Internet of Things (IoT) has been highly based on the advances on wireless communications and sensing capabilities of smart devices, along with a, still increasing, number of applications that are being developed which manage to cover various small and more important aspects of every people's life. This chapter aims at presenting the wireless technologies and protocols that are used for the IoT communications, along with the main architectures and middleware that have been proposed to serve and enhance the IoT capabilities and increase its efficiency. Finally, since the generated data that are spread in an IoT ecosystem might include sensitive information (e.g., personal medical data by sensors), we will also discuss the security and privacy hazards that are introduced from the advances in the development and application of an IoT environment.

Keywords Wireless communications · Internet of Things (IoT)
Architecture · Middleware · Privacy · Security

D. G. Kogias (✉) · E. T. Michailidis
Department of Electrical & Electronics Engineering, University of West Attica,
12244 Egaleo, Greece
e-mail: dimikog@puas.gr

E. T. Michailidis
e-mail: emichail@puas.gr

G. Tuna
Department of Computer Programming, Trakya University, 22020 Edirne, Turkey
e-mail: gurkantuna@trakya.edu.tr

V. C. Gungor (✉)
Department of Computer Engineering, Abdullah Gul University, 38039 Kayseri, Turkey
e-mail: cagri.gungor@agu.edu.tr

© Springer Nature Singapore Pte Ltd. 2018
K. V. Arya et al. (eds.), *Emerging Wireless Communication and Network Technologies*,
https://doi.org/10.1007/978-981-13-0396-8_10

1 Introduction of IoT with Emerging Wireless Technologies

In recent years, as the number of smart devices, wireless technologies, networking protocols, and sensors rapidly grows, the IoT has emerged as the vision of a global infrastructure of networked real-world smart objects with little or no human intervention [1–5]. The IoT constitutes an interoperable, energy-efficient, and secure network architecture of connected heterogeneous devices and objects and has gained the attention in the academia and industry as an integrated part of the so-called Future Internet. According to forecasts, the IoT will consist of billions connected things, i.e., IoT devices, including televisions, cars, kitchen appliances, surveillance cameras, smartphones, utility meters, cardiac monitors, thermostats, etc.

Among the most notable challenges, wireless and mobile technologies are the underlying technologies for realizing the IoT [6, 7]. Both long-range and short-range wireless communication technologies can be supported with some advantages and weaknesses and different characteristics in terms of the data range and rate, network size, channels, bandwidth, and power consumption. This capability can lead to the provision of numerous services. This section attempts to shed a light on the different types of wireless communication infrastructure within the IoT, underline the corresponding potential applications, and highlight the privacy issues considering both the technological aspects and their implications on business models and strategies.

1.1 Types of Communication Within the IoT and Challenges

The IoT may utilize a wide range of communications at various radio frequencies and proprietary protocols with different implementations at the physical, data link, and network layers, in order to successfully and uninterruptedly interconnect the devices [6–8].

The communications in IoT can be classified as Device-to-Device (D2D), device to human and vice versa, and device to distributed storage. Depending on the number of relay nodes within the networks, single-hop or multiple-hop connections may be present. The former includes an access point or a base station, whereas the latter considers that multiple devices relay information for each other to attain a seamless end-to-end communication. In addition, the communication may take place with, or without, human intervention depending on the requirements for human decision making. Inherent hardware and software constraints may exist (such as low computation capabilities, low transmission power, low-power consumption, insufficient memory, limited battery life, reduced implementation and operational cost, wide coverage range, simplicity), which should be explicitly adhered. In addition, the wireless radio channel suffers from much impairment, such as fading, shadowing, and interferences, which may degrade the system performance.

Consequently, the choice of technology is actually a major challenge to be faced for the applicability of the IoT and is limited to particular scenarios. Possible modifications of the available technologies may also be requisite.

1.2 Potential Applications

To improve the quality of life in our society, the emerging IoT intends to positively affect a huge number of applications depending on the type of network availability, coverage, scale, heterogeneity, repeatability, user involvement and impact. In particular, the potential applications of IoT can be classified into the following domains, which are currently characterized with only primitive intelligence and elementary communication capabilities [9–11]:

- **Personal, home, and social applications**: A typical paradigm of such applications is the home monitoring system for health care, which involves both medical staff and patients and allows for the measuring and monitoring of physiological parameters. To succeed in the healthcare domain, implantable, wearable, and ambient sensors should be used to store health records. Moreover, IoT enables the vision of the smart home/office and optimizes the automation, security, and energy management by controlling the corresponding home equipment, e.g., heating devices, lighting, alarm systems, via distributed sensors and actuators.
- **Enterprise applications for the advancement of communities**: This type of applications mainly refers to large-scale work environments, where the information data are exclusively used by the owners and are selectively provided to other users. IoT will, also, make possible the monitoring of vehicle emissions and air quality, the collection of recyclable materials, and the reuse of packaging resources. In addition, the mobile ticketing transportation services will allow the users to buy tickets via mobile phones using *Near Field Communication (NFC)* tags located on posters or panels, whereas driving services will improve the navigation and safety by employing collision avoidance systems and monitoring of transportation of hazardous materials.
- **Applications at a national or regional scale**: This application domain corresponds to service optimization via extended networks rather than consumer consumption and includes smart resource management and efficient energy consumption, in order to increase the profit, as well as smart grid, and smart metering.
- **Smart transportation and smart logistics**: IoT is capable of providing several advantages in the automotive industry and the smart transportation, where vehicles, e.g., cars, trains, buses, and bicycles, are equipped with wireless sensors and actuators with increased processing powers. Specifically, exploiting the IoT, monitoring of various critical mechanical and electronic parameters, vehicle travel times on motorway and streets, queue lengths, transport network state, air pollution, and noise emissions can be realized.

1.3 Privacy Hazards

With the IoT functionalities and services covering a large variety of a human's life, as has been described above, the most important feature is the information that is generated, shared and, even, being processed to test whether certain conditions are met to reach for decisions. For example, in health care, the data generated from a patient are being processed to alert the doctor when hazardous conditions are close to emerge and, even, prevent them when early signs are recognized or, in home automation, the temperature of the room is measured and when it reaches certain threshold values, then the air condition or the heat (depending on the value) is turned on.

Since, often, the generated data include very sensitive, even private, information the handling of these data should be following specific and strict rules. Therefore, in an IoT ecosystem, reserving the privacy of the users data is of ultimate importance and, along with security, they are in the center of attention from the architecture view (see Sect. 3 for more information) to the functionality and performance of the system. In fact, the importance (both regarding the collection and processing of the user's data) and the need to ensure the users that their data are being used for well-specified reasons (stated explicitly in each application scenario and service demanding user consent), is such that recently the European Parliament regulated accordingly [12]. In Sect. 3, a more detail discussion on IoT security and privacy hazards is presented.

2 Basic Elements for Forming IoT and Addressing the Issues Associated with It

2.1 Components of IoT

To efficiently and effectually set up and construct IoT-based wireless networks and succeed in accomplishing complex tasks and providing advanced services, all the components of these networks should uninterruptedly interoperate and cooperate. The major components of the IoT can be summarized as follows [9]:

- **Identification (ID)**: A particular method of object identification (ID) is indispensable. Hence, an ID should be assigned to each object based on the MAC ID, IPv6 ID, a universal product code or some other custom method.
- **Security controls**: To manage and limit the risks associated with unauthorized connections among the devices, restrictions on the types of devices that can be connected should be designated.
- **Service discovery**: The devices of IoT networks should be aware about the capabilities of other devices and the types of services that they can provide.

- **Relationship management**: It is necessary to manage the relationships among the devices and correspond each device with another device based on the similarity of the service they can provide.

2.2 Application Protocols

The IoT aims at covering a huge range of applications that may involve massive cross-platform deployments of embedded technologies, cloud systems, and devices connecting in real-time [4, 9]. The protocols intend to act as a common language and a set of rules and instructions that are used by two or more devices, especially distributed devices in different locations, to interact with each other and exchange information.

The future of IoT lies in standardizing the protocols used across the network stack. The choice of a protocol depends on the underlying scenario, the reliability of the network, and the QoS. Besides, multiple protocols may be occupied in complex scenarios. Since the IoT involves the connection of physical objects to the Internet, these objects should operate in an Internet Protocol (IP) manner, in order to be capable to provide data exchange among different devices and servers over the Internet. However, the devices of a local network may exploit non-IP protocols to communicate. Then, an Internet gateway is necessary to enable communication with an Internet service or external devices.

The *Hypertext Transfer Protocol (HTTP)* is the typical protocol for distributed, collaborative, and hypermedia information systems. However, in resource constrained environments, other alternate protocols with limited parsing overhead are required. Paradigms of open protocols for IoT environments, which provide mechanisms for asynchronous communication, run on IP, and have a range of implementations are briefly described below:

- **CoAP**: This protocol is actually an alternative to HTTP and it is optimized for constrained application environments, especially for Machine-to-Machine (M2M) applications, such as smart energy and building automation. CoAP uses the Efficient XML Interchanges (EXI) binary data format, which is more efficient in terms of space. CoAP also supports built in header compression, resource discovery, auto-configuration, asynchronous message exchange, congestion control, and support for multicast messages.
- **MQTT**: This protocol is a lightweight publish/subscribe (pub-sub) protocol that runs over the Transmission Control Protocol (TCP) and is suitable for the majority of IoT applications, especially in situations characterized by low-bandwidth, limited processing capabilities, small memory capacities, and high latency. Considering low-power and low-cost devices, the MQTT-S/MQTT-SN extension of MQTT is suggested.
- **AMQP**: This enterprise application oriented protocol is a message-centric middleware based on queuing and topic-based publish-and-subscribe messaging. It

is mainly used for building large-scale, reliable, resilient, or clustered messaging infrastructures.

- **RESTful HTTP**: This protocol leverages the HTTP, which stands for the service interface, and is usually used in mobile applications and distributed systems, such as social networking, mashup tools, and automated business processes. In this protocol, the devices are able to create, read, update, and delete data on a server. In particular, the data are uniquely referenced by a designated Web page that contains an eXtensible Markup Language (XML) file containing the desired content and can be acted upon using HTTP operations.
- **XMPP**: This protocol is based on TCP and XML and facilitates the exchange of structured data between multiple connected entities. Among the positive aspects of XMPP are its application to publish–subscribe systems, the decentralization, and the capability of providing authentication using a centralized XMPP server. However, XMPP lacks end-to-end encryption and cannot ensure QoS.

2.3 Frequency Bands and Regulations

To realize the wireless IoT, specific frequency bands should be licensed on a global, regional, or national basis and for each frequency band, appropriate regulatory provisions should be established based on technical, regulatory, and operational studies. At the same time, the biggest challenge in IoT, which should be addressed by the standards organizations, is the interoperability among devices from different vendors to exchange data.

Since the IoT is heterogeneous and encompasses different types of applications and services with various technical and operational requirements, a wide variety of spectrum solutions for access to spectrum that fits all the scenarios is required [3, 4]. Frequencies ranging between 100 MHz and 5.8 GHz are, in general, suggested. Although high frequencies allow for sufficient data rate, lower frequencies have an extended coverage range. The frequency bands that can be used for the IoT include:

- **Frequencies with no requirements for individual authorization**: These frequencies are mainly identified for consumer IoT services, low-cost and low-power sensing and data-collecting devices, e.g., health and fitness monitoring and smart home devices, and short-range links, in the following bands: 169, 433, 863–870, 2,400–2483.5, 5,150–5,350 and 5,470–5,725 MHz.
- **Frequencies with particular requirements for individual authorization**: These frequencies should be allocated for the implementation and operation of public cellular networks (current 2G, 3G, and 4G and upcoming 5G networks in frequencies between about 1 and 4 GHz depending on the location and the generation of cellular technology), professional mobile radio networks and fixed-service frequencies (e.g., wireless local loop, microwave).

3 General Architecture and Factors Affecting IoT Centric Features

3.1 General Architecture and Reference Ones

An IoT ecosystem is a very heterogeneous environment that consists of many different parts and devices that include, but are not limited to, the following:

- significant number of the various sensors and actuators which generate the data,
- physical devices (e.g., aggregators or gateways) that are capable to provide preliminary processing of the data and that can deal with the interconnection between them and/or the outside world (e.g., the Internet), and
- server-side (probably cloud based) components that enhance the performance by supporting the system's functionalities, gather the data, visualize them, and provide services to the end-users through applications. Here, a platform is usually found that enhances the system performance and usability.

Each of the aforementioned parts can be found in a generic, three-layer IoT architecture [2, 9] that can be seen in Fig. 1, along with the interconnections between them. The three depicted layers are the *Sensing, Transport,* and *Application* layer that generally include the respective parts described above. In more details:

- The *Sensing* layer includes all the sensors and actuators that are in the system, sensing for specific physical parameters or for the existence of other smart devices in the close environment,
- The *Transport* layer that includes both the gateways (located at the edge [13]) that forward the information from the sensors to the core of the system and the routing between the devices (and/or the platform) for the dissemination of the unprocessed (or raw) information. This layer uses communication technologies like 3G or Wi-Fi as discussed in the previous section,

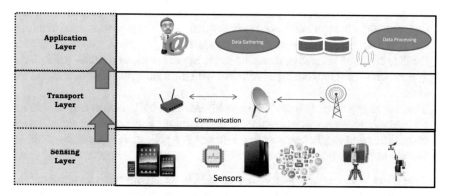

Fig. 1 A generic three-layer IoT architecture

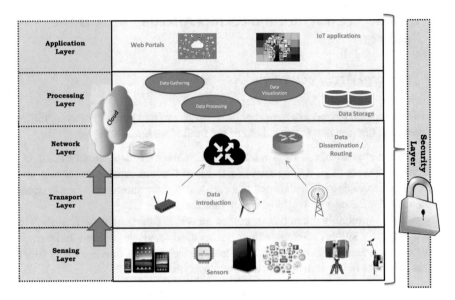

Fig. 2 Extended IoT architecture (five layers + security overall)

- The *Application* layer that includes the data gathering and data processing functionality which allows for the delivery of those data, in a desired form, to the end-user by using applications to visualize and manage it.

An extension to this generic architecture can expand some of the existing modules to create new ones that emphasize on the management, processing and networking features that can be found in an IoT ecosystem. To this end, the expanded IoT architecture might include five layers: *Sensing, Transport, Network, Processing, and Application* layer as this can be seen in Fig. 2. The two new layers can possess the following characteristics:

- The *Network* layer could be responsible for the dissemination of the generated data that have already been introduced to the system by the Transport layer. Therefore, the Transport layer will be, mainly, responsible for the gateway node to forward the data from the sensors or actuators to the upper layers, as an entry point to the system while the Network layer will deal with the routing of those data inside the system,
- The *Processing* layer, that could be considered as a middleware, is responsible for the analysis, storage, and processing of the data, along with providing visualization capabilities that can be used by the, following, Application layer as a service to the end-user.

One more layer can be considered vertically in the architecture, covering all the aforementioned layers and interacting with them, the layer of *Security*, where the needed mechanisms or techniques at each level should be described and applied.

In addition, the Network and Processing layers could be located in the *Cloud*, where the servers will be responsible for the described actions in these levels, forming a centralized IoT architecture. If multiple cloud solutions would be used, then a federation of those clouds should be considered [14].

Finally, lately, due to the large volume of data and the increased processing capabilities of the gateways (with a sharp decrease in their cost), there is an effort to move part of the data analysis and processing features at the sensors and gateways, which will be, therefore, inserting processed data in the system, allowing them to be more easily handled and shared among the end-users. This new architecture is called *Fog computing* [15] and, also, inserts storage, monitoring and security layers [16] in the generic three-layer architecture discussed above and seen in Fig. 1.

In an effort to standardize the structure of an IoT architecture, many initiatives took place to provide their ideas for a reference architecture [13, 17–19] that, not only could be followed by all manufacturers and interested parties, but could also provide solutions to many issues found in an IoT environment. Those issues usually concern power consumption, connectivity and communications, security and scalability, and other key performance indicators that will be discussed later in this section.

To this end, in [18] a reference architecture for IoT is presented by proposing a reference model that has five different views in the architecture (functional, system, communications, information, and usage view). In [20], ITU-T Y.2060, another reference model is presented. This model consists of four layers: application, service and application support, network and device layer along with the security and management requirements that should be met by each one of these layers, highlighting their importance throughout the proposed structure. Finally, in [21], the described architecture consists of four layers: the Sensors and Actuators, the Aggregator/Gateway, the Internet and the Service layer for single administrative domains, while for multiple administrative domains, the view changes including layers for data harmonization, distribution, and virtualization.

3.2 IoT Key Performance Indicators

In order to be able to measure the performance of an IoT system, there are several metrics that can be used as Key Performance Indicators (KPIs). In this subsection, we will discuss the most important ones that can be considered for all the IoT systems and that are affected by the aforementioned reference architectures. Those metrics are:

- *Scalability*: With the number of devices that are included in an IoT ecosystem rising rapidly, along with the volume of the generated data, IoT architecture should be able to follow this growth in numbers and manage, store and process the huge size of data using modern data analytics process in order to reach the expected levels of performance.

- *Reliability*: An IoT system should be reliable both in the sense of intermittent communication and in the sense of data validation and Quality of Information (QoI) that is shared among the IoT devices. This is very important in an IoT system due to the highly heterogeneous environment that is created and in an effort to allow the data to be used more efficiently by the interested parties or the end-users.
- *Low power and latency*: For the system to be able to expand its lifetime, the sensing devices should be able to work with low-power demands, therefore saving their energy, especially considering that these devices might be scattered in distant and difficult to approach places. In addition, the intermittent communications should be enhanced with low latency results, because often the transmitted data are of great importance (e.g., life critical data from a patient are transmitted to the doctor to monitor his/her health or data from a home's alarm system regarding a possible intrusion).
- *Large coverage*: Coverage capabilities must be extended to meet the needs, depending on the application scenario.
- *Low module cost*: The cost for creating and maintaining the IoT ecosystem is important for its performance and its life cycle. Recent advances in technology permit the creation of many powerful sensors in a small cost that can be part of such an ecosystem.
- *Mobility and roaming support*: The IoT enabled devices, very often, are not static but they are moving around, generating and sending data to the (potential) IoT platform to analyze and process. Therefore, the system should be able to support such a behavior providing all the communication and operational functionalities to this end.
- *SLA support*: The Service Level Agreement (SLA) is a characteristic to preserve the privacy of the generated data in an IoT ecosystem. It is an agreement between an end-user and a customer (i.e., company) that will handle, store, and process the data of the individual according to the agreed rules and laws.

One more very important KPI for an IoT ecosystem, and one that has been under consideration from, almost, all the described reference architectures is *security*. But, because of its great importance, it will be covered in more details in the following subsection.

To conclude, the aforementioned KPIs have not only been presented to characterize the performance of an IoT ecosystem but also to be seen as requirements that should be met in an IoT system.

3.3 Security Threats and Countermeasures

Security is of ultimate importance in an IoT ecosystem and is a characteristic that is considered in all the layers of an IoT architecture. The reasons for this are related with the nature of the information that can be collected by the many sensors and which is disseminated in the system. In addition, the fact that these data travel through the Internet increases the hazards that they might encounter and the need to pay special attention for their integrity and security.

In an IoT ecosystem, there are different types of security that could be described:

- *Device security* that deals with the sensors and any attacks on them either to cause a system failure or in an attempt to possibly take control of a number of those devices and possibly misuse them.
- *Edge security* that, mainly, deals with the security in the communications in the system and aims at altering the integrity of the generated data.
- *Network security* that deals with possible cyber-attacks on the system that are used to either steal data or to break down the system and affect its functionality and performance.

Due to its intrinsic characteristics, an IoT system not only deals with classic network security hazards (e.g., man-in-the-middle, Denial of Service, Spoofing, Injection and Routing Protocol attacks [22]) but also with hazards specific to the environment. Those hazards will not be limited to the physical access of the attackers to the IoT nodes, but can also include cyber-attacks, originated by sophisticated intruders (e.g., invalid input data by injecting exploits, unauthorized access due to misconfigured security parameters, hacking and installation of unauthorized software [22]). The communication between the devices (also known as M2M communication) is usually targeted.

To address those issues, an IoT ecosystem should be, mainly, interested in securing the integrity of the data by meeting the following requirements:

- *Encryption* in the communication and storage of data, especially at the devices that are capable (a lot of sensors might not have the necessary power to apply this requirement),
- *User authentication* and *identity management*, preferable by more sophisticated solutions that are based on tokens rather than the traditional user and password combination.
- Use of *firewalls* and *special intrusion detection systems* to support and enhance the network security in the system,
- *Smooth management of keys and tokens* throughout the layers of the architecture, and
- *Access Control* that is based on specific policy rules and, often, user-managed based on XAMCL [23].

4   Middleware Technology for Grooming IoT Essence

4.1   Middleware for Internet of Things: Challenges

Essentially, IoT middleware is software which operates as an interface between components of an IoT ecosystem and enables to join the heterogeneous domains of applications that communicate over different interfaces. It creates a very ubiquitous environment for different and already existing programs, and makes communication possible among Things that would not otherwise be capable of [24].

Although IoT middleware eases many troubles and worries related to installing, configuring and using IoT devices from different vendors, the process of designing and developing middleware is not free of hassles. There are many challenges that developers possibly face while creating middleware including: *scalability, edge computing, efficient communication architectures, protocol support, and cognitive computing*. In particular, edge computing points to the reevaluation of centralized intelligence and control topologies in favor of distributed architectures, with intelligence pushed into each edge device or data endpoint [9]. On the other hand, efficient communication architectures, scalability and efficient protocol support are needed to take advantage of the inherent characteristics of a modern IoT system.

The most critical point of most IoT solutions is the capability to consume data from any end point in the IoT chain. However, the devices in the IoT chain may come from all over the world with different size, shape, and standards. Hence, the middleware should offer the capability to consume data immediately with the help of plug-ins or Application Programming Interfaces (APIs).

4.2   Classification of IoT Middleware

IoT middleware manages the structure, format, and encoding of the information that is being exchanged between different layers, devices, and sensors. It acts as a common standard among the diversity of devices, sensors, operating systems, and applications that make up the IoT ecosystem architecture. API management is, also, a critical function of IoT middleware since it tries to call APIs from disparate systems, taking care of the interconnection. IoT middleware has also a role to play in security.

Although a customized and fully configurable IoT middleware is desired, such middleware needs time for development and testing before deployment, resulting in cost and time overrun in many projects. At the same time, the middleware must be adaptive and optimized for scalability in an agile fashion. In this regard, the use of open middleware may be ideal for most of the practitioners.

If IoT middleware systems in the literature are reviewed, these systems can be broadly classified into *service* and *agent-oriented* systems. Hence, IoT middleware systems can be designed either as an ecosystem of services or as an ecosystem of agents. Importantly, both of these approaches can successfully overcome heterogeneity issues.

On the other hand, if the existing middleware solutions are classified based on their design techniques, they can be grouped into seven categories [25–27], namely: *event-based, service-oriented, virtual machine-based, agent-based, tuple-spaces, database-oriented,* and *application-specific*. In event-based middleware solutions, components, applications, and all the other participants interact through events and each event has a type and a set of parameters [25]. In service-oriented middleware solutions, software, or applications are built in the form of services and their characteristics such as service reusability, service discoverability, service composability, loose coupling, and technology neutrality suit well to the demands of IoT applications [25]. Virtual machine-based middleware solutions provide programming support for a safe execution environment for user applications by the virtualization of the infrastructure [25].

4.3   Popular IoT Middleware

As we have seen, IoT platforms and middleware exist between physical devices or data endpoints, and higher level software applications like artificial intelligence, predictive analytics, and cognitive computing.

Accordingly, in recent years, well-known software companies such as Oracle, RedHat, Mulesoft, and WSO2 have started to offer IoT middleware solutions that provide API management in addition to basic messaging, routing and message transformation [28]. APIs enable to address IoT devices and the services which they create; nevertheless, they do not enable the connectivity. For IoT solutions, it is critical to ensure that the IoT solution is addressable via APIs. On the other hand, MachineShop, one of the well-known IoT platforms designed as a Platform of Services instead of a Platform as a Service (PaS), relies on an API-first approach and, without the limitations of a proprietary platform, enables all processes to be expressed as discrete services [29]. Therefore, it is able to transform the way developers build, integrate, and manage solutions. There are many commercial middleware solutions. For instance, snapIoT is an IoT middleware and application enablement platform. It allows creating end-to-end IoT applications which handle and manage the information provided by any smart sensor, smart phones, or IoT cloud, on any communication protocol or any operating system/device.

Despite the many well-known commercial middleware solutions, it is expected that open-source middleware solutions will rule the IoT market by offering users benefits that proprietary middleware solutions fail to deliver. For example, ownership is a prominent benefit of open-source software. Proprietary middleware solutions cannot, also, offer the flexibility offered by open-source solutions. Finally, as in the realm of IoT in which data security becomes a more important issue, transparency gives open-source middleware solutions a definite edge.

5 The Device-to-Device (D2D) Communications Protocol Stack

We have seen that in an IoT ecosystem many different communications take place. Here, we will discuss the probably most prominent one, D2D communications presenting the prominent technologies, their characteristics and how routing is addressed in them.

5.1 Device-to-Device (D2D) Communication Technologies

Hosted in the Sensing layer of the architecture (see Figs. 1 and 2), the devices responsible for the gathering of the produced, often sensitive, data use mainly wireless communications either to interact between them, or with a base station, an aggregator or a network gateway. The wireless protocols that are used for this dissemination of information vary from well-known solutions to new emerging technologies.

In more details, the most important communication technologies that are used for D2D communications in an IoT system are the following [30]:

- **Bluetooth**: One of the most popular short-range (or personal area) communication technologies that has been used extensively from smartphones and wearables. Especially, the recent version of *Bluetooth Low Energy* (*BLE* [27]) has been IoT enabled by providing excellent performance for small data propagation in small areas (from 50 to 150 m long) and using high speeds (around 1 Mbps). An important advantage is the use of the existing IP structure and the presence in many mobile devices that facilitates its use.
- **6LowPAN**: IPv6 Low-power wireless Personal Area Network (6LowPAN) is a network protocol that uses the IPv6 addressing, along with encapsulation and header compression to be enabled for IoT D2D communications [31]. Apart from this, 6LowPAN can be used in many different frequencies depending on the wireless technology that needs to interact with (e.g., Bluetooth, Wi-Fi or any low-power RF communication platform).
- **Radio Frequency Identification (RFID)**: The IoT was initially inspired by the success of the non-IP RFID technology and considered uniquely identifiable interoperable connected objects. This low-power technology is suitable for many of the devices involved in the IoT, especially devices that are lightweight and operate on batteries. The RFID is based on tags, i.e., small chips with antennas, which transmit data via radio waves to a RFID reader. This identification technology supports ranges up to hundreds of meters depending on the frequency and does not require Line-of-Sight (LoS) communication between the tag and the reader and human intervention.
- **Near Field Communication (NFC)**: A common communication technology for very-short-range low-power communications is also the NFC, which is based on the RFID. Although communication links in distances of few centimeters can be

accomplished, the requirement of close proximity between devices ensures secure transactions, e.g., payments.

- **Z-Wave**: Z-Wave is a developing communication technology from Sigma Designs which manages to operate efficiently in the sub-1 GHz band [32]. Its main use is in Home Automation providing solutions for smart lamps controllers or motion detectors, among others. Its main characteristic is that it is a lightweight, reliable protocol, whose performance is not affected by the presence in the area of other well-used wireless communication technologies, like Wi-Fi or ZigBee. It has a range of 30 m and is scalable enough, able to support up to 232 devices in the area. The achieved data rates are around 9.6/40/400 kbps.
- **ZigBee:** It is one of the most popular and widely used communication technologies that are suitable for low-power operation, small-speed (e.g., around 250 kbps) transmissions at infrequent time intervals [33]. ZigBee is highly scalable, managing to support a large number of devices, secure and robust solution that is mainly applied in the industrial domain. It is based on the IEEE802.15.4 protocol which is an industry standard solution that transmits at 2.4 GHz and has a range from 10 to 100 m.
- **Wi-Fi**: It can be considered as the most popular communication technology that is based in the IEEE 802.11n standard and transmits in two bands at 2.4 and 5 GHz [34].
- **Cellular**: This is a very popular solution for wide area transmissions of small rate data over small period of time. The use of a GSM/3G/4G (now moving slowly to 5G [35]) communication is ideal for sensors that comply with the previously described characteristics, especially when they are placed in difficult to approach areas that do have network coverage and provide for cellular communication.
- **NFC**: Near-Field Communication (NFC) is a wireless technology that encourages a simple two-way contactless communication between electronic devices in a safe manner [36]. Real-life scenarios that take advantage of NFC include the use of a smartphone for payments, or the binding between smartphones to exchange data and information. Following the ISO/IEC 18000-3 standard, NFC transmits at 13.56 MHz with data rates up to 100–420 kbps in a range of 10 cm, for security reasons.
- **LoRaWAN**: This is a solution for Wide Area Networks (WANs) as the name suggests [37]. It provides characteristics for low power (e.g., Low-Power WAN, LPWAN), bidirectional communication specifically between IoT enabled devices, for industrial and smart cities applications. One of its advantages is the high scalability properties that it offers in a range of 2–5 km for an urban environment, while it manages to extend to around 15 km for suburban ones. Supporting many frequencies, the achieved data rates vary from 0.3 to 50 kbps.

Other popular D2D communication technologies include *Neul*, *Sigfox* and *Thread* that have started lately to gain in popularity and use.

5.2 *Characteristics of D2D Communication Within IoT*

D2D communications have to take under consideration the many inherent characteristics of an IoT environment in order to be effective and increase the system's performance. The aforementioned solutions manage to efficiently adapt to the demands
and, therefore, gain in popularity. More specifically, the conditions that they need to
address include, among others:

- *Heterogeneity and cooperation of devices*: The many different smart devices that
  work in the Sensing layer not only might be used to measure different things, but
  they could also use different technologies for communication and have different
  energy consumption functionality.
- *Device collaboration*: In an IoT system, the devices should be able to cooperate
  between them in order to find the desired information and serve a certain need. For
  example, a user's vacuum cleaner needs to be able to check whether the alarm in
  the house is activated in order to perform cleaning without any problems. To this
  end, the two devices could be connected and collaborate to achieve the scheduled
  or desired result. These collaboration and communication between the devices are
  also considered under the notion of Social Internet of Things (SIoT) [38]. The use
  of middleware software is also preferred in IoT solutions.
- *Diverse networks and networking standards*: The various different sensors and
  aggregators are connected in many different networking topologies, others using
  an aggregator node as a gateway to the deeper layers of the architecture, or they
  may possess more ad hoc characteristics, relaying the information through them
  to the network. For their functionality, they may use different protocols depending
  on their application (e.g., MQTT, REST, HTTP) but this should not affect the
  overall performance and communication of the system. A discussion on the many
  different routing protocols used in IoT follows in the next subsection.
- *Device limitations*: The smart devices that coexist in today's IoT systems meet high
  capacity requirements and include many communication technologies (e.g., on a
  smartphone you can find available Bluetooth, NFC, Wi-FI and Cellular communication technologies), while operating at low cost with small power consumption
  aiming at extending the life cycle of the device, before been (re)charged or stop
  operating and in need of replacement. The harvesting of energy with the use of
  renewable energy solutions would highly contribute to bend those limitations and
  would significantly boost the performance of an IoT system.
- *Self-configuration, self-organization and autonomy*: The nodes in an IoT system
  need to be able to adapt to any environmental hazards and to any network changes
  that affect its functionality and operation. The self-configuration characteristic is
  of great importance, also, against any potential privacy risks since it would deter
  the possible intruders and would help the system to update and defend against any
  hacking threats.
- *Unpredictable mobility pattern*: As has already been explained, the various sensors could be mobile or static, a characteristic that should not affect the systems
  performance. Geolocation data received by an IoT application to inform a human

user for available parking spaces in his were-abouts, do not follow a predefined direction but rather dynamically move in the area as the user searches the urban environment for a possible end to his/her parking request.

- *Multihop communication*: Based on the network topology and the selected IoT architecture, either the sensors have to relay the data to an aggregator node that then forwards them to the gateway or they might need to communicate with other sensors that are not located near them, therefore they should use the Internet for their request.

5.3   Existing Routing Algorithms and Protocols

Due to its intrinsic characteristics, IoT needs a different approach for routing, dealing not only with the IP address problem that leads to incorporate IPv6-based solutions (the proliferation of devices with Internet access has urged the use of IPv6), but also with solutions that demand low power and have small memory demands. With this in mind, the most popular IoT algorithms and protocols are presented below [39, 40]:

- *IPv6*: As stated above, for IP-based networks, the selection of IPv6 solutions (like 6LowPAN) is rather straightforward but is recommended only for devices that have high process capabilities and memory.
- *CoAP*: The most important feature is the translation of this routing protocol to an HTTP message that can be integrated with web services. It can, also, support multicast functionality with very small overhead.
- Ad Hoc *on Demand Multipath Distance Vector Routing protocol for IoT (AOMDV-IoT)*: A connection between connected regular nodes (without internet access) and internet nodes takes place with this routing protocol. Unfortunately, there are no security mechanisms deployed and since, it is not a context-aware routing protocol (routing is rather based on the number of hops from the destination as with most distance vector solutions) and, therefore, there is no mechanism to change the route based on the energy consumption of the nodes on the selected path.
- *Secure Multihop Routing Protocol (SMRP)*: It is a routing protocol that focuses on security and for this reason, each IoT network must register its applications, network addresses, and data link addresses to an authorized Service Provider (SP). Then, the SP is responsible to create an Encrypted File (EF) that should be installed on every individual device that will be used from them in order to authenticate any future communications between the nodes. SMRP is not, also, a context-aware protocol and this affects the network's lifetime and needs more memory usage.
- *IPv6 Routing Protocols for Low-Power and Lossy Network (RPL)*: As the name suggests, this is a routing protocol that is better used for routing between low-power devices with constrained process and memory capabilities in noisy environments that cause losses in the network functionality.

Other well-known solutions include *Energy aware Ant Routing Algorithm (EARA), Pruned Adaptive IoT Routing (PAIR)*, and *Multiparent routing in RPL*.

6 Conclusions

With the evolution of the IoT being closely related to the advances in the sensing and power capabilities of the smart devices, along with the wireless capabilities that the smart devices possess and use, this paper tried to emphasize on the challenges and characteristics of the use of wireless technologies in this environment. From their role in the proposed architectures to the study of the D2D communications, one of the most important features of the IoT, wireless technologies emerge as the force behind the wheel of IoT. At the same time, special care and attention should be given on the advanced security features needed from the inherent characteristic of such a heterogeneous, rapidly developing, scalable, and large size system. We also discuss the security hazards and possibly solutions for a complete study of the effect of wireless technologies in the IoT environment.

References

1. Gubbi J, Buyya R, Marusic S and Palaniswami M (2013) Internet of Things (IoT): A vision, architectural elements, and future directions. Future Generation Computer Systems 29(7): 1645–1660.
2. Yaqoob I, Ahmed E, Hashem I, Ahmed A, Gani A, Imran M and Guizani M (2017) Internet of Things Architecture: Recent Advances, Taxonomy, Requirements, and Open Challenges. IEEE Wireless Communications 24(3): 10–16.
3. Atzori L, Iera A and Morabito G (2010) The Internet of Things: A survey. Computer Networks 54(15): 2787–2805.
4. Al-Fuqaha A, Guizani M, Mohammadi M, Aledhari M and Ayyash M (2015) Internet of Things: A Survey on Enabling Technologies, Protocols, and Applications. IEEE Communications Surveys & Tutorials 17(4): 2347–2376.
5. Xu L, He W and Li S (2014) Internet of Things in Industries: A Survey. IEEE Transactions on Industrial Informatics 10(4): 2233–2243.
6. Elkhodr M, Shahrestani S and Cheung H (2016) Emerging Wireless Technologies in the Internet of Things: A Comparative Study. International Journal of Wireless & Mobile Networks 8(5): 67–82.
7. Palattella M, Dohler M, Grieco A, Rizzo G, Torsner J, Engel T and Ladid L (2016) Internet of Things in the 5G Era: Enablers, Architecture, and Business Models. IEEE Journal on Selected Areas in Communications 34(3): 510–527.
8. Bello O and Zeadally S (2016) Intelligent Device-to-Device Communication in the Internet of Things. IEEE Systems Journal 10(3): 1172–1182.
9. Sethi P and Sarangi S (2017) Internet of Things: Architectures, Protocols, and Applications. Journal of Electrical and Computer Engineering 2017: 1–25.
10. Bandyopadhyay D and Sen J (2011) Internet of Things: Applications and Challenges in Technology and Standardization. Wireless Personal Communications 58(1): 49–69.
11. Miorandi D, Sicari S, De Pellegrini F and Chlamtac I (2012) Internet of things: Vision, applications and research challenges. Ad Hoc Networks 10(7): 1497–1516.
12. European Parliament (2017) General Data Protection Regulation 2016/79. Available at: http://eur-lex.europa.eu/legal-content/EN/TXT/?uri=uriserv:OJ.L_.2016.119.01.0001.01. ENG&toc=OJ:L:2016:119:TOC, 2017. Accessed 25 July 2017.
13. IEC (2016) IoT 2020: Smart and secure IoT platform. White paper.

14. Kogias D, Xevgenis M and Patrikakis C (2016) Cloud Federation and the Evolution of Cloud Computing. Computer 49(11): 96–99.
15. Puliafito C, Mingozzi E and Anastasi G (2017) Fog Computing for the Internet of Mobile Things: Issues and Challenges. In: IEEE International Conference on Smart Computing, SMARTCOMP 2017, pp. 1–6.
16. ZVEI – German Electrical and Electronic Manufacturers' Association (2015) Industrie 4.0: The Reference Architectural Model Industrie 4.0 (RAMI 4.0).
17. Weyrich M and Ebert C (2016) Reference Architectures for the Internet of Things. IEEE Software 33(1): 112–116.
18. The International Electrotechnical Commission (2016) Standards on Information Technology - "Internet of Things Reference Architecture (IoT RA)", ISO/IEC CD 30141.
19. Fremantle P (2015) A Reference Architecture for the Internet of Things. White Paper.
20. International Telecommunication Union (2012) ITU-T Study Group 13, Next Generation Networks – Frameworks and Functional Models: Overview of the Internet of Things.
21. Minerva R (2016) IEEE IoT Initiative Chair – Tim Lab, "Toward a Definition of Internet of Things".
22. Tuna G, Kogias D, Gungor V, Gezer C, Taşkın E and Ayday E (2017) A survey on information security threats and solutions for Machine to Machine (M2M) communications. Journal of Parallel and Distributed Computing 109: 142–154.
23. OASIS (2017) OASIS eXtensible Access Control Markup Language (XACML) TC. *OASIS eXtensible Access Control Markup Language (XACML) TC | OASIS*. Available at: https://www.oasis-open.org/committees/tc_home.php?wg_abbrev=xacml. Accessed 25 July 2017.
24. Bandyopadhyay S., Sengupta M., Maiti S., Dutta S. (2011) A Survey of Middleware for Internet of Things. In: Özcan A., Zizka J., Nagamalai D. (eds) Recent Trends in Wireless and Mobile Networks. Communications in Computer and Information Science, vol 162. Springer, Berlin, Heidelberg.
25. Razzaque M, Milojevic-Jevric M, Palade A and Clarke S (2016) Middleware for Internet of Things: A Survey. IEEE Internet of Things Journal 3(1): 70–95.
26. Zhou H (2012) The Internet of Things in the Cloud: A Middleware Perspective (1st ed.). Boca Raton, FL: CRC Press.
27. Bisdikian C (2001) An overview of the Bluetooth wireless technology. IEEE Communications Magazine 39(12): 86–94.
28. Cavalcante E, Pitanga Alves M, Batista T, Coimbra Delicato F and Pires P F (2015) An Analysis of Reference Architectures for the Internet of Things. In: Proceedings of the 1st International Workshop on Exploring Component-based Techniques for Constructing Reference Architectures (CobRA '15), pp. 13–16.
29. Inc., M.S., MachineShop | The Edge Computing Company (2017) MachineShop | The Edge Computing Company. Available at: http://www.machineshop.io/. Accessed 25 July 2017.
30. RS Components (2017) 11 Internet of Things (IoT) Protocols You Need to Know About. Available at: https://www.rs-online.com/designspark/eleven-internet-of-things-iot-protocols-you-need-to-know-about. Accessed 25 July 2017.
31. Bormann C and Shelby Z (2013) 6lowpan. Hoboken, N.J.: Wiley.
32. Z-Wave (2017) Z-Wave. Available at: http://www.z-wave.com/. Accessed 25 July 2017.
33. ZigBee Alliance (2017) ZigBee. Available at: http://www.zigbee.org/. Accessed 25 July 2017.
34. Wi-Fi Alliance (2017) Wi-Fi. Available at: http://www.wi-fi.org/. Accessed 25 July 2017.
35. Gupta A and Jha R (2015) A Survey of 5G Network: Architecture and Emerging Technologies. IEEE Access 3: 1206–1232.
36. NFC Forum (2017) NFC Forum. Available at: http://nfc-forum org/, last accessed at 25 July 2017.
37. lora-alliance (2017) lora-alliance. Available at: https://www.lora-alliance.org/. Accessed 25 July 2017.
38. Kasnesis P, Patrikakis C, Kogias D, Toumanidis L and Venieris I (2017) Cognitive friendship and goal management for the social IoT. Computers & Electrical Engineering 58: 412–428.

39. Dhumane A, Prasad R, Prasad J (2016) Routing Issues in Internet of Things: A Survey. In Proceedings of the International MultiConference of Engineers and Computer Scientists 2016, vol I, pp. 404–412.
40. Sutaria R (2014) Understanding Wireless Routing For IoT Networks. *Electronic Design*. Available at: http://www.electronicdesign.com/communications/understanding-wireless-routing-iot-networks. Accessed 25 July 2017.

Fast and Flexible Initial Uplink Synchronization for Long-Term Evolution

Md. Mashud Hyder and Kaushik Mahata

Abstract Orthogonal frequency-division multiple access has been widely adopted by the modern wireless networking standards. These use initial uplink synchronization (IUS) process to detect and uplink-synchronize with new user equipments (UEs) (3rd Generation Partnership Project; technical specification group radio access network; evolved universal terrestrial radio access (E-UTRA); physical channels and modulation (release 10), (2011) [1]). IUS is a random access process where a UE intending to start communication transmits a code during an "IUS opportunity". The code is chosen uniformly at random from a predefined codebook. The eNodeB uses the received signal to detect the codes, and estimate the uplink channel parameters associated with each detected code. This detection and estimation problem is known to be quite challenging, particularly when the number of UEs transmitting during an IUS opportunity is not small. We discuss some recent sparse signal processing methods to address this problem in the context of long-term evolution (LTE) standards. This research does not only give some new directions to solve the detection and estimation problem but also provides guidelines for designing the codebook. In addition, the key ideas are applicable to other OFDMA systems.

1 Background

1.1 Initial Uplink Synchronization (IUS)

The orthogonal frequency-division multiple access (OFDMA) is the most successful wireless access technology in mitigating the adverse effects of multi-access interference and wireless channel fading. Most modern wireless communication standards

Research is supported by the Australian Research Council.

Md. M. Hyder (✉) · K. Mahata
The University of Newcastle, Callaghan, NSW 2308, Australia
e-mail: md.hyder@uon.edu.au

K. Mahata
e-mail: kaushik.mahata@newcastle.edu.au

use OFDMA [1, 20]. However for OFDMA to work properly, the OFDM symbols from different UEs must arrive at the eNodeB at the same time with similar power levels. Since different UEs are located at different points within the cell, and experience different degrees of wireless connectivity, they must adjust their transmission power levels and delay their transmission appropriately to ensure their symbols arrive at the eNodeB with similar power levels. Hence, these delay and power level values must be estimated for each individual UE before it can start uplink transmissions. To facilitate this estimation process, OFDMA systems require the new UEs to follow a network entry procedure. It is called initial ranging (IR) in IEEE 802.16, and random access (RA) in long-term evolution (LTE) [1, 20, 27]. In this scheme, a new UE must downlink synchronize itself, and wait for an IUS opportunity. The eNodeB periodically broadcasts the resource allocations of IUS opportunities in the downlink channel. These allocations consist of a number of IUS subcarriers in some time slots. The downlink synchronized user equipment (UEs) willing to commence communication can use this opportunity by transmitting certain code via the IUS subcarriers during an IUS time slot. The code must be chosen at random from a predefined codebook. If multiple UEs transmit their IUS signal simultaneously, they are allowed to share the same IUS subcarriers. Since UEs are located at different positions, their signal arrival time delay at the eNodeB will be different. In addition, the uplink wireless channel also varies widely for different UEs. The eNodeB receives the signal resulted from the transmission of all the UEs. It is eNodeB's task to use the received signal to detect the IUS codes transmitted along with the channel parameters and the delays corresponding to each code [3, 25, 26, 30, 31]. Subsequently, the eNodeB broadcasts a response message indicating the detected codes and the corresponding timing and power adjustment parameters. From this, a UE can infer its IUS code has been detected or not. If the eNodeB fails to detect its code then the UE must make a transmission in the next IUS opportunity [26].

1.2 Related Works

The correlation-based approach proposed in [21] is working based on the principle that a time delay can be represented by a phase shift in the frequency domain. In [33], it is shown that the frequency-domain correlation-based approach outperforms its time-domain counterpart. Lee [22] replaces the IUS codes by a set of generalized chirp-like polyphase sequences to get a more accurate timing estimate. The IUS resource allocation scheme in [14] requires the UEs to transmit their codes on disjoint sets of subcarriers, resulting in a minimum level of MAI. This scheme reduces the number of effective subcarriers for each user, resulting in some degradation of timing estimation performance [25]. The IUS scheme in [4] needs that the uplink signals from UEs are transmitted over disjoint subcarriers, and the receivers use filter banks to separate multiuser codes. The generalized likelihood ratio test (GLRT) based methods have been proposed in [26, 27], which are used in LTE and IEEE 802.16-based networks, respectively, estimate the timing offset and the channel response jointly via the maximum-likelihood (ML) method. GLRT performs very well with one or two UEs. However, its performance appears to degrade quickly with an increase in the number of UEs.

The iterative maximum-likelihood algorithm in [29, 31] applied an expectation-maximization (EM) type technique to mitigate MAI. Successive interference cancellation (SIC) algorithms [23, 25, 26] are very popular in IUS. In its most basic form, the algorithm works in an iterative fashion where the strongest path of each active RT is detected and removed from the received signal, and the resulting signal is used in succeeding iterations. The complexity of SIC algorithms are generally low and have efficient user detection capability.

1.3   Emerging Challenges and Our Contributions

The user detection and channel estimation problem become very challenging in multiuser environment due to large multi-access interference (MAI). The state-of-the-art methods are still unable to reliably solve the underlying detection and estimation problem if the number of synchronizing terminals is not small. In addition, the computation times of the more accurate algorithms are often quite high. This limits the practical utility of these algorithms. Third Generation Partnership Project [1] deploys a number of collision resolution algorithms, such as binary exponential back-off, to keep the number of IUS terminals to a manageable level. But in the emerging M2M networks with a vast number of devices, it will be hard to limit the number of simultaneous connection requests to a small number [32].

In this chapter, we describe a new signal processing framework for addressing above challenges. We show that IUS can be cast as a sparse signal representation problem. Subsequently, we use some analytic and computational techniques from the compressive sensing literature to address the issues described above. The resulting algorithms can reliably detect more codes than the commonly used methods. It is possible to accelerate the computations significantly via some tricks employing the fast Fourier transform (FFT). In addition, this framework allows us to use the theoretical results in compressive sensing to quantify the desirable characteristics of a codebook. This makes it possible to cast the codebook design problem as an optimization problem. The solution to the optimization problem gives an optimal codebook which ensures the best possible performance of a code detection-estimation algorithm. This code design method is also extended to handle carrier frequency offsets. Typically, it is very difficult to detect code of fast-moving IUS terminal due to Doppler drift of its carrier frequency. The proposed framework allows us to design the codebook such that the effect of this carrier frequency drift on the detection-estimation performance is minimized.

2   IUS Signal Representation

Single RT

Consider an OFDMA LTE system with N data samples N_p cyclic prefix samples per OFDM frame, and M *adjacent* IUS subcarriers (also known as physical random access channel (PRACH)) [1]. Suppose the IUS codebook consists of G codes, where each code is an M-dimensional vector of complex numbers. LTE employs Zadoff–Chu (ZC) sequences to generate codes. Given a root u, the ZC sequence is generated as [11, 13]

$$Z^u(s) = e^{-i\pi us(s+1)/M}, \quad s \in \mathcal{K}, \tag{1}$$

where $\mathcal{K} = \{0, 1, 2, \ldots, M - 1\}$. The $(s + 1)$ th element of the $k + 1$ th code $\hat{\mathbf{c}}_{k+1}$ in the codebook is given by

$$\hat{c}_{s+1,k+1} = Z^u \{(s + k\, n_{cs}) \mod M\}, \quad s \in \mathcal{K}. \tag{2}$$

Here, "mod" is the mathematical modulo operator. The value of n_{cs} can be chosen from [1, Table 5.7.2-2]. Total $\lfloor M/n_{cs} \rfloor$ codes can be generated from a single root [30]. We need multiple roots for generating codes more than $\lfloor M/n_{cs} \rfloor$.

At an IUS opportunity, an IUS terminal picks a code $\hat{\mathbf{c}}_k$ randomly from G available codes, calculate its M point discrete Fourier transform (DFT) $\mathbf{c}_k$ and transmit to the eNodeB. Being M a prime number in LTE, it can be shown that [5, 24]

$$c_{s,k} = \tilde{Z}^u(s - 1)e^{i2\pi(s-1)(k-1)n_{cs}/M}, \tag{3}$$

$$\text{with,} \qquad \tilde{Z}^u(s) = \frac{e^{i\pi uu^+ s(u^+ s+1)/M}}{\sqrt{M}} \sum_{n=0}^{M-1} Z^u(n), \tag{4}$$

where u^+ is such that $(uu^+) \mod M = 1$.

The received signal at eNodeB is given by [17, Eq. (15)]

$$\mathbf{v} = \mathbf{R}\text{diag}(\tilde{Z}^u)\mathbf{U}_k \Theta \mathbf{F}\tilde{\mathbf{h}} \tag{5}$$

$$\mathbf{U}_k = \text{diag}\{1, e^{i2\pi(k-1)n_{cs}/M}, \ldots, e^{i2\pi(M-1)(k-1)n_{cs}/M}\} \tag{6}$$

$$\mathbf{R} = \Theta \mathbf{F} \mathbf{D} \mathbf{F}^* \Theta^\mathsf{T}, \tag{7}$$

$$\tilde{\mathbf{h}} := [\mathbf{0}_{1\times d}\, h(0)\, \ldots\, h(P - 1)\, \mathbf{0}_{1\times(N-P-d)}]^\mathsf{T} \tag{8}$$

$$\text{where } \mathbf{D} = \text{diag}\{1, e^{i2\pi\epsilon/N}, \ldots e^{i2\pi(N-1)\epsilon/N}\}. \tag{9}$$

Here, $\mathbf{F}$ is the DFT matrix where $[\mathbf{F}]_{s,\ell} = \exp\{-i2\pi(s - 1)(\ell - 1)/N\}/\sqrt{N}$, $s, \ell \in \{1, 2, \ldots N\}$, $\Theta \in \mathbb{R}^{M \times N}$ whose ℓ-th row is the j_ℓ-th row of the $N \times N$ identity matrix, ϵ is the normalized carrier frequency offset (CFO). Equation (5) expresses the received signal vector $\mathbf{v}$ in a convenient manner where $\mathbf{R}$ represents the effect of carrier frequency offset, $\text{diag}(\tilde{Z}^u)\mathbf{U}_k$ depends on the code transmitted, and $\tilde{\mathbf{h}}$ represents the channel impulse response (CIR) $h(n)$, $n \in \{0, 1, 2, \ldots, P-1\}$ delayed by d where the value of d depends on the distance between eNodeB and the RT.

In [17, Appendix A], it has been shown that

$$\mathbf{R} = \mathbf{I} + \mathbf{W}\epsilon + O(\epsilon^2),$$

$$\text{where} \quad \mathbf{W}_{(s,n)} = \begin{cases} i\pi(1 - 1/N), & s = n, \\ -\dfrac{\pi e^{i\pi(s-n)/N}}{N\sin(\pi(s-n)/N)}, & s \neq n \end{cases} \tag{10}$$

and $\mathbf{I}$ is an identity matrix. Let $P_{\max}$ and $\widehat{D}$ are the maximum value of P and d, respectively. In practical systems, $P_{\max}$ is known from the statistical characteristics of the channel, and $\widehat{D}$ is known because the cell radius is known. Hence, $N_1 = P_{\max} + \widehat{D}$ is also known. Now $d + P \leq N_1$. By construction (see (8)), all components of $\tilde{\mathbf{h}}$ with indices larger than N_1 are zeros. Therefore, we can rewrite (5) as

$$\mathbf{v} = \mathbf{R}\tilde{\mathbf{E}}_\ell\widehat{\mathbf{h}}, \tag{11}$$

$$\text{where,} \quad \tilde{\mathbf{E}}_\ell = \text{diag}(\tilde{Z}^u)\mathbf{U}_\ell\Theta\mathbf{F}_{(:,1:N_1)}$$

$$\widehat{\mathbf{h}} = \tilde{\mathbf{h}}_{(1:N_1)}$$

$\mathbf{F}_{(:,1:N_1)}$ is constructed from $\mathbf{F}$ by taking its first N_1 columns. Note that $\tilde{\mathbf{E}}_\ell$ is known $\forall \ell$. But $\{h(n)\}$, d and ϵ are unknown.

Multiple RTs

So far, we considered only one IUS terminal. In practice, there are multiple IUS terminals sending multiple codes. We assume that one RA code does not transmit by multiple UEs (see [19] for a justification). Hence, the received data $\mathbf{y}$ can be expressed as

$$\mathbf{y} = \mathbf{A}\mathbf{x} + \mathbf{e}, \quad \mathbf{A} = [\ \mathbf{R}_1\tilde{\mathbf{E}}_1 \ \ \mathbf{R}_2\tilde{\mathbf{E}}_2 \ \ \ldots \ \ \mathbf{R}_G\tilde{\mathbf{E}}_G\ ],$$

$$\mathbf{x} := [\ \widehat{\mathbf{h}}_1^\mathsf{T} \ \ \widehat{\mathbf{h}}_2^\mathsf{T} \ \ \ldots \ \ \widehat{\mathbf{h}}_G^\mathsf{T}\ ]^\mathsf{T}, \tag{12}$$

where $\mathbf{e}$ is the noise term, and $\widehat{\mathbf{h}}_\ell$ is the CIR (see (11)) of RT transmitting $\mathbf{c}_\ell$. Note that $\widehat{\mathbf{h}}_\ell = 0$ if no RT sends $\mathbf{c}_\ell$. Finally, $\mathbf{R}_\ell$ embedded the CFO contribution of that RT.

With this background the IUS problem can be stated as follows: *Given* $\mathbf{y}$ *the eNodeB needs to (a) find the set* $\mathcal{L} = \{\ell : \|\widehat{\mathbf{h}}_\ell\|_2 \neq 0\}$, *(b) timing offsets associated to each user, i.e.,* d_ℓ *for all* $\ell \in \mathcal{L}$, *and (c) power associated with the channel impulse response vectors* $\widehat{\mathbf{h}}_\ell$ *for all* $\ell \in \mathcal{L}$. Recall that the first d components of $\tilde{\mathbf{h}}$ are zero, see (8). Hence by construction of $\widehat{\mathbf{h}}_\ell$ in (11), the index of the first nonzero component of $\widehat{\mathbf{h}}_\ell$ is $1 + d_\ell$.

3 A Sparse Recovery Method

For now, we consider CFO is negligible [25, 27]. Hence, $\mathbf{R} = \mathbf{I}$, see (10). Later we see how the ZC codes can be selected to mitigate the effects of CFO. When we use these specially selected codes, the algorithms designed by setting $\mathbf{R} = \mathbf{I}$ become immune to the adverse effects of CFO.

With $\mathbf{R} = \mathbf{I}$, (12) becomes

$$\mathbf{y} = \mathbf{A}\mathbf{x} + \mathbf{e}, \quad \mathbf{A} = [\ \tilde{\mathbf{E}}_1 \ \tilde{\mathbf{E}}_2 \ \dots \ \tilde{\mathbf{E}}_G \ ]. \tag{13}$$

Since $\tilde{\mathbf{E}}_\ell$ is known $\forall \ell$, we can construct $\mathbf{A}$. In practice, the number of RTs is much smaller than G. Hence, $\widehat{\mathbf{h}}_\ell = 0$ for most of $\ell \in \{1, 2, \dots, G\}$. Even if $\widehat{\mathbf{h}}_\ell \neq 0$ for some ℓ, most components of that $\widehat{\mathbf{h}}_\ell$ have very small magnitude. As a result, $\widehat{\mathbf{h}}_\ell$ is sparse [23, 25, 26], and hence, $\mathbf{x}$ is very sparse. Therefore, recovering $\mathbf{x}$ from $\mathbf{y}$ can be cast as a sparse recovery problem.

From a sparse estimate $\check{\mathbf{x}}$ of $\mathbf{x}$, BS can extract the IUS information as follows. Partition $\check{\mathbf{x}}$ into G sub-vectors:

$$\check{\mathbf{x}} = [\ \grave{\mathbf{h}}_1^\mathsf{T} \ \grave{\mathbf{h}}_2^\mathsf{T} \ \dots \ \grave{\mathbf{h}}_G^\mathsf{T} \ ]^\mathsf{T},$$

where each $\grave{\mathbf{h}}_\ell$ is of length N_1. Then we declare $\ell \in \mathcal{L}$ only if $\|\grave{\mathbf{h}}_\ell\| \neq 0$ and the index of the first nonzero component of $\grave{\mathbf{h}}_\ell$ leads to an estimate of d_ℓ. In the following, we consider ways to estimate $\mathbf{x}$ from (13).

Since $\mathbf{A}$ is an $M \times G.N_1$ matrix with $M < G.N_1$, (13) is underdetermined, and has infinitely many solutions of $\mathbf{x}$, even when $\mathbf{e} = 0$. With $\mathbf{e} \neq 0$ we have more possible combinations of $\mathbf{x}$ and $\mathbf{e}$ satisfying (13). To solve $\mathbf{x}$ we need prior knowledge to identify the desired solution from the solution set. This in our case is that $\mathbf{x}$ is sparse. The most popular way to estimate a sparse $\mathbf{x}$ is to solve

$$\underset{\mathbf{x}}{\text{minimize}} \ \|\mathbf{x}\|_p, \quad \text{subject to } \mathbf{y} = \mathbf{A}\mathbf{x}, \tag{14}$$

with $p \leq 1$ [9]. The solution for $p = 0$ (commonly referred to as the ℓ_0 optimization method) has the smallest number of nonzero components. But this is a combinatorial problem [10]. Hence, researchers either suggest using $0 < p \leq 1$ [15], or use some smoothed approximation of the zero norm [18]. The later is often referred to as the ℓ_0 approximation algorithms. Taking $p = 1$ gives a convex problem popularly known as basis pursuit (BP) [8, 10]. ℓ_0 approximation algorithms are resilient to noise. But being non-convex these require some good initialization. Basis pursuit, on the other hand, is not that noise tolerant.

Our algorithm in [19] starts with an iterative primal-dual interior point algorithm for basis pursuit and generates a good approximate solution in a few iterations. This approximate solution is then used to initialize a smoothed ℓ_0 approximation algorithm called ISL0 [18]. The resulting algorithm can be viewed as an iterative re-weighted least square (IRLS) method, which is known to offer super-quadratic convergence when initialized sufficiently close to the final solution [15]. The algorithm developed in the following sections will be called "Handover" algorithm.

3.1 ℓ_1 Optimization Strategy

Let $\mathbf{a}_i$ be the ith column of $\mathbf{A}$, and $\mathbf{1}$ denote the $G.N_1$-dimensional vector of all ones. In [19, Theorem 1], it is shown that BP can be solved by solving

$$\text{maximize}_{\mathbf{g}} \; (\mathbf{g}^*\mathbf{y} + \mathbf{y}^*\mathbf{g})/2 \tag{15a}$$

$$\text{subject to } \mathbf{g}^*\mathbf{a}_i\mathbf{a}_i^*\mathbf{g} \leq 1, \quad i = 1, \ldots G.N_1, \tag{15b}$$

jointly with its Lagrangian dual

$$\text{minimize}_{\mathbf{z}} \; \{\mathbf{1}^\mathsf{T}\mathbf{z} + \mathbf{y}^*[\mathbf{A}\,\text{diag}(\mathbf{z})\mathbf{A}^*]^{-1}\mathbf{y}\}/2 \tag{16a}$$

$$\text{subject to } \mathbf{z} \geq 0. \tag{16b}$$

Table 1 lists the primal-dual algorithm of [19] to solve the primal-dual pair (15) and (16) concurrently. This is based on the path following the method outlined in [7]. In Table 1 $f(\mathbf{g}) = [f_1(\mathbf{g})\; f_2(\mathbf{g}) \ldots f_{G.N_1}(\mathbf{g})]^\mathsf{T}$, with

$$f_i(\mathbf{g}) = \mathbf{g}^*\mathbf{a}_i\mathbf{a}_i^*\mathbf{g} - 1, \text{ for } i = 1, 2, \ldots, G.N_1. \tag{17}$$

In addition, we define the vector $\mathbf{b}$ such that

$$[\mathbf{b}]_i = 1/f_i(\mathbf{g}), \text{ for } i = 1, \ldots, G.N_1. \tag{18}$$

It can be shown that the Jacobean matrix of $f(\mathbf{g})$ is

$$\mathbf{J} = [\, \mathbf{a}_1\mathbf{a}_1^*\mathbf{g} \; \mathbf{a}_2\mathbf{a}_2^*\mathbf{g} \; \ldots \; \mathbf{a}_{G.N_1}\mathbf{a}_{G.N_1}^*\mathbf{g} \,]^\mathsf{T}. \tag{19}$$

In Table 1, μ controls the surrogate duality gap, and α controls the step length in backtracking line search. As suggested in [7], we take $\mu = 0.5$ and $\alpha = 0.1$. The algorithm is run until a sparse enough $\mathbf{x}$ is produced. The energy of the $M/2$ most significant components of $\mathbf{x}$ expressed as a fraction of the total energy of $\mathbf{x}$ is used as the measure of sparsity [15]. When this fraction is above a threshold κ we terminate the algorithm. The choice of κ is analyzed in detail in [19].

Table 1 Primal-dual algorithm for initialization ℓ_1 (INL-1)

Initialization: Set $\mathbf{g} = \mathbf{0}, \mathbf{z} = \mathbf{0}$, and $\mu, 0 < \alpha \leq 1$.

repeat

1. Compute primal-dual search directions:

$\Delta\mathbf{g} = (\mathbf{A} \operatorname{diag}(\mathbf{z})\mathbf{A}^* - \mathbf{J}^*\mathbf{S}\mathbf{J})^{-1}(-\mathbf{y} + \mu\mathbf{J}^*\mathbf{b})$

$\Delta\mathbf{z} = -(\mathbf{z} + \mu\mathbf{b} + \mathbf{S}\mathbf{J}\Delta\mathbf{g})$; where, $\mathbf{S} = [\operatorname{diag}\{f(\mathbf{g})\}]^{-1}\operatorname{diag}(\mathbf{z})$.

2. Find $0 < s \leq 1$ such that:

$f_i(\mathbf{g} + s\Delta\mathbf{g}) \leq 0, \mathbf{z}(i) + s\Delta\mathbf{z}(i) \geq 0; \forall i$.

3. Set $\mathbf{g} = \mathbf{g} + s\Delta\mathbf{g}, \mathbf{z} = \mathbf{z} + s\Delta\mathbf{z}, \mu = \alpha\mu$.

4. Compute $\mathbf{x} = \operatorname{diag}(\mathbf{z})\mathbf{A}^*\mathbf{g}$.

until (*A rough estimate of optimal* $\mathbf{x}$ *is obtained*)

Output: $\mathbf{x}^{(1)} = \mathbf{x}$.

3.2 Smoothed ℓ_0 Norm Minimization [18]

The solution produced by the primal-dual method serves as the initial point in ISL0 algorithm. [16, 18]. ISL0 approximates $||\mathbf{x}||_0$ by the sum of Gaussian functions [16, 18]. In effect, it minimizes

$$L_\sigma(\mathbf{x}) := -\sum_{j=1}^{G.N_1} e^{-\frac{||\mathbf{x}||_j^2}{2\sigma^2}} + \frac{\lambda}{2}||\mathbf{y} - \mathbf{A}\mathbf{x}||_2^2 \tag{20}$$

to estimate $\mathbf{x}$ from (13). Here, σ is a small real number and $\lambda > 0$. Taking $\sigma \to 0$ approximates $||\mathbf{x}||_0$ closely. But L_σ highly non-smooth when $\sigma \to 0$. ISL0 implements a the graduated non-convexity (GNC) strategy [6], where σ controls the degree of non-convexity. ISL0 constructs a sequence $\sigma_n > \sigma_{n-1} > \cdots > \sigma_0$ with $\sigma_j - \sigma_{j-1}$ being a small positive number for each j. Thereby, the minimizer of $L_{\sigma_{j-1}}$ is quite close to the minimizer of L_{σ_j}. If ISL0 finds out the minimizer of L_{σ_j}, then it can be used to initialize the solver for minimizing $L_{\sigma_{j-1}}$. This procedure of successive optimization continues until the terminating value of σ, i.e., σ_0 is reached. The value of σ_0 can be chosen using a procedure outlined in [18], which turns out to be 0.001 in our simulations.

ISL0 employs Gauss–Newton-based convex–concave procedure to minimize L_σ for a fixed σ. This algorithm uses the fact that $L_\sigma(\mathbf{x})$ is decreasing along $\zeta_\sigma(\mathbf{x}) - \mathbf{x}$ [16], where

$$\zeta_\sigma(\mathbf{x}) = \lambda\left[W_\sigma(\mathbf{x})/\sigma^2 + \lambda\mathbf{A}^*\mathbf{A}\right]^{-1}\mathbf{A}^*\mathbf{y}, \tag{21}$$

$$\text{and } W_\sigma(\mathbf{x}) = \operatorname{diag}\left\{e^{-\frac{||\mathbf{x}||_1^2}{2\sigma^2}}, \ldots, e^{-\frac{||\mathbf{x}||_{G.N_1}|^2}{2\sigma^2}}\right\}. \tag{22}$$

Table 2 ISL0 Algorithm

Input: $\mathbf{x}, \sigma_{st}, \lambda, \sigma_0$.

Initialization: $\sigma = \sigma_{st}$, and $\rho, \eta, \gamma \in [0, 1), i = 0, \beta = 1$.

repeat

  1. while $L_\sigma\{\beta\zeta_\sigma(\mathbf{x}) + (1 - \beta)\mathbf{x}\} > L_\sigma(\mathbf{x})$

    $\beta = \gamma\beta$.

  end

  2. $\mathbf{x}_o = \mathbf{x}$.

  3. $\mathbf{x} = \beta\zeta_\sigma(\mathbf{x}) + (1 - \beta)\mathbf{x}$. Set $\beta = 1$.

  4. If $||\mathbf{x} - \mathbf{x}_o||_2 < \eta\sigma$ then $\sigma = \rho\sigma$.

while $\sigma \geq \sigma_0$.

Output: $\bar{\mathbf{x}} = \mathbf{x}$.

Furthermore, at the minimum point $\mathbf{x}_*$ of $L_\sigma(\mathbf{x})$ it holds that

$$\mathbf{x}_* = \zeta_\sigma(\mathbf{x}_*). \tag{23}$$

The ISL0 algorithm is given in Table 2. The value of λ depends on the noise level. Following the recommendations in [28], we take $\lambda = c/\sqrt{2\sigma_e^2 M \log(G.N_1)}$, where c depends on $\mathbf{A}$ [28]. We follow the recommendations in [16, 18] and set $\rho = 0.3, \eta = 0.5, \gamma = 0.5$.

The initial value of σ depends on the output $\mathbf{x}^{(1)}$ of the ℓ_1-optimization routine in Table 1, which is used to initialize ISL0. For the GNC strategy of ISL0 to work well, we must choose the starting value of σ_{st} carefully. We cannot allow $\mathbf{x}^{(1)}$ to be far from the minimum point of $L_{\sigma_{st}}$. In particular, if we like to ensure that $\mathbf{x}^{(1)}$ is the minimizer of $L_{\sigma_{st}}$, then (21) and (23) require

$$\mathbf{x}^{(1)} = \lambda \left[W_{\sigma_{st}}(\mathbf{x}^{(1)})/\sigma_{st}^2 + \lambda \mathbf{A}^*\mathbf{A} \right]^{-1} \mathbf{A}^*\mathbf{y}. \tag{24}$$

We take σ_{st} as the solution to the least squares problem

$$\sigma_{st}^2 = \arg\min_{\sigma^2} \left\| \frac{W_\sigma(\mathbf{x}^{(1)})}{\sigma^2}\mathbf{x}^{(1)} - \lambda \mathbf{A}^*(\mathbf{y} - \mathbf{A}\mathbf{x}^{(1)}) \right\|_2^2, \tag{25}$$

induced by (24). The problem (25) being one-dimensional can be solved reliably by an interior trust region algorithm [12].

The major computational steps in the algorithms in Tables 1 and 2 can be accelerated significantly by using FFT algorithm. This possible because $\mathbf{A}$ depends on the DFT matrix $\mathbf{F}$. We refer the readers to [19] for details.

4 Codebook Design

Recall that LTE uses ZC codes in its codebook. In fact, LTE uses some restricted set of ZC codes in the sense that only a few values of the ZC root u and the parameter n_{cs} are allowed in LTE, see [1, Table 5.7.2-2]. This is because the codes used in the codebook has a profound effect on the performance of the code detection algorithms. To illustrate this, we simulate three active IUS terminals with low CFO. The LTE system model used for the simulation is detailed in Sect. 6. The user detection performances of three IUS algorithms are given in Table 3 as functions of n_{cs}. According to LTE standards, we can pick any n_{cs} from [1, Table 5.7.2-2]. However, the results in Table 3 reveal that the probability of successful code detection (P_s) depends on n_{cs}. When $r = 2.1$ km, P_s is poor for $n_{cs} \leq 15$. In contrast, P_s are above 0.9 for $n_{cs} \geq 18$. Again, $n_{cs} \geq 26$ is required for $r = 3.2$ km. This result demonstrates the importance of choosing appropriate n_{cs}.

In this section, we show how the sparse recovery framework described in the previous section helps us in the codebook design process. In particular, the code design method aims to produce a matrix $\mathbf{A}$ for which we get the best possible sparse recovery performance. For that, we apply the sparse signal recovery theory which shows that the performance of sparse recovery algorithm generally depends on a factor called "matrix coherence". We establish a connection between n_{cs} and matrix coherence of the IUS problem. We propose an efficient RA code matrix design procedure by using the theory. The code matrix can also avoid the effect of CFO for code detection. It turns out that there are better codes than those suggested by the LTE standard, and these codes are particularly useful to mitigate the effects of larger carrier frequency offsets.

Table 3 Code detection probabilities for three IUS users. SNR of the users is distributed in [0, 15] dB

Algorithm	$n_{cs} = 13$	15	18	22	26	32
P_s with $r = 2.1$ km						
SMUD [25]	0.086	0.374	0.980	0.986	0.988	0.998
SRMD [23]	0.340	0.620	0.950	0.966	0.974	0.996
Handover (proposed)	0.375	0.670	0.973	0.99	0.999	0.999
P_s with $r = 3.2$ km						
SMUD [25]	0.006	0.014	0.080	0.460	0.978	0.996
SRMD [23]	0.040	0.106	0.246	0.760	0.974	0.994
Handover (proposed)	0.07	0.160	0.513	0.98	0.99	0.999

4.1 The Mutual Coherence

Being a sparse recovery problem, the IUS code detection performance can be enhanced by minimizing the mutual coherence $\mu(\mathbf{A})$ of $\mathbf{A}$ [?]

$$\mu(\mathbf{A}) = \max_{m \neq n} \frac{|[\mathbf{A}]^*_{(:,m)}[\mathbf{A}]_{(:,n)}|}{\|[\mathbf{A}]_{(:,m)}\|_2 \|[\mathbf{A}]_{(:,n)}\|_2}. \tag{26}$$

The angular distance between columns of $\mathbf{A}$ increases with decreasing $\mu(\mathbf{A})$. Therefore, for smaller $\mu(\mathbf{A})$, it is possible to identify a sparse signal efficiently by applying sparse recovery algorithms [?, ?]. In [17], it is shown how $\mu(\mathbf{A})$ influences the performances of some IUS algorithms, which were not originally proposed under the sparse recovery framework. It was found that several popular IUS methods implicitly relies on $\mu(\mathbf{A})$ to be high to yield good detection performance. In addition, we found that another related measure called block coherence has a significant impact on the estimation performance. The block coherence between two blocks $\tilde{\mathbf{E}}_m$ and $\tilde{\mathbf{E}}_\ell$ of $\mathbf{A}$, see (12), is defined as

$$\hat{\mu}(\tilde{\mathbf{E}}_m, \tilde{\mathbf{E}}_\ell) = \max_{s,n} \frac{|[\tilde{\mathbf{E}}_m]^*_{(:,s)}[\tilde{\mathbf{E}}_\ell]_{(:,n)}|}{\|[\tilde{\mathbf{E}}_m]_{(:,s)}\|_2 \|[\tilde{\mathbf{E}}_\ell]_{(:,n)}\|_2}. \tag{27}$$

Using (26) and (27), it can be shown that

$$\mu(\mathbf{A}) = \max\{\mu(\tilde{\mathbf{E}}_1), \max_{k \neq p} \hat{\mu}(\tilde{\mathbf{E}}_p, \tilde{\mathbf{E}}_k)\}. \tag{28}$$

5 Zadoff–Chu Sequence Selection

5.1 Small CFO

The following results describe the ZC code design procedure at low CFO.
  Single root case: Define

$$\mathcal{G}(n) = M^{-1}|\sin(\pi n M/N)/\{\sin(\pi n/N)\}|. \tag{29}$$

The following theory establishes the relation of n_{cs} with the mutual coherence of the code matrix $\mathbf{A}$.

Theorem 1 *If* $\tilde{\mathbf{E}}_k = \text{diag}(\tilde{Z}^u)\mathbf{U}_k\Theta\mathbf{F}_{(:,1:N_1)}$, *then* $\mu(\tilde{\mathbf{E}}_k) = \mathcal{G}(1)$; $\forall k$. *In addition, if* $k, n \in \{1, 2, \ldots, G\}$, *and*

$$MN_1/N \leq n_{cs} \leq M(N - N_1)/\{N(G - 1)\} \tag{30}$$

then $\hat{\mu}(\tilde{\mathbf{E}}_k, \tilde{\mathbf{E}}_n) \leq \mathcal{G}(1)$ for any $k \neq n$. Define

$$\zeta(n_{cs}) = \frac{n_{cs}N}{M} - (N_1 - 1). \tag{31}$$

In general, if $n_{cs} \leq \frac{M}{G}$ for some n_{cs} satisfying

$$0.25N(M-1)^{-1} \leq \zeta(n_{cs}) \leq 0.5N \tag{32}$$

then $\hat{\mu}(\tilde{\mathbf{E}}_k, \tilde{\mathbf{E}}_n) \leq \mathcal{H}(\zeta(n_{cs})/N)$, where $\mathcal{H}$ is a monotonically decreasing function over $0.25(M-1)^{-1} \leq \zeta(n_{cs})/N \leq 0.5$:

$$\mathcal{H}(v) = \begin{cases} \frac{12}{\pi(6-\pi^2v^2)(M+1)v}, & \frac{1}{4(M-1)} \leq v \leq \sqrt{2}/\pi, \\ \frac{1}{(M+1)v}, & \sqrt{2}/\pi \leq v \leq 1/2. \end{cases} \tag{33}$$

See [17] for a proof. Since $\mu(\tilde{\mathbf{E}}_k) = \mathcal{G}(1)$, $\forall k$, the value of $\mu([\tilde{\mathbf{E}}_1 \ \ldots \ \tilde{\mathbf{E}}_G]) \geq \mathcal{G}(1)$. The lower bound is possible if (i) $n_{cs} \geq MN_1/N$ (see (30)), and (ii) $G \leq n_{cs}^{-1}M(1 - N_1/N) + 1$. Thus, we can maintain $\mu([\tilde{\mathbf{E}}_1 \ \ldots \ \tilde{\mathbf{E}}_G]) = \mathcal{G}(1)$ by setting $n_{cs} = \lceil MN_1/N \rceil$, while we maximize G. Next, (32)–(33) show that we can lower the block coherence values $\hat{\mu}(\tilde{\mathbf{E}}_k, \tilde{\mathbf{E}}_n)$, $k \neq n$ by increasing n_{cs} which results in some performance improvement. However, Eq. (33) shows that the value of block coherence decreases slowly at larger n_{cs}. Therefore, a significantly larger n_{cs} is not very effective. Furthermore, as we always need to satisfy (30) for smallest $\mu(\mathbf{A})$, a larger n_{cs} enforces smaller G.

Multiple root case: Suppose in a random access environment we need total $\bar{G}$ codes. If $G < \bar{G}$, we must use $v = \lceil \bar{G}/G \rceil$ roots. Denote $\mathbf{E}_\ell^u = \text{diag}(\tilde{Z}^u)\mathbf{U}_\ell \Theta \mathbf{F}_{(:,1:N_1)}$ and $\mathbf{B}_u = [\mathbf{E}_1^u \ \ldots \ \mathbf{E}_G^u]$. We wish to design the root set $\mathcal{U} := \{u_1, u_2, \ldots u_v\}$ which makes $\mu(\mathbf{A})$ minimal, where $\mathbf{A} = [\mathbf{B}_{u_1} \ \ldots \ \mathbf{B}_{u_v}]$. Using (28), we get

$$\mu(\mathbf{A}) = \max\{\max_{v \in \mathcal{U}} \mu(\mathbf{B}_u), \max_{\substack{v_r \neq v_t; \\ v_r, v_t \in \mathcal{U}}} \hat{\mu}(\mathbf{B}_{v_r}, \mathbf{B}_{v_t})\}. \tag{34}$$

Choosing n_{cs} and G by using the procedure described in the previous section ensures $\mu(\mathbf{B}_u) = \mathcal{G}(1)$, $\forall u$. For $\hat{\mu}(\mathbf{B}_{u_r}, \mathbf{B}_{u_t})$, we use the result from [17, Lemma 1] which shows that for any value of ℓ, m, p, k if $u_r \neq u_t$, then $\hat{\mu}(\mathbf{E}_\ell^{u_r}, \mathbf{E}_m^{u_t})$ is approximately equal to $1/\sqrt{M}$. Furthermore, in a typical LTE system $1/\sqrt{M} \ll \mathcal{G}(1)$. Hence, finding $\mathcal{U}$ such that

$$\hat{\mu}(\mathbf{E}_\ell^{u_r}, \mathbf{E}_m^{u_t}) \leq \mathcal{G}(1), \ \forall r, m, \ell, t. \tag{35}$$

is not hard. Therefore, we can choose $\{u_k\}_{k=1}^v$ such that

$$\mu([\mathbf{B}_{u_1} \ \mathbf{B}_{u_2} \ \ldots \ \mathbf{B}_{u_v}]) = \mathcal{G}(1),$$

for which a sufficient condition is (35).

5.2   Mitigating the Adverse Effects of CFO

For simplicity, we assume single root cause. Let ϵ_ℓ be the CFO associated with the code $\mathbf{c}_\ell$. Write $\mathbf{W}$ in (12) as $\mathbf{W} = i\pi(1 - 1/N)\mathbf{I} + \widehat{\mathbf{W}}$, where

$$\widehat{\mathbf{W}}_{(s,n)} = \begin{cases} 0, & s = n, \\ -\dfrac{\pi\,e^{i\pi(s-n)/N}}{N\sin(\pi(s-n)/N)}, & s \neq n. \end{cases} \tag{36}$$

Denoting $\breve{\mathbf{h}}_k = \mathbf{h}_k + i\pi(1 - 1/N)\mathbf{h}_k\epsilon_k$, (12) reduces to

$$\mathbf{y} = \sum_{k=1}^{G}\{\tilde{\mathbf{E}}_k\breve{\mathbf{h}}_k + \widehat{\mathbf{W}}\tilde{\mathbf{E}}_k\mathbf{h}_k\epsilon_k\} + \mathbf{e}, \tag{37}$$

where the second-order term of ϵ is absorbed in $\mathbf{e}$.[1] For an efficient code detection, we want to select the ZC sequences such that $\mu(\mathbf{A}) = \mathcal{G}(1)$ and at the same time it can reduce the effect of the second term in (37) on user detection. This approach is useful for generic methods outside the class of sparse recovery methods. Some illustrative examples are available in [17].

The ZC code selection procedure described in the previous section ensures $\max_{\ell \neq m}\hat{\mu}(\tilde{\mathbf{E}}_m, \tilde{\mathbf{E}}_\ell)$ is the smallest. Now, we want to make

$$\widehat{\mu} = \max_{\substack{s \neq n \\ p,q}} \left| [\tilde{\mathbf{E}}_n]^*_{(:,q)}[\widehat{\mathbf{W}}\tilde{\mathbf{E}}_s]_{(:,p)} \right|$$

small, where $p, q \in \{1, 2, \ldots N_1\}$ and $s, n \in \{1, 2, \ldots G\}$. Note that $\widehat{\mu}$ is monotonic in G. Therefore, to minimize $\widehat{\mu}$, we fixed a threshold ξ first. Then, we select a G so that $\tilde{\mu} \leq \xi$, which can be done easily in a computer.

6   Simulation Results

Table 4 shows LTE simulation parameters. The signal-to-noise ratio (SNR) of each RT is independent and uniformly distributed in $[0, Q]$ dB. The SNR for ℓ-th RT is defined as SNR $= 20\log_{10}(\|\mathbf{v}_\ell\|_2/\|\mathbf{e}\|_2)$ where $\mathbf{v}_\ell$ is defined in (11). The wireless channels are modeled according to Extended Pedestrian A model (EPA) [2] if mobile speed is less than 5 m/s and Extended Vehicular A model (EVA) otherwise. The value of N_1 equal to 506 and 731 for cell radius $r = 2.1$ km and 3.2 km, respectively. $\mathcal{L}$ denotes the set of active RA code indices, while $\hat{\mathcal{L}}$ denotes the code indices detected by an algorithm. The probability that $\mathcal{L} = \hat{\mathcal{L}}$, denoted by P_s, is used as the detection performance measure. We also evaluate the false detection probability (P_f). A false

---

[1]Because the practical values of $|\epsilon_\ell| \leq 0.5$.

Table 4 Simulation parameters

Parameters	Notation	Values
Carrier frequency	f_c	2.5 GHz
Sampling frequency	f_n	30.72 MHz
Subcarrier spacing	Δf	1.25 KHz
No. of RA samples	N	24,576
Cyclic prefix (CP) samples	N_p	3168
Total PRACH subcarriers	M	839

detection occurs when the algorithm includes the index of an inactive code into $\hat{\mathcal{L}}$. RA codes are generated using two LTE standards i) unrestricted set standard and ii) restricted set standards (see [1, Sect.-5.7.2]). The algorithm developed in Sect. 3 will be called "Handover". The simulation results obtained by applying Handover on the unrestricted and restricted set will be denoted by "Handover: unres" and "Handover: res", respectively.

6.1 Performance Evaluation at Low CFO

The value of CFO of every UE is uniformly distributed in $[-0.015, 0.015]$. At first, we evaluate IUS parameter estimation performance by different algorithms for a fixed codebook where $n_{cs} = 22$. Figure 1a compares the value of P_s achieved by different algorithms as a function of K at different values of SNR. Handover performs the best. For instance, the value of P_s for Handover at $[0, 20]$ dB SNR is significantly higher than that achieved by other algorithms. Thus, the gain in detection performance offered by Handover is significant when compared to the other algorithms. Also note that Handover is robust to noise. Its performance at $[0, 20]$ dB SNR is just slightly worse than that at $[0, 15]$ dB SNR with as many as 10 users. As can be noted that the user detection performance of all algorithms degrades at high SNR variation. This is due to the fact that at high SNR variation some users have very high channel power compared to some other users. Consequently, the interference of high energy users affects significantly the low power users.

In Fig. 1c, we plot the root mean squared error (RMSE) associated with the estimate of the timing as a function of K for different values of SNR. As expected, the MSE of timing estimate increases with K and decrease in SNR, while the Handover algorithm outperforms the other algorithms by a respectable margin. Figure 1d shows the RMSE of channel power estimation. Here, SRMD and Handover perform almost similarly. SMUD performs well for $[0, 15]$ dB, but cannot perform similarly at higher SNR variation $[0, 20]$ dB.

We now investigate the effect of codebook design on IUS parameter estimation. Here, we validate the following results:

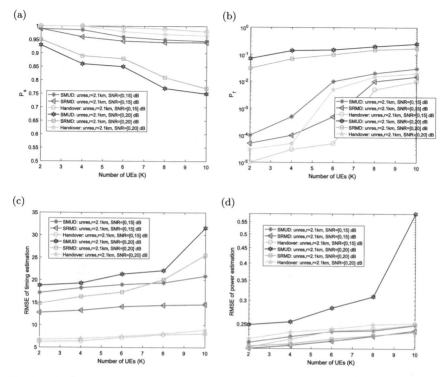

Fig. 1 IUS parameter estimation by different algorithms as a function of UEs. The mean value of CFO $\in [-0.015, 0.015]$ and wireless cell radius $r = 2.1$ km with $n_{cs} = 22$. **a** Probability of successful code detection, **b** probability of false code detection, **c** RMSE of estimated timing offset, and **d** RMSE of estimated channel power

R1

We need to choose n_{cs} that minimize $\mu(\mathbf{A})$ to enhance user detection performance. Applying (30) we see that the smallest n_{cs} from [1, Table 5.7.2-2] that minimize $\mu(\mathbf{A})$ are 18 and 26 for $N_1 = 506$ and 731, respectively.

R2

The discussion after Theorem 1 claims that we can enhance the user detection performance by increasing n_{cs}.

IUS parameters estimation performance for different values of n_{cs} is illustrated in Fig. 2. Figure 2a shows that with $r = 2.1$ km and $n_{cs} \leq 15$, the P_s value of all algorithms is very low. The value goes above 0.86 for $n_{cs} \geq 18$. After that, P_s increases slowly with increasing n_{cs}. We get a similar result for $r = 3.2$ km (Fig. 2c). The timing offset estimation results are shown in Fig. 2b. The timing estimation performance of all algorithms is improved for larger value of n_{cs}. We also observe that n_{cs} has a similar effect on both unrestricted and restricted type RA codes.

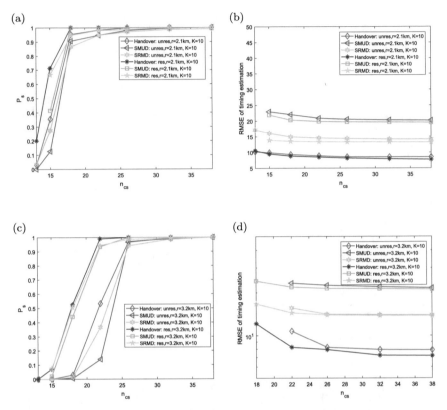

Fig. 2 IUS parameters estimation performances as a function of n_{cs}. User CFO is low with SNR = $[0, 15]$ dB. **a** Successful user detection rate for $r = 2.1$ km, **b** timing offset estimation accuracy for $r = 2.1$ km, **c** successful user detection rate for $r = 3.2$ km, and **d** timing offset estimation accuracy for $r = 3.2$ km

6.2 Performance Evaluation at Larger CFO

The value of CFO for ℓ-th active user is $\epsilon_\ell \sim \mathcal{N}(\mathfrak{s} \cdot \tau, 10^{-4})$, where $\mathfrak{s}$ is the sign bit takes values in $\{+1, -1\}$ with equal probability. Code detection performance of different algorithms in varying CFO environment is tabulated in Table 5. By "SMUD:proposed", we denote the code detection performance of SMUD when the RA codes are designed by using our proposed method (see Sect. 5.2). We set $n_{cs} = 32$ for generating the codes. As can be seen, the code detection performance of all algorithms are significantly better when we design codes by the proposed method, specially at higher CFO. Algorithms with unrestricted LTE codes cannot performs equally when $\tau \geq 0.15$. Restricted LTE codes exhibits some improved performance until $\tau \leq 0.3$. In contrast, P_s value of "SMUD:proposed" remains above 0.85 for $n_{cs} \leq 0.5$.

Table 5 Success rate of user detection for different values of CFO mean (τ). SNR of the users is in [0, 15] **dB** and $K = 10$

Algorithm	P_s with $r = 2.1$ km								
	$\tau = 0.01$	0.05	0.1	0.15	0.2	0.25	0.3	0.4	0.5
SMUD:unres	0.9960	0.8740	0.5960	0.4800	0.4000	0.3400	0.2700	0.1180	0.0060
SMUD:unres	0.9860	0.9200	0.7000	0.5300	0.4400	0.3800	0.3000	0.1600	0.0140
SMUD:res	0.999	0.9980	0.9980	0.9940	0.9860	0.9740	0.7860	0.3260	0.0980
SMUD:res	0.998	0.9940	0.9900	0.9860	0.9820	0.9800	0.8660	0.3660	0.1260
SMUD:proposed	0.9999	0.9980	0.9980	0.9960	0.9940	0.9940	0.9940	0.9540	0.8600
SMUD:proposed	0.9999	0.9980	0.9975	0.9960	0.9945	0.9860	0.9780	0.8740	0.6860

7 Need of Advancement of IUS Algorithms for Future Wireless Communication

In this work, IUS problem is solved by using a sparse signal recovery framework. Comparing to the state-of-the-art works, the proposed algorithm shows a clear improvement of IUS parameters estimation performance. Nevertheless, there are still some areas which need to improve for future system:

- Current IUS algorithms can efficiently detect up to 10 users. In near future, a huge number of devices are expected to be conceded to the same network [32]. Consequently, the number of active IUS users will increase further. More efficient IUS algorithms will be necessary for a large number of IUS users.
- Figure 2 shows that performance of all algorithms degrades at larger SNR variation. Because at large SNR variation the received signal energy at the eNodeB from some users are very high compared to other low SNR users. As a result, high SNR users act as interference source to low SNR users making it difficult to detect low SNR users. This problem will increase with increasing the number of active users.
- Section 6.2 shows that the state-of-the-art algorithms can detect users efficiently at high CFO environment by using our designed RA code matrix. However, the algorithms can not estimate the CFO associated with each user. If CFO of a user is high, then it can degrade the data transmission rate. Therefore, a suitable algorithm is necessary with can detect user and estimate CFO simultaneously.

8 Conclusion

This chapter describes a new direction for solving the initial uplink synchronization problem in LTE systems. Unlike conventional methods, we show IUS as a sparse signal recovery problem. Section 2 develops a new data model which matched perfectly with the standard sparse recovery problem. This representation allows us to develop a time efficient algorithm by combining $\ell_1 - \ell_0$ norm minimization approach. The sparse signal recovery theory also gives a direction of designing RA code matrix that can enhance user detection performance both in low and high CFO environments.

References

1. 3rd Generation Partnership Project; technical specification group radio access network; evolved universal terrestrial radio access (E-UTRA); physical channels and modulation (release 10) (2011)
2. 3GPP-TS-36.101: User equipment (UE) radio transmission and reception. 3rd Generation Partnership Project; Technical Specification Group Radio Access Network; Evolved Universal Terrestrial Radio Access (E-UTRA) (2012)
3. Bao, P., Guan, Q., Guan, M.: A multiuser detection algorithm in the uplink SC-FDMA system for green communication network. IEEE Access **4**, 5982–5989 (2016). https://doi.org/10.1109/ACCESS.2016.2556279
4. Barbarossa, S., Pompili, M., Giannakis, G.: Channel-independent synchronization of orthogonal frequency division multiple access systems. Selected Areas in Communications, IEEE Journal on **20**(2), 474–486 (2002)
5. Beyme, S., Leung, C.: Efficient computation of DFT of Zadoff-Chu sequences. Electronics Letters **45**(9), 461–463 (2009). https://doi.org/10.1049/el.2009.3330
6. Blake, A., Zisserman, A.: Visual Reconstruction. MIT Press, Cambridge, MA (1987)
7. Boyd, S., Vandenberghe, L.: Convex Optimization. Cambridge Univ. Press, Cambridge, U.K. (2004)
8. Candés, E.J., Romberg, J., Tao, T.: Robust uncertainty principles: exact signal reconstruction from highly incomplete frequency information. IEEE Transactions on Information Theory **52**, 489–509 (2006)
9. Candés, E.J., Wakin, M.B.: An introduction to compressive sampling. IEEE Signal Processing Magazine **25**, 21–30 (2008)
10. Chen, S.S., Donoho, D.L., Saunders, M.A.: Atomic decomposition by basis pursuit. SIAM Journal on Scientific Computing **20**, 33–61 (1999)
11. Chu, D.: Polyphase codes with good periodic correlation properties (corresp.). Information Theory, IEEE Transactions on **18**(4), 531–532 (1972). https://doi.org/10.1109/TIT.1972.1054840
12. Coleman, T., Li, Y.: An interior trust region approach for nonlinear minimization subject to bounds. SIAM Journal on Optimization **6**(2), 418–445 (1996)
13. Frank, R., Zadoff, S.: Phase shift pulse codes with good periodic correlation properties (corresp.). Information Theory, IRE Transactions on **8**(6), 381–382 (1962). https://doi.org/10.1109/TIT.1962.1057786
14. Fu, X., Li, Y., Minn, H.: A new ranging method for OFDMA systems. Wireless Communications, IEEE Transactions on **6**(2), 659–669 (2007)
15. Gorodnitsky, I., Rao, B.: Sparse signal reconstruction from limited data using FOCUSS: a reweighted minimum norm algorithm. IEEE Transactions on Signal Processing **45**(3), 600–616 (1997)
16. Hyder, M., Mahata, K.: Direction-of-arrival estimation using a mixed $\ell_{2,0}$ norm approximation. IEEE Transactions on Signal Processing **58**(9), 4646–4655 (2010)
17. Hyder, M., Mahata, K.: Zadoff-Chu sequence design for random access initial uplink synchronization in LTE-like systems. IEEE Transactions on Wireless Communications **16**(1), 503–511 (2017). https://doi.org/10.1109/TWC.2016.2625319
18. Hyder, M.M., Mahata, K.: An improved smoothed ℓ^0 approximation algorithm for sparse representation. IEEE transactions on Signal Processing **58**(4), 2194–2205 (2010)

19. Hyder, M.M., Mahata, K.: A sparse recovery method for initial uplink synchronization in OFDMA systems. IEEE Transactions on Communications **64**(1), 377–386 (2016). https://doi.org/10.1109/TCOMM.2015.2497232
20. IEEE standard for local and metropolitan area networks part 16: Air interface for broadband wireless access systems. IEEE Std 802.16-2009 (Revision of IEEE Std 802.16-2004) pp. 1–2080 (2009)
21. Lee, C.C., Krinock, J., Singh, M., Paff, M.: Comments on OFDMA ranging scheme described in IEEE 802.16ab-01/01r1. IEEE 802.16abc-01/24 (2001)
22. Lee, D.H.: OFDMA uplink ranging for IEEE 802.16e using modified generalized chirp-like polyphase sequences. In: Internet, 2005. The First IEEE and IFIP International Conference in Central Asia on (2005)
23. Lin, C.L., Su, S.L.: A robust ranging detection with MAI cancellation for OFDMA systems. In: Advanced Communication Technology (ICACT), 2011 13th International Conference on, pp. 937–941 (2011)
24. Popovic, B.: Efficient DFT of Zadoff-Chu sequences. Electronics Letters **46**(7), 502–503 (2010). https://doi.org/10.1049/el.2010.3510
25. Ruan, M., Reed, M., Shi, Z.: Successive multiuser detection and interference cancelation for contention based OFDMA ranging channel. Wireless Communications, IEEE Transactions on **9**(2), 481–487 (2010)
26. Sanguinetti, L., Morelli, M.: An initial ranging scheme for the IEEE 802.16 OFDMA uplink. Wireless Communications, IEEE Transactions on **11**(9), 3204–3215 (2012)
27. Sanguinetti, L., Morelli, M., Marchetti, L.: A random access algorithm for LTE systems. Transactions on Emerging Telecommunications Technologies **24**(1), 49–58 (2013). https://doi.org/10.1002/ett.2575
28. Wainwright, M.: Sharp thresholds for high-dimensional and noisy sparsity recovery using ℓ_1-constrained quadratic programming (LASSO). Information Theory, IEEE Transactions on **55**(5), 2183–2202 (2009)
29. Wang, Q., Ren, G.: Iterative maximum likelihood detection for initial ranging process in 802.16 OFDMA systems. Wireless Communications, IEEE Transactions on **14**(5), 2778–2787 (2015)
30. Wang, Q., Ren, G., Wu, J.: A multiuser detection algorithm for random access procedure with the presence of carrier frequency offsets in LTE systems. IEEE Transactions on Communications **63**(9), 3299–3312 (2015)
31. Wang, Q., Ren, G., Wu, J.: A centralized preamble detection-based random access scheme for LTE CoMP transmission. IEEE Transactions on Vehicular Technology **65**(7), 5200–5211 (2016)
32. White Paper, E.: More than 50 billion connected devices. Ericsson Tech Report **284 23-3149 Uen** (2011)
33. Zhou, Y., Zhang, Z., Zhou, X.: OFDMA initial ranging for IEEE 802.16e based on time-domain and frequency-domain approaches. In: Communication Technology, 2006. ICCT '06. International Conference on, pp. 1–5 (2006)

Part III
Wireless Technology and Communications—Advancement & Future Scope

Toward a "Green" Generation
of Wireless Communications Systems

Fernando Gregorio and Juan Cousseau

Abstract Green mobile networks are essential to enable sustainable growth of future communications systems. Communication transceivers with reduced-power consumption and high-spectral efficiency will enable suitable connectivity while maintaining user access costs at accessible levels. The term "green" in our case refers to devices of low implementation cost but that is efficient from two points of view: energy consumption and use of radio spectrum. Based on the premise of spectral efficiency, this chapter describes spectrum access techniques, in particular full-duplex and massive communication techniques. It is worth to mention that these techniques are presented in a realistic scenario considering RF front-end imperfections always present in low-cost units. Furthermore, a performance study of these systems is also presented. On the other hand, in order to improve energy capabilities, the more challenging context of massive MIMO systems is studied.

1 Introduction

The demand of high data rates, motivated by the new generation of wireless systems that requires huge data transfers while keeping the transceiver cost and power consumption at reasonable levels, is a challenging issue [1].

As illustrated in Fig. 1, a wireless transceiver is generically composed of two fundamental blocks: (a) the digital section and, (b) the RF analog block. The receiver front-end amplifies the incoming signal and downconvert it to baseband. The baseband signal is digitized and processed by the digital block in order to recover the transmitted symbols. On the other hand, the baseband section of the transmitter generates modulated signals. These signals are upconverted to RF, amplified and transmitted. The RF front-end consists of several components such as: oscillators, mixer for down/up conversion operation, low-noise amplifier, filters, and power amplifiers. These elements need to be carefully chosen to ensure a proper system operation.

F. Gregorio (✉) · J. Cousseau
IIIE-CONICET, Dpto, Ing. Electrica y Computadoras, Universidad Nacional del Sur, Bahía Blanca, Argentina
e-mail: fernando.gregorio@uns.edu.ar

© Springer Nature Singapore Pte Ltd. 2018
K. V. Arya et al. (eds.), *Emerging Wireless Communication and Network Technologies*,
https://doi.org/10.1007/978-981-13-0396-8_12

Fig. 1 OFDM transceiver
model

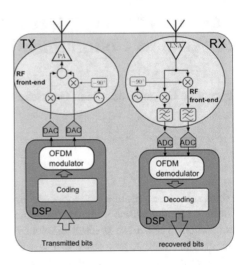

The number of components employed in the RF front-end is small if compared
to the components required to implement the digital part. However, especially for
the case of base stations and mobile devices [2], the RF components are usually
responsible for a significant part of the cost, performance, and power consumption
of the complete radio system. Moreover, the analog-to-digital (ADC) and digital-
to-analog (DAC) converters working as interface between the analog and digital
"worlds" and vice versa are also critical components when the overall system power
consumption and implementation cost need to be optimized [3].

The modulation scheme, the operation bandwidth, and the power budget of the
transceiver are significant parameters to be considered for the election of the ADC.
The increasing use of multicarrier modulation techniques, as orthogonal frequency
division multiplexing (OFDM), imposes severe requirements over the quality of the
RF front-end. OFDM has been chosen by the majority of wireless communication
standards due to its high-spectral efficiency and robustness against multipath channel
[4]. However, its performance is jeopardized by the distortions generated at the analog
front-end. Particularly, the high peak-to-average-power ratio (PAPR) associated to
the OFDM signals is an important obstacle when power efficiency is a requirement.
The performance of the whole communication system is affected by its RF front-end
and needs to be carefully designed to avoid performance degradation and waste of
significant resources [5].

This chapter is focused on the challenging issue of energy efficiency in novel
communication systems based on OFDM modulation. The chapter begins with an
overview of RF impairment models in an OFDM context. Afterwards, full-duplex
techniques are addressed. Finally, implementation issues of massive MIMO systems
are introduced in the last section.

2 RF Front-End Imperfection Models

In the following, we present an overview of the more common RF impairments that affect the performance of wireless communication systems.

From the transmitter side, power amplifier (PA) distortion, phase noise, and mixer imperfections are the more significant contributors to degrade the system performance [6]. PA is also one the most power demanding devices. At least for medium or high power transmitters, PA governs the consumption of the whole system. Considering the receiver front-end, local oscillator, downconverter, and ADC are the critical components to be optimized in order to increase the receiver sensitivity and minimize its power consumption.

When the energy efficiency is the main priority and the challenge topic, the energy dissipated in baseband processing and analog components need to be carefully analyzed. It is assumed, generically, that the analog block dominates the power consumption of the system. However, in low-power applications, the digital block starts to dominate the consumption of the transceiver [7]. Typical scenarios in the upcoming 5G are the ultra-dense small-cell networks, where the distance between base station and users is small, requiring low-power transmission. In this context, the processing power dominates over the analog power. The processing energy cannot be ignored in short-range link systems [8].

2.1 Power Amplifier Nonlinear Distortion

The power amplifier (PA) is a key component in present and future broadband communication systems. In order to maximize its power efficiency, the PA is operated close to the saturation point where a nonlinear behavior is actually present, generating nonlinear distortion (NLD). The nonlinear response is further emphasized when it is driven by high PAPR OFDM signals. Nonlinear behavior creates in-band and out-of-band distortions that degrade the system performance and also create unwanted emission on neighborhood bands.

If the PA has a flat frequency response characteristic over its entire working frequency range, it can be modeled as a memoryless system fully characterized by their AM/AM (amplitude to amplitude) and AM/PM (amplitude to phase) conversions [9], which depend only on the current input signal magnitude [10].

In broadband applications, PA frequency response is no longer flat over the frequency operation region. Several models can be employed to characterize the behavior of broadband power amplifiers [11, 12].

Based on the Bussgang's theorem [13], the output of a nonlinear memoryless PA driven by a Gaussian distributed signal $x(n)$ can be represented by $x_{pa}(n) = k_{pa}x(n) + d(n)$, where the distortion term $d(n)$ is uncorrelated with $x(n)$ and k_{pa} is the scaling factor.

2.2   Phase Noise

The noise that originates from oscillator nonidealities appears in the baseband signal as additional phase and amplitude modulation. Assuming that there is no deterministic offset in the system, the phase noise can be modeled as a zero-mean random process, characterized by its power spectral density (PSD). The PSD of ideal LO is represented by an impulse placed at the carrier frequency. The PSD of practical oscillators is represented by side bands.

The LO signal quality is specified by the *integrated single-side band phase noise* (IPN), given by $IPN = \int_0^\infty S(f)df$ [dBc].

The contribution of the phase noise can be viewed as an additional multiplicative effect of the channel, like fast and slow fading. In case of OFDM systems, the multiplication effect results in intercarrier interference (ICI) [14]. Considering an OFDM transmitter, the signal at subcarrier k can be expressed as

$$X_{up}(k) = \Phi_0 X(k) + \sum_{l=0, l \neq k}^{N-1} \Phi(k-l) X(k-l)$$

where $\Phi(k)$ is the frequency-domain representation of the phase noise process, given by $\Phi(k) = \frac{1}{N} \sum_{m=0}^{N-1} j\phi(n)t e^{-j\frac{2\pi}{N}km}$

Two effects can be observed, the common phase error (CPE), $\Phi_0 X(k)$, and the intercarrier interference (ICI), $\sum_{l=0, l \neq k}^{N-1} \Phi(k-l) X(k-l)$. The phase noise induced ICI power depends on the bandwidth of the phase noise process and the subcarrier spacing $B_S = 1/NTs$. The ICI power can be expressed as

$$\sigma_{ici}^2(k) = \sigma_{ici}^2 \approx \int_{-\infty}^{+\infty} \left(1 - sinc^2\left(\frac{f}{B_s}\right)\right) S(f)df$$

where $S(f)$ is the PSD of the local oscillator.

2.3   Analog-to-Digital Converter Quantization Noise

An ADC with high resolution reduces the quantization noise increasing the achievable data rate. On the other hand, the power dissipated at the ADC is increased reducing the power efficiency of the receiver. There is a trade-off between the allowed quantization distortion and the consumed power. Moreover, the use of large bandwidth signal requires high sampling rate that inexorably increases the ADC consumption. From the perspective of the future communication networks, the situation is even worse due to the expected use of large bandwidth, high constellation size, and energy constraints.

Several parameters can be taken into account to evaluate the performance of an ADC:

Quantization noise: Following the Bussgang theorem, the output of an ADC can be expressed by $x_{adc} = Q(x) = \alpha x + n_q$, where $Q(x)$ denotes the quantization function, α is the scaling factor, and n_q is the quantization noise that

$$E[n_q] = (1 - \alpha)E[x]$$

$$\sigma_{n_q}^2 = (1 - \alpha)\alpha\sigma_x^2$$

$$\alpha = \left(1 - \frac{\sigma_x^2}{\sigma_{eq}^2}\right) = \left(1 - \frac{\pi\sqrt[2]{3}}{2b^2}\right)$$

where b is the number of bits. The scaling factor is calculated considering a Gaussian input signal with a nonuniform quantizer.

Power consumption: The power dissipation of ADC shows a dependence with the sampling rate W and the number of quantization bits b and can be modeled as $P_{ADC} = cW2^b$, where c is the energy consumption per conversion step [15, 16]. Magnitude of c is a function of the implementation technology, the kind of converter, and the resolution. Energy consumption per conversion step of 1 pJ/step is reported in [15]. The state of the art is approaching 0.1 pJ/step [17]. Nowadays, due to the large signal bandwidth and high constellation size (which demand ADC with high resolution), ADC becomes a power hungry device that governs the power consumption of the complete receiver.

2.4 IQ Imbalances

To reduce the implementation cost and size, the chip area and the number of components need to be minimized; toward that objective a direct conversion (DC) transceiver is a good choice. Unfortunately, the DC implementation is more sensitive to amplitude and phase mismatch between I and Q branches and carrier frequency offset (CFO) [18]. The performance degradation observed due to these imperfections motivates the implementation of digital domain compensation techniques in a cost-effective manner. IQ imbalance appears in both, upconverters (TX) and down-converters (RX). The distortion introduced by the transmitter can be described by

$$X_{up(k)} = \mu_{tx}X(k) + \upsilon_{tx}X^\#(k)$$

where $X^\#(k) = X^*(N - k)$ represents the mirrored OFDM symbol. μ_{tx} and υ_{tx} are given by

$$\mu_{tx} = \cos(\Delta_\theta) + j\varrho\,\sin(\Delta_\theta)$$

Table 1 Simulation parameters for RF impairments evaluation

Parameter	Value
No subcarriers	512
Bandwidth	5, 10, 20 MHz
Upconverter/downconverter	$\mu = 5\%$ $\varrho = 2°$ IPN $= -20$ dBc
Power amplifier	SSPA model, IBO $= 0$ to -6 dB
ADC resolution	8–12 bits

$$v_{tx} = \varrho \cos(\Delta_\theta) - j \sin(\Delta_\theta)$$

with Δ_θ and ϱ being the phase and amplitude imbalances between the I and Q branches. Identical effects appear at the receiver side due to imbalances in the downconverter. The mirror signal attenuation can be quantified by the image rejection ratio defined by $IRR = 10 \log_{10}\left(\frac{\mu}{v}\right)$.

2.5 Quantification of the RF Front-End Quality

The contribution of the different RF impairments to the system degradation can be quantified by using the error vector magnitude (EVM), as is widely adopted in communication standards. EVM provides adequate information regarding distortions generated by each RF component. On the other hand, the SINR is employed to define the specifications at the receiver side. SINR evaluation includes the effects of the receiver RF impairments.

For illustration purposes, the degradation of the different RF impairments is evaluated in a typical OFDM transceiver (with specifications in Table 1), as described in Fig. 2.

The constellations obtained in different parts of the transceiver are illustrated in Fig. 3. Phase rotation and ICI plus mirror interference are observed in TP1. In TP2,

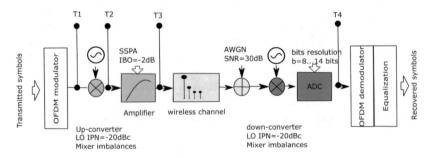

Fig. 2 OFDM transceiver simulation model

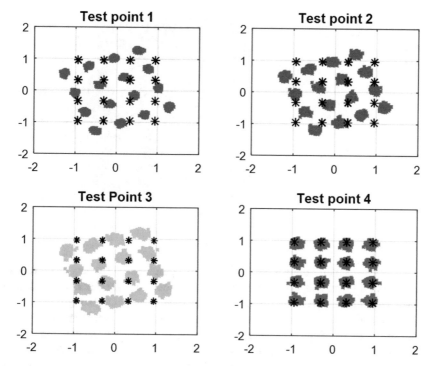

Fig. 3 Constellation at several test points

is depicted the NLD added by the PA. Downconverter distortion and quantization noise are plotted in TP3. At the final point of the chain, TP4, the recovered symbols after channel equalization are shown.

CPE effects can be easily removed by equalization. On the other hand, ICI terms due to phase noise, quantization noise, and NLD, is modeled as an additive Gaussian noise and its removal requires sophisticated algorithms.

Figure 4 shows EVM curves as a function of the PA operation point (back-off) and the ADC resolution, with $IRR = 20$ dB and IPN $= -20$ dBc. Ideal channel model is considered in this evaluation. These results indicate that in order to operate with large constellation size, a back-off large than 5 dB and ADC resolution larger than 9 bits are required.

3 Spectral Efficiency Using Full-Duplex

The full-duplex technology has emerged as an interesting option to reach the required link capacity increment by allowing transmitting and receiving simultaneously in the same frequency band. Full-duplex transceivers enjoy doubled transmission rates in

Fig. 4 EVM in function of
PA input back-off and ADC
resolution

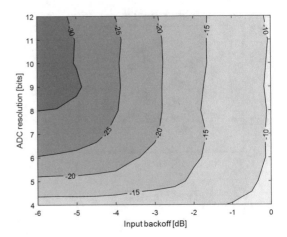

comparison with their conventional half-duplex counterparts. However, in order to
obtain that improvement, one needs to mitigate strong self-interference (SI) due to
its own transmitted signal that is coupled to the receive antenna [19].

Passive and active antenna cancelation techniques can be employed to mitigate
the SI at the input of the receiver chain. After that, analog RF cancelation is also
included to suppress the SI signal in order to keep the input signal at an adequate
level in the downconversion stage and avoid operating the mixer in the nonlinear
region of its transfer curve. This also reduces the saturation of the ADC and allows
to operate it with an adequate dynamic range enabling the implementation of DSP
for SI cancelation [20, 21].

There are two factors that determine the required amount of SI cancelation: (a)
the dynamic range of the ADC which defines the required level of cancelation at the
analog domain to avoid large quantization noise [22], and (b) the level of the desired
signal at the receiver. Ideally, the SI should be mitigated so that its power goes down
to the level of the receiver noise floor.

As discussed in the previous section, in the case of full-duplex systems, hardware
imperfections significantly limit the self-interference suppression capability.

3.1 Self-interference Reduction Techniques: Antenna, RF, and Digital Cancelation with "Real-Life" Components

3.1.1 Antenna Cancelation

Due to the short distance between TX and RX antennas, the SI is typically stronger
than the signal of interest and its cancelation/suppression is mandatory. From
the antenna point of view, increasing the TX-RX antenna separation increases
the attenuation and reduces the self-interference level. However, due to practical

reasons (cost and physical dimensions), the amount of attenuation is restricted to 20–30 dBs [23].

To achieve sufficient cancelation, it is necessary to increase SI attenuation by active antenna cancelation techniques [24]. Several options can be considered to improve the antenna isolation: (a) directional antennas, where the TX antenna gain is minimized in direction of the receive antenna and vice versa, and (b) cross-polarization techniques that employ transmit and receive antennas in orthogonal polarization states.

A simple design and implementation of a patch antenna array for full-duplex (FD) applications was presented in [25]. The antenna array operates in the 2.4–2.6 GHz band and is designed to provide large isolation between its TX and RX ports. The feeding circuit for the TX antennas is a *rat race* coupler used to create 180° phase change between the outputs. The antenna prototype and rate race coupler are illustrated in Fig. 5.

Measurements of the isolation between TX-RX antennas are plotted in Fig. 6 where an isolation between 52 and 57 dB is verified in the 2.5–2.55 GHz band.

The far-field pattern is shown in Fig. 7 (azimuth and elevation). The obtained patterns are very well aligned with those obtained in the simulation. In the radiation pattern, we can identify the nulls regions where the RX antennas should be placed to get maximum isolation.

3.1.2   RF Cancelation Techniques

The RF canceler operates in the analog domain by adding a canceling signal to the received signal at the victim antenna [26]. The canceling signal is generated by phase and amplitude alignment of a reference interference signal to match the interference

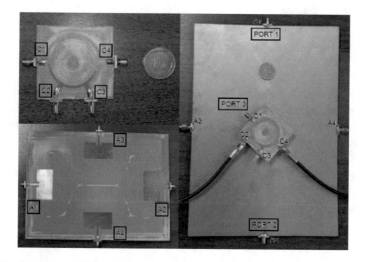

Fig. 5 Patch antenna design with rate race coupler

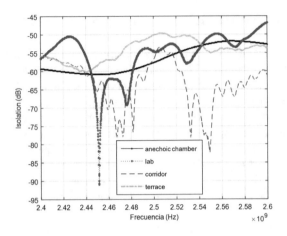

Fig. 6 TX-RX antennas isolation

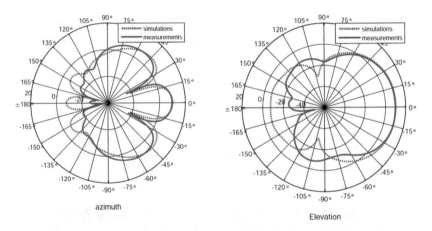

Fig. 7 Radiation pattern azimuth and elevation

at the RX antenna. Generically, the RF cancelers are designed assuming a single-tap SI channel, which introduces attenuation and phase delay as main parameters. If it were the case of a multipath SI channel, the level of residual SI would be increased deteriorating the performance. However, typical measurements show that the delay spread of the SI channel is small compared with the OFDM sub-symbol duration in the digital baseband domain and the SI channel typically manifests itself as a flat channel for baseband processing. Despite that, the reflected paths should be considered in the implementation of RF cancelers [27].

3.1.3 Baseband Cancelation

After the implementation of antenna and RF cancelation techniques, the residual SI is removed at baseband using digital domain techniques by subtracting the transmitted baseband waveform from the received signal. The SI signal samples are generated by the linear convolution between the transmitted symbols and the residual SI channel estimate. The residual channel is composed by the channel coupling and RX-TX front-ends.

3.2 RF Imperfections on Full-Duplex Relays

The link model comprises a source node, a full-duplex relay, and a destination node, as is illustrated in Fig. 8. First, the received signal at the RN is given by

$$y_{R(n)} = h_{SR(n)} * x(n) + h_{SI}(n) * \beta y_R(n - D) + w_R(n)$$

where $x(n)$ is the transmitted time-domain signal generated by an OFDM modulator for N subcarriers, $h_{SR(n)}$ is the source-relay channel, $h_{SI}(n)$ models the loopback coupling channel, and β is a scalar (amplify-and-forward relay). Parameter D is the processing delay of the repeater and $w_R(n)$ denotes the channel noise.

The forwarded signal that is received at destination can be expressed as

$$y_D(n) = h_{RD}(n) * \beta y_R(n - D) + h_{SD}(n) * x(n) + w_D(n)$$

where $h_{RD}(n)$ and $h_{SD}(n)$ represent the relay-destination and source–destination channels, respectively.

Front-end imperfections limit the performance of full-duplex relays. In the following, the effects of typical RF impairments are addressed [28, 29]. A full-duplex relay diagram is illustrated in Fig. 9.

Fig. 8 Relay link model

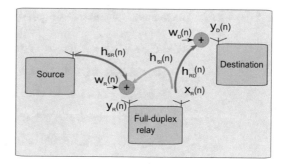

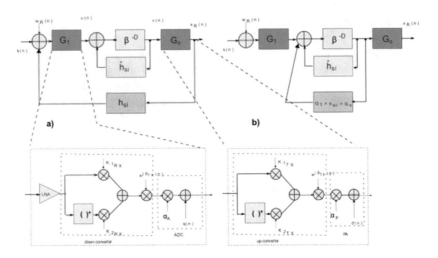

Fig. 9 **a** Full-duplex relay model, **b** equivalent model

To simplify the analysis, we define an equivalent SI channel, $h_{eq}(n) = G_o(n) * h_{si}(n) * G_1(n)$, where $G_o(n)$ denotes the TX front-end, and $G_1(n)$ models the RX front-end that involves the LNA, mixer, and ADC.

The output of up/downconverter mixers including phase noise are given by

$$x_{rf}(n) = (K1_x x(n) + K2_x x(n))e^{-j\theta x(n)} = x_{iq}(n)e^{-j\theta r(n)}$$

where $K1_x$ and $K2_x$ denote the imbalance coefficients associated to IQ mismatches in the transmitter $(x = t)$ or receiver $(x = r)$ assuming that upconverter and downconverter share the LO (single local oscillator, SLO), i.e., $e^{j\theta t(n)} = e^{-j\theta r(n)}$. The baseband representation of the frequency-domain signal at k-th subcarrier (dropping the block index n) is given by [6] $X_{rf}(k) = \Psi(0)X_{iq}(k) + \gamma(k)$, where $\Psi(0)$ is the CPE term and $\gamma(k)$ is ICI induced by the phase noise.

On the other hand, the ADC output is modeled by $x_Q(n) = \alpha_A x(n) + Q(n)$,

A linearized model is also employed to characterize the PA response, i.e., $x_{pa}(n) = \alpha_p x(n) + d(n)$. The feedback signal by using the equivalent relay model is given by

$$
\begin{aligned}
z(n) = {} & G_o(n)r(n) * \hat{h}_{si}(n) * G_1(n) \\
= {} & \alpha_a \alpha_p G_L k1_r k1_t e^{-j\theta r(n)} * \hat{h}_{si}(n)e^{j\theta t(n)}r(n) \\
& + \alpha_a \alpha_p G_L k2_t k1_r e^{-j\theta r(n)} * \hat{h}_{si}(n)e^{j\theta t(n)}r^*(n) \\
& + \alpha_a \alpha_p G_L k1_t k2_r e^{-j\theta r(n)} * \widehat{h^*}_{si}(n)e^{j\theta t(n)}r^*(n) \\
& + \alpha_a \alpha_p G_L k2_t^* k2_r^* e^{j\theta r(n)} * \widehat{h^*}_{si}(n)e^{j\theta t(n)}r(n) \\
& + \alpha_a \alpha_p G_L (k1_r d(n) + k2_r d^*(n))e^{-j\theta r(n)} + Q(n)
\end{aligned}
$$

where $r(n)$ and $z(n)$ denote the input and output signal of the equivalent feedback channel, respectively. From this equation, several terms can be identified:

(a) **Dominant term**: $\alpha_a \alpha_p G_L k1_r k1_t e^{-j\theta r(n)} * \hat{h}_{si}(n) e^{j\theta t(n)} r(n)$.
(b) **Mirror terms**: Due to SLO, the second term can be written as $\alpha_a \alpha_p G_L \ k2_t k1_r \widehat{H}_{si}(k) r(N-k)$. In the case of the third term, the expression at subcarrier k is given by $\alpha_a \alpha_p G_L k1_t k2_r \widehat{H}_{si}(k) r(N-k) \Psi_{tr}(0) + \gamma_{tr}(k)$. For practical mixers, the fourth-term can be neglected.
(c) **Additive term**: this term comes from the distortion generated by the PA and the quantization noise from the ADC.

- **Self-interference channel estimate**: The estimation of the SI channel is required for the digital cancelation technique. If we employ a set of pilots symbols defined to eliminate the effect of mirror subcarriers [30], a noisy version of the dominant term is obtained $\widehat{H}_{si_{eq}}(k) = \alpha_a \alpha_p G_L k1_r k1_t \widetilde{H}_{si}(k)$.

  The channel estimation error is given by $\epsilon_H = H_{si\_eq}(k) - \widehat{H}_{si_{eq}}(k)$, where $H_{si\_eq}(k) = G_o(n) \widetilde{H}_{si}(k) G_1(k)$, and composed by the dominant term, the mirror components, and the additive terms previously described.
- **Effective signal-to-interference-plus-noise ratio**
  The SINR at the output of the relay at any subcarrier can be expressed as

$$SINR_o = \frac{P_{soi}}{P_{ici} + P_{iq} + P_Q + P_{rsi} + P_{pa} + P_{feed} + P_n}$$

where

- P_{ici} is the ICI: $P_{ici} = |\alpha_a|^2 |\alpha_p|^2 |\beta|^2 |G_L|^2 |k1_r|^2 |k1_t|^2 \sigma_\gamma^2(k)$
- P_{iq} is the interference from mirror subcarriers:

$$P_{iq} = |\alpha_a|^2 |\alpha_p|^2 |\beta|^2 |G_L|^2 \left(|k1_t|^2 |k2_r|^2 + |k2_t|^2 |k1_r|^2\right) E_s$$

- P_Q is associated to the ADC quantization noise:

$$P_Q = |\beta|^2 \left(|k1_t|^2 + |k2_t|^2\right) \sigma_Q^2(k)$$

- P_{rsi} is the residual self-interference due to imperfect channel estimation: $P_{rsi} = |\beta|^2 \left(|k1_t|^2 + |k2_t|^2\right) \sigma_\epsilon^2(k) P_{YR}$, where P_{YR} is the power transmitted by the relay.
- $P_{pa} = \sigma_d^2(k)$ is the power of PA nonlinear distortion.
- P_{feed} is the distortion due to the residual self-interference channel:

$$\begin{aligned}
P_{feed} = |\alpha_a|^2 |\beta|^2 |G_L|^2 [&\left(|k1_r|^2 + |k2_r|^2\right)\left(|k1_t|^2 + |k2_t|^2\right)\sigma_d^2 \\
&+ |\alpha_p|^2 (|k1_t|^2 |k2_t|^2 |k1_r|^2 + |k1_t|^2 |k2_t|^2 |k2_r|^2 \\
&+ |k2_t|^2 |k2_r|^2 |k1_r|^2 + |k2_t|^2 |k1_t|^2 |k2_r|^2) P_{YR}] \sigma_{\widetilde{H}}^2
\end{aligned}$$

Table 2 Simulation parameters (DRR denotes directed to reflected paths ratio)

Parameter	Value
OFDM	N = 512, B = 5 MHz
Relay Tx power	$P_{y_R} = 0$–20 dBm
Antenna + RF cancelation	$A_c = 30$ dB $A_{rf} = 30$ dB
Mixer imbalances and phase noise	$IRR = -35$ dB, $f_{3dB} = 100$
ADC resolution	8–12 bits
Power amplifier	Third-order polynomial model $c_1 = 1.1136 + 0.184j, c_3 = -0.3807 - 0.0705j$
SI channel	Ricean channel DRR = 30 dB

where $\sigma_{\tilde{H}}^2 = \sigma_{\tilde{H}}^2 A_c^2 A_{rf}^2$
- P_n is the S-R channel noise:

$$P_n = |\alpha_a|^2 |\alpha_p|^2 |\beta|^2 |G_L|^2 \left(|k1_t|^2 + |k2_r|^2\right)\left(|k2_t|^2 + |k1_r|^2\right)\sigma_w^2$$

3.2.1 Numerical Evaluations of FD Relay Performance

Link capacity and SINR at the relay output are evaluated to quantify the performance of FD relay affected by RF impairments. Simulation parameters are summarized in Table 2.

System capacity is illustrated in Figs. 10 and 11. The capacity for a HD relay is also included for comparison purpose. From Fig. 10, it can be observed that FD link outperforms HD for DSP cancelation levels larger than 30 dB even using a low-resolution ADC. In this figure, a noise floor due to quantization noise appears. Noise floor can be also appreciated in Fig. 11 due to nonlinear distortion and mixer imbalances.

Figure 12 illustrates the dependence of system capacity with the relay transmitted power (the SNR at destination is normalized in 30 dB). The effects of SI increment (due to large transmitted power) over the overall system capacity are plotted in Fig. 13. In this case, the dynamic range of the ADC input signal is large, and the ADC resolution needs to be increased.

The effective SINR at the output of the relay node is shown in Fig. 14. Considering that the SNR at the relay input is 30 dB, we can observe that this value is reached only for large levels of DSP cancelation with ideal front-end. From practical implementations, a SINR degradation larger than 10 dB is obtained. To alleviate this problem, nonlinear and widely linear DSP cancelers, predistorters, and ADC compensation are the options to be explored.

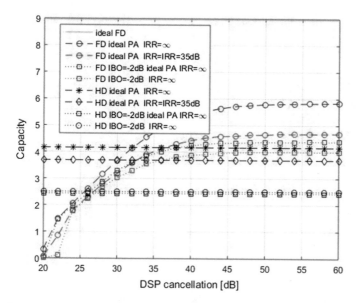

Fig. 10 Capacity versus DSP cancelation. FD and HD links are evaluated considering IQ imbalances and nonlinear PA distortion. ADC resolution 8 bits

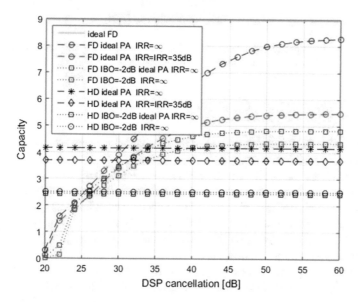

Fig. 11 Capacity versus DSP cancelation. FD and HD links are evaluated considering IQ imbalances and nonlinear PA distortion. ADC resolution 14 bits

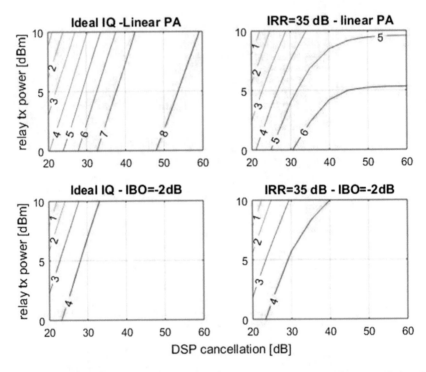

Fig. 12 FD relay. Capacity dependence with relay transmitted power and DSP cancelation. IQ imbalances and PA distortion is included. ADC resolution 14 bits

3.3 Self-interference Channel Models

The knowledge of the characteristics of the SI channel is a key issue to design self-interference cancelation techniques

We investigate the equivalent SI channel model considering the proposed patch FD antenna. The frequency response of a SI channel was measured using a vector network analyzer (VNA) in four scenarios: anechoic chamber, laboratory, corridor, and terrace. The S parameters at $N = 2000$ uniformly spaced frequency components in 2.4–2.6 GHz ($B_v = 200$ MHz) band were obtained. Each frequency sample value was averaged over 30 measurements.

The time resolution is given by $\tau_{min} = 1/B_v = 5$ ns and a maximum time delay of $\tau_{max} = \frac{N-1}{B_v} = 15\,\mu s$. Several parameters are employed to characterize a wireless cannel:

- **Average isolation**: $I = \frac{1}{N}\sum_{f_l}^{f_u}|S_{13}(f)|^2$ where f_l and f_u are lower and upper frequency limits of the measurement bandwidth B_v.
- **Power delay profile**: is obtained by taking the inverse discrete-time Fourier transform (IDFT) of the measured frequency response, $p(t) = 20log_{10}|s_{13}(t)|$ where

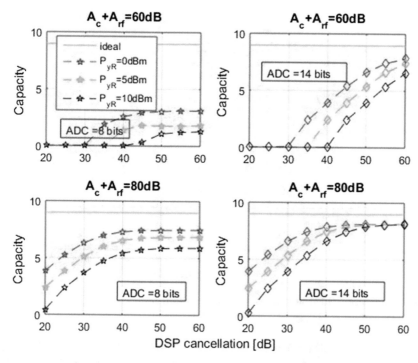

Fig. 13 FD relay. Capacity dependence with antenna + RF cancelation, ADC resolution, relay transmitted power, and DSP cancelation

Table 3 Self-interference channel metrics

	Isolation I (dB)	Rms delay spread σ_τ (ns)	Coherence bandwidth B_c (MHz)
Terrace	54.3	70	2.84
Corridor	63.1	110	1.80
Laboratory	55.3	84	2.36
Anechoic chamber	57.0	19	10.5

$s_{13}(t) = F^{-1}[S_{13}(f)]$, and $F^{-1}[\cdot]$ is the inverse discrete-time Fourier transform (IDFT).

- **Delay spread**: It is given by, $\sigma_\tau = \sqrt{\overline{\tau^2} - \overline{\tau}^2}$ where $\overline{\tau^2} = \sum_{l=1}^{N} \tau_l^2 p(\tau_l)$ and $\overline{\tau} = \sum_{l=1}^{N} \tau_l p(\tau_l)$.
- **Coherence bandwidth**: it quantifies the frequency selectivity of a wireless channel, $B_c = \frac{0.02}{\sigma_\tau}$.

A synthesis of the measured parameters is listed in Table 3.

From these results, we can infer that only in case of nonreflective scenarios, the SI can be considered as frequency flat requiring a single-tap DSP canceler. However, for baseband processing, the single or multipath characteristic of a SI channel is a

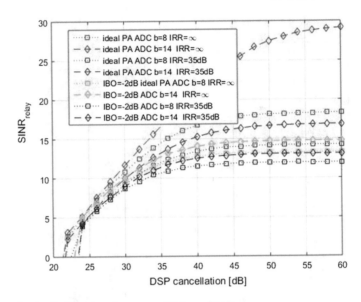

Fig. 14 Effective SINR at the relay output. $SNR_R = 30$ dB

function of the sampling frequency. Considering an OFDM system operating with n sampling frequency $f_s = \frac{1}{T_s}$, each baseband channel tap can be obtained as the combination of T_s/τ_{min} samples of $p(t)$.

Figure 15 shows the power delay profile of power normalized SI channel considering OFDM bandwidths of $B_{ofdm} = 5$, 10, and 40 MHz. In case of lower sampling frequency, the energy of the channel is concentrated in a dominant path. When the sampling frequency is increased, secondary paths have also significant levels of energy in reflective scenarios.

4 Massive MIMO Systems

The development of massive multiple-input multiple-output (MaMIMO) was motivated by the necessity of a large spectral efficiency and reduced power consumption. The implementation of a large number of antennas offers a large spectral efficiency and link reliability. Moreover, the use of MaMIMO allows scaling down the transmitted power proportionally to the number of antennas which potentially leads to a significant improvement in terms of energy efficiency [31].

MaMIMO systems have been developed to provide very high data rate, without increasing the spectrum bandwidth and energy consumption [32, 33]. Massive MIMO usually consists of tens to hundreds of antenna elements, placed in one or some of the involved communication devices. A base station with a massive antenna array serving several single-antenna user terminals is a typical example of massive antenna

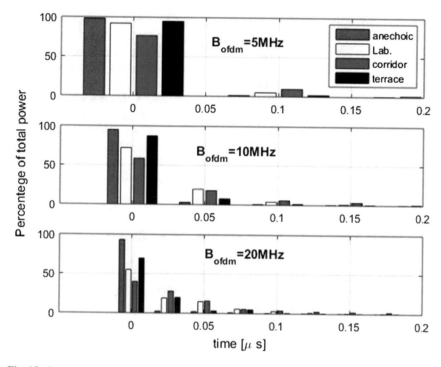

Fig. 15 Sampled channel impulse response of SI coupling. Energy distribution per tap for an OFDM system with bandwidth of 5, 10, and 20 MHz

application. The use of large number of antennas at the base station is essential to enhance the capacity without extra spectral resources [34].

OFDM and MaMIMO present a huge potential to obtain very high data rates and high quality of service (QoS). However, MaMIMO systems requires a mobile equipped with multiple antennas at the transmitter that is a challenging issue in mobile devices due mostly to their size, cost, and computing power limitations. These technological issues constrain the use of large number of antennas only to the base station. In the mobile device, only single or a couple of antennas are allowed [35].

4.1 Low Complexity RF Front-End Design

The radiated power per antenna decreases linearly with the number of antennas. Moreover, the effects of small-scale fading, noncoherent interference and receiver noise are minimized. However, a massive number of antennas require a separate transceiver chain and power amplifier (PA) for each antenna. In this situation, the size and costs of the RF front-end (including ADC and DAC) and PA become a critical issue. The optimization of cost and size implies the use of low-cost components which creates several imperfections/impairments that degrade the system performance. For large number of base station (BS) antennas, the system capacity is only limited by the impairments at the user equipment. That result implies that RF imperfection at BS can be supported with low performance degradation [36, 37].

A critical component of MaMIMO front-ends is the ADC. In case of large number of antennas, a MaMIMO front-end implies in many ADCs, i.e., cost and power consumption prohibitive for practical implementations. Low-resolution ADCs are proposed to minimize the cost and power consumption of each RF chain [38–40]. The trade-off between quantization noise and power consumption need to be investigated in order to find an optimal resolution that minimizes the SNR degradation.

4.2 Power Efficiency and Linear Precoding Techniques

The MaMIMO system model based on OFDM illustrated in Fig. 16 is composed of a single-cell scenario that operates in a TDD mode. It includes a base station with M antennas with a total power constraint P_{tr} that serve N single—antenna users ($M \times N$). It is assumed that the channel is uncorrelated without coupling between antenna elements. The vector of the received signal on subcarrier k, y_k, with ($k = 1, 2, \ldots, N_{FFT}$), for an arbitrary OFDM symbol is given by

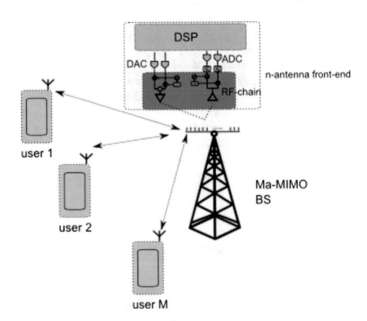

Fig. 16 Massive MIMO system

$$y_k = \rho_k H_k Q_k s_k + w_k$$

where $y_k = [y_k^1, \ldots, y_k^N]^T \in C^{N \times 1}$, the response of the wireless fading channel at subcarrier k is $H_k = [h_k^1, \ldots, h_k^N] \in C^{N \times M}$, with elements $h_k^{n \cdot m}$ with $n = 1, \ldots, N$ and $m = 1, \ldots, M$). $Q_k \in C^{M \times N}$ is a precoding matrix with elements $q_k^{m,n}$ and $s_k = [s_k^1, \ldots, s_k^N]^T \in C^{N \times 1}$ is the data symbols vector, and $w_k = [w_k^1, \ldots, w_k^N]^T \in C^{N \times 1}$, with w_k^n are sampled of independent and identically distributed (i.i.d.) zero-mean complex white Gaussian with variance σ_w^2. The power transmitter is subject to a constraint given through $\rho_k = \sqrt{p_k}$ and it is assumed that power per user is normalized $[E|x^n|^2] = 1$. The precoding matrix is obtained assuming that the BS has perfect knowledge of the channel. The precoded signal is given by $x_k = Q_k s_k$. Considering a linear precoder, the transmitted signal can be written as

$$x_k^n = \sum_{i=1}^{N} q_k^{n,i} s_k^i$$

And the received signal by the user n is

$$y_k^n = \rho_k^n h_k^n q_k^n s_k^n + \rho_k^n \sum_{i=1,i \neq n}^{N} h_k^n q_k^i s_k^i + w_k^n$$

where the desired signal, the interference, and the noise can be easily identified. The precoding matrix is designed to minimize the multiuser interference, improving the signal-to-interference-plus-noise ratio (SINR). That SINR can be expressed by

$$SINR_k^n = \frac{\rho_k^n E[|h_k^n q_k^n|^2]}{p_k^n \sum_{i=1,i \neq n}^{N} E[|h_k^n q_k^n|^2] + \sigma_w^2}$$

where $p_k = \rho_k^2$. The achievable rate for user n can be approximated by

$$R^n \cong log_2(1 + SINR^n)$$

Complexity issues motivate the use of linear precoding techniques, which present a good performance and lower implementation complexity. Particularly, two linear precoding techniques will be explored: Least square (LS) and maximum ratio transmitter (MRT).

LS or zero-forcing is designed to eliminate multiuser interference (MUI). The precoding matrix for each subcarrier k is given by

$$Q_{k,LS} = \beta_{LS}^{-1} H_k^H (H_k H_k^H)^{-1}$$

where $\beta_{LS} = \sqrt{\left\| H_k^H \left(H_k H_k^H \right)^{-1} \right\|_F^2}$. LS precoding avoids the MUI. However, the channel noise is amplified when the channel has large deeps in its frequency response. **Maximum ratio transmitter (MRT)** precoding maximizes the power of the desired signal disregarding the interference. The precoding is given by

$$Q_{k,MRT} = \beta_{MRT}^{-1} H_k^H$$

where $\beta_{MRT} = \sqrt{\left\| H_k^H \right\|_F^2}$. The MRT scheme avoids matrix inversion leading to the simplest precoding solution.

4.2.1   Channel State Information

CSI is estimated at the BS using a training sequence allocated to each user. During the training, all N users transmit non-orthogonal sequences and the BS estimates each channel. By considering uplink–downlink channel reciprocity, the estimates are employed to make the precoding matrix.

In case that lower resolution ADC is employed in each RF chain of the BS, the estimate will be affected by quantization noise and the downlink performance will be degraded.

4.3   Numerical Evaluation

In order to check the performance of Ma-MIMO systems with low-resolution ADCs, we define a simulation scenario that consists of a BS equipped with N BS and M single-antenna users. MRT and LS precoders were evaluated.

The recovered symbol constellation with 1 and 2 bits ADC at the BS is illustrated in Fig. 17 for MRT precoding. Two active users are considered. The BS is equipped with 40 and 250 antennas. Severe quantization noise can be observed when 1 bit ADC is employed. Using the LS precoder and 40 antennas, the MUI is severe and cannot be suppressed. The results are improved for the case of 250 antenna elements. When an ADC of 2 bits is adopted, a significant noise/interference reduction is obtained. For large number of antennas, the quantization noise can be considered negligible even for the case of 2 bits ADC.

The capacity for user n is illustrated in Fig. 18 that is evaluated for MRT and LS precoding varying the number of antenna elements and the ADC resolution. LS precoding obtains the best performance in the high SNR region. When 100 antennas and 2 bits ADC are employed, the performance is almost identical to the ideal with a low penalty. On the other hand, it can be seen that the rate of the MRT quantized

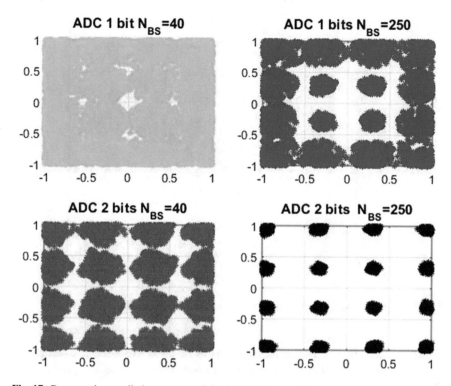

Fig. 17 Recovered constellation at user n (BS using MRT precoding)

systems is limited by the MUI. The implementation complexity and the associate power consumption of both precoders need to be carefully evaluated in order to decide which the best option for Ma-MIMO is. The trade-off between cost in terms of power consumption and the penalty in terms of SINR are the variables to be optimized.

5 Summary of the Chapter

- RF components are fundamental elements that govern the performance of communication devices.
- The power consumption of long range communication systems is a function of power amplifier consumption. In this case, the implementation of linearization techniques is mandatory. However, when the transmitted power is lower (short-range communication), DSP and ADC converters appear as significant contributors to the power driven.

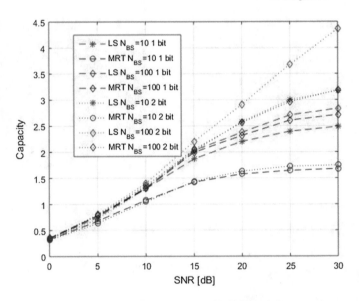

Fig. 18 Capacity versus number of antenna elements and ADC resolution

- The consumption of digital blocks needs to be considered in low-mean power devices. ADCs and digital processing block need to be carefully designed in massive MIMO applications.
- Full-duplex systems allow increasing the system capacity and spectral efficiency. Practical implementation of self-interference cancelation techniques is an open issue where several items are still unsolved.

References

1. Mamta Agiwal, Abhishek Roy, Navrati Saxena, "Next Generation 5G Wireless Networks: A Comprehensive Survey", Communications Surveys & Tutorials IEEE, pp. 1617–1655, 2016.
2. G. Auer et al., "How much energy is needed to run a wireless network?," in *IEEE Wireless Communications*, vol. 18, no. 5, pp. 40–49, October 2011.
3. O. Orhan, E. Erkip and S. Rangan, "Low power analog-to-digital conversion in millimeter wave systems: Impact of resolution and bandwidth on performance," *2015 Information Theory and Applications Workshop (ITA)*, San Diego, CA, 2015, pp. 191–198.
4. G. Andrews et al., "What Will 5G Be? *IEEE JSAC*, vol. 32, no. 6, pp. 1065–82, 2014.
5. Tim Schenk, RF Imperfections in High-rate Wireless Systems-Impact and Digital Compensation, Springer, 2008.
6. F. Gregorio, J. Cousseau, S. Werner, T. Riihonen and R. Wichman, "EVM Analysis for Broadband OFDM Direct-Conversion Transmitters," in *IEEE Transactions on Vehicular Technology*, vol. 62, no. 7, pp. 3443–3451, Sept. 2013.
7. A. Mammela and A. Anttonen, "Why Will Computing Power Need Particular Attention in Future Wireless Devices?" in *IEEE Circuits and Systems Magazine*, vol. 17, no. 1, pp. 12–26, 2017.

8. A. Mezghani and J. A. Nossek, "Modeling and minimization of transceiver power consumption in wireless networks," *2011 International ITG Workshop on Smart Antennas*, Aachen, 2011, pp. 1–8.
9. J. Minkoff, "The role of AM-to-PM conversion in memoryless nonlinear systems," *IEEE Trans. Commun.*, vol. 33, no. 2, pp. 139–144, Feb. 1985.
10. F. Gregorio, Analysis and Compensation of Nonlinear Power Amplifier Effects in Multi-Antenna OFDM Systems. Ph.d. Thesis, Helsinki University of technology, Finland, 2007.
11. J. Cousseau, J. Figueroa, S. Werner, and T. Laakso, "Efficient nonlinear Wiener model identification using a complex-valued simplicial canonical piecewise linear filter," *IEEE Trans. Signal Process. Part 1*, vol. 55, no. 5, pp. 1780–1792, May 2007.
12. D.R. Morgan, Z. Ma, J. Kim, M.G. Zierdt and J. Pastalan, "A Generalized Memory Polynomial Model for Digital Predistortion of RF Power Amplifiers", IEEE Trans. Signal Process., vol. 54, no. 10, pp. 3852–3860, Oct. 2006.
13. J. J. Bussgang, "Cross correlation function of amplitude-distorted Gaussian input signals," Res. Lab Electron., M.I.T., Cambridge, MA, Tech. Rep. 216, vol. 3, Mar. 1952.
14. D. Petrovic, W. Rave, G. Fettweis, "Effects of phase noise on OFDM systems with and without PLL: Characterization and compensation", *IEEE Trans. Commun.*, vol. 55, no. 8, pp. 1607–1616, Aug. 2007.
15. H. S. Lee and C. G. Sodini, "Analog-to-Digital Converters: Digitizing the Analog World," in *Proceedings of the IEEE*, vol. 96, no. 2, pp. 323–334, Feb. 2008.
16. Q. Bai, A. Mezghani and J. A. Nossek, "On the Optimization of ADC Resolution in Multi-antenna Systems," *ISWCS 2013*, Ilmenau, Germany, 2013, pp. 1–5.
17. B. Murmann, "ADC Performance Survey 1997–2016," [Online]. Available: http://web.stanford.edu/~murmann/adcsurvey.html.
18. M. Windisch and G. Fettweis, Standard-Independent I/Q imbalance compensation in OFDM direct-conversion receivers, *Proc. 9th Intl. OFDM Workshop (InOWo)*, (*Dresden*), pp. 57–61.
19. M. Duarte, C. Dick, and A. Sabharwal, "Experiment-driven characterization of full-duplex wireless systems," IEEE Trans. Wireless Commun., vol. 11, no. 12, pp. 4296–4307, Dec. 2012.
20. A. Raghavan, E. Gebara, E. M. Tentzeris and J. Laskar, "Analysis and design of an interference canceller for collocated radios," in *IEEE Transactions on Microwave Theory and Techniques*, vol. 53, no. 11, pp. 3498–3508, Nov. 2005.
21. D. Korpi, T. Huusari, Y. S. Choi, L. Anttila, S. Talwar and M. Valkama, "Digital self-interference cancellation under nonideal RF components: Advanced algorithms and measured performance," *2015 IEEE 16th International Workshop on Signal Processing Advances in Wireless Communications (SPAWC)*, Stockholm, 2015, pp. 286–290.
22. T. Riihonen and R. Wichman, "Analog and digital self-interference cancellation in full-duplex MIMO-OFDM transceivers with limited resolution in A/D conversion," in Proc. Asilomar Conference on Signals, Systems and Computers, Nov. 2012.
23. Choi, J.I., Jain, M., Srinivasan, K., et al.: 'Achieving single channel, full duplex wireless communication'. Mobicom'10, IL, USA, September 2010, pp. 1–12.
24. A. Batgerel and S. Y. Eom, "High-isolation microstrip patch array antenna for single channel full duplex communications," in IET Microwaves, Antennas & Propagation, vol. 9, no. 11, 2015.
25. J. M. Laco, F. H. Gregorio, G. González, J. E. Cousseau, T. Riihonen and R. Wichman, "Patch antenna design for full-duplex transceivers," *2017 European Conference on Networks and Communications (EuCNC)*, Oulu, Finland, 2017, pp. 1–5.
26. J. Tamminen et al., "Digitally-controlled RF self-interference canceller for full-duplex radios," *2016 24th European Signal Processing Conference, Budapest, 2016, pp. 783–787.
27. Björn Debaillie, DUPLO deliverable 2.1, D2.1—Design and measurement report for RF and antenna solutions for self-interference cancellation, 2014.
28. G. González, F. Gregorio, J. Cousseau, T. Riihonen, and R. Wichman, "Performance analysis of full-duplex AF relaying with transceiver hardware impairments," in European Wireless (EW) conference, Oulu, Finland, May 2016.

29. F. Gregorio, G. González, J. Cousseau, T. Riihonen, and R. Wichman, "RF front-end implementation challenges of in-band full-duplex relay transceivers," in European Wireless (EW) conference, Oulu, Finland, May 2016.
30. F. Gregorio, J. Cousseau, S. Werner, R. Wichman and T. Riihonen, "Sequential Compensation of RF Impairments in OFDM Systems," *2010 IEEE Wireless Communication and Networking Conference*, Sydney, NSW, 2010, pp. 1–6.
31. Massive MIMO for efficient transmission, FP7 Project MAMMOET, http://mammoet-project.eu/.
32. C. Desset, B. Debaillie, and F. Louagie. Modeling the hardware power consumption of large scale antenna systems. In Proc. IEEE OnlineGreenComm, 2014.
33. H. Yang and T.L. Marzetta. Total energy efficiency of cellular large scale antenna system multiple access mobile networks. In Proc. IEEE Online Green Comm, 2013.
34. F. Rusek et al., "Scaling Up MIMO: Opportunities and Challenges with Very Large Arrays," in *IEEE Signal Processing Magazine*, vol. 30, no. 1, pp. 40–60, Jan. 2013.
35. Thomas L. Marzetta, Erik G. Larsson, Hong Yang, Hien Quoc Ngo, Fundamentals of Massive MIMO, Cambridge University Press 978-1-107-17557-0.
36. E. Björnson, J. Hoydis, M. Kountouris, and M. Debbah, "Massive MIMO systems with non-ideal hardware: Energy efficiency, estimation, and capacity limits," IEEE Trans. Inf. Theory, vol. 60, no. 11, pp. 7112–7139, Nov. 2014.
37. Y. Zou et al., "Impact of Power Amplifier Nonlinearities in Multi-User Massive MIMO Downlink," *2015 IEEE Globecom Workshops* (*GC Wkshps*), San Diego, CA, 2015, pp. 1–7.
38. Y. Li, C. Tao, G. Seco-Granados, A. Mezghani, A. L. Swindlehurst and L. Liu, "Channel Estimation and Performance Analysis of One-Bit Massive MIMO Systems," in *IEEE Transactions on Signal Processing*, vol. 65, no. 15, pp. 4075–4089, Aug. 1, 1 2017.
39. C. Mollén, J. Choi, E. G. Larsson and R. W. Heath, "Uplink Performance of Wideband Massive MIMO With One-Bit ADCs," in *IEEE Transactions on Wireless Communications*, vol. 16, no. 1, pp. 87–100, Jan. 2017.
40. C. Studer, G. Durisi, "Quantized massive MU-MIMO-OFDM uplink", *IEEE Trans. Commun.*, vol. 64, no. 6, pp. 2387–2399, Jun. 2016.

Security Attacks on Wireless Networks and Their Detection Techniques

Rizwan Ur Rahman and Deepak Singh Tomar

Abstract The security issues of wireless networks have been a continuous research area in the recent years. With the advancement of wireless networking systems, building secured and reliable communication is particularly important. Security of wireless systems is to prevent unauthorized access or harm to computers from the attackers. Since wireless networks are open in nature, it is possible for attacker to launch numerous types of attacks on wireless network. Hence, the security of wireless network remains a serious and challenging issue. In this chapter, security issues in context to wireless network are presented. In the first section, vulnerabilities such as unknown network boundary and no physical access required are explored. The next section is dedicated to attacks in wireless networks, attacks classification such as active attack versus passive attack, and the other types of attack in wireless networks. Special section is dedicated to emerging attacks in bitcoin peer-to-peer network. The last section gives the security mechanism of wireless network and attack detection techniques including active attack detection and passive attack detection techniques.

Keywords Wireless security · Vulnerabilities · Attacks · Active attacks
Passive attacks · Cryptocurrency · Bitcoin · Blockchain · Bitcoin attacks · WIDS
WIPS · Firewall · Anomaly detection

1 Introduction

Wireless technology is becoming the most exhilarating field of communication and networking. The rapid development of cell phone use, a variety of satellite services, and at the moment the wireless local area networks wireless Internet are making

R. U. Rahman (✉) · D. S. Tomar
Department of Computer Science and Engineering,
Maulana Azad National Institute of Technology, Bhopal, India
e-mail: rizwan.143212007@manit.ac.in; rizwan.rahman12@gmail.com

D. S. Tomar
e-mail: deepaktomar@manit.ac.in

© Springer Nature Singapore Pte Ltd. 2018
K. V. Arya et al. (eds.), *Emerging Wireless Communication and Network Technologies*,
https://doi.org/10.1007/978-981-13-0396-8_13

incredible changes in communication and networking. This chapter explores the core areas in the field of security in wireless network:

- Wireless networking introduction,
- Vulnerabilities in wireless networks,
- Attacks in wireless networks, and
- Attack detection in wireless networks.

1.1 Introduction to Wireless Networking

A wireless network is a computer network that allows computing devices to communicate with each other without being connected via a physical communication medium, such as networking cable. Modern wireless networks typically rely upon radio communications which take place in the band of frequencies beyond infrared light in the electromagnetic spectrum [1]. Other means of wireless signaling which rely upon higher energy frequencies within the electromagnetic spectrum are also possible within the broader framework of network communications. Wireless networks are implemented at the physical layer that is layer 1 of the OSI (Open System Interconnection) reference model. A typical wireless system is shown in Fig. 1.

1.2 Wireless Access Points

Wireless access point is used to provide wireless entrance to a device that does not have wireless access. It can be thought as a converter between regular local area network and a wireless world. Wireless access point is employed to accomplish other tasks, for instance, increase a wireless network as well [2].

Wireless access points are typically physically connected to a wired network via a network router. One of the primary components of a wireless access point is a radio transceiver which is a device that allows for both the transmission and reception of radio signals client. Devices wishing to join a wireless network also have a transceiver and in this way, two-way wireless communication between a client and a WEP becomes possible. Wireless access points commonly use the Wi-Fi set of communication standards which are alternatively known as the IEEE 802.11 protocol suite.

1.3 Advantages of Using Wireless Network

Wireless networks are advantageous in many different ways and one good way of understanding the advantages that wireless networks provide is to divide them into

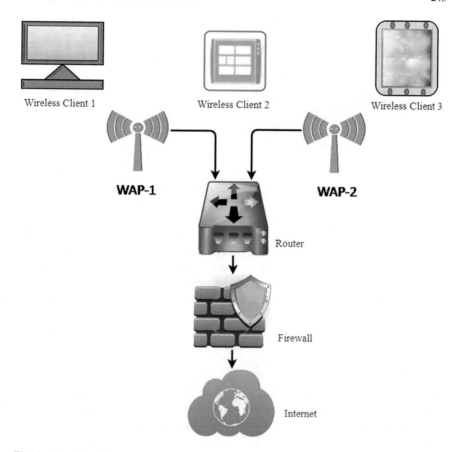

Fig. 1 A typical wireless system

two groups. The first is the advantages for network users and the second is the advantages for network providers.

1.3.1 Advantages for Wireless Network Users

For network users, wireless networks provide both convenience and mobility. Wireless networks are often more convenient to users than wired networks because they allow user to connect to a network without needing to attach a network cable to computing device and without needing to specify any cumbersome network configuration settings such as IP addresses, subnet masks, gateways, DNS servers, and so forth. Wireless networks also provide users with mobility rather than being forced to stay in one physical location as with a wired network. Users of wireless networks are free to roam to any location in the world which provides wireless access and can still be connected to the network [3].

Table 1 Summary of wireless network advantages

S. no.	Advantages for wireless network users	Advantages for wireless network providers
1	Work with multiple devices	Network cables are not required
2	Guest use	Fewer points of failure
3	Convenience	Fewer points of attacks
4	Increased mobility	Lower maintenance and installation cost
5	Safety	Scalability

Another advantage of using wireless networks is that it works with numerous devices. Whether the device of end user is a PC, tablet, laptop, or even cell phone, it can be connected to the network with a trouble-free click. This is suitable for visiting clients, plus employees in an office who want to complete different tasks during the day, such as making notes or checking emails during meetings. So, in this way, it gives a convenience for guests visiting the office or any other place. Last but not least, it provides safety to the end users. As there are no cables used in a wireless connection, the possible risk of stumbling over any trailing wires can be avoided in general.

1.3.2 Advantages for Wireless Network Providers

Wireless networks also offer many advantages for network providers. First, with wireless networks, a network provider needs to run much less network cable that would be necessary if it used an exclusively wired network of the same capacity. This concept also extends into mobile telephony. Developed countries, for example, spent more than a century laying millions of kilometers of telephone wires. In the modern era, developing and emerging countries can simply deploy wireless telephone networks in order to allow their citizens to communicate with each other.

The summary of wireless network advantages is presented in Table 1.

1.4 Communiqué Using Wireless Networks

An end user who wishes to employ a wireless network must possess a wireless network interface card (NIC) which supports the communications protocols. In the early period of computer networking, network interface cards (NICs) were often separate devices that had to be installed into a computer's expansion slot. But now NICs have become standard components that are mostly a permanent part of computing devices in primary circuitry. It should be noted that each NIC has a media access control (MAC) address which in theory provides a mean through which the network interface card can be uniquely identified. The capability of each computing device to be uniquely identifiable and addressable is essential for computer networking including wireless networking [4].

Most modern wireless networks rely upon the IEEE 802.11 protocol suite of communications protocols. The 802.11 protocol suite is a family of related communications protocols that provide detailed specifications for implementing local area wireless computer networks. For instance, IEEE 802.11 based wireless protocols include 802.11a and 802.11b. The usable range of 802.11 wireless networks varies widely from approximately 25 m to approximately half kilometers depending upon the local environmental conditions and the specific protocol being used.

1.5 Data Frames in Wireless Networking

Wireless data are being transmitted in blocks of data known as wireless data frames [5]. Each wireless data frame is made up of three key parts. A typical wireless data frame is shown in Fig. 2:

1. Frame header
2. Frame payload
3. Frame checksum sequence.

Frame Header:
The frame header contains five numbers of different attributes that are required in order to make wireless networking realistic in a multiuser environment.

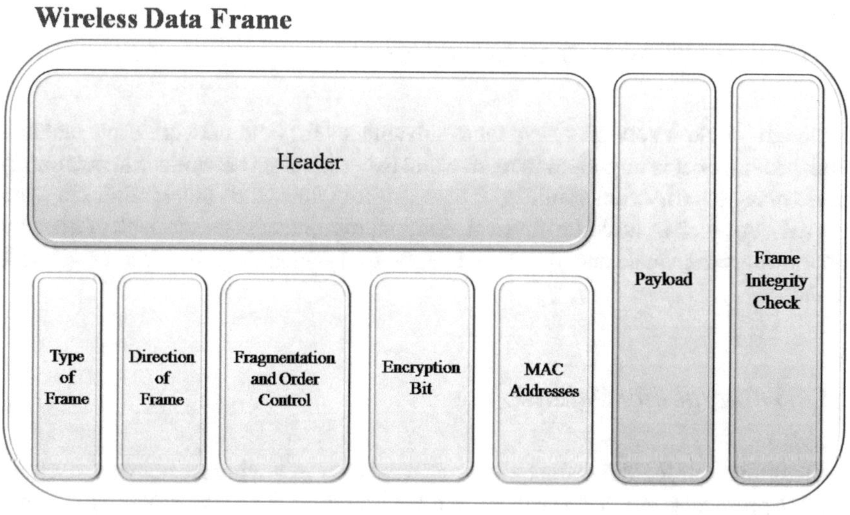

Wireless Data Frame

Fig. 2 Wireless data frame

1. **Type of Frame**: The first one among these is the frame type which could be a usual data frame or some other types of frame such as an authentication frame or beacon frame or an association frame.
2. **Direction of Frame**: The next attribute in the frame header specifies the direction of the frame that is whether the frame is sent from a wireless access point to a client or a client to the wireless access point.
3. **Order Control and Fragmentation**: The third part of the frame header deals with fragmentation and order control as wireless messages are subdivided into frames; it is essential to maintain the order and sequencing of each frame within a message.
4. **Bit of Encryption**: The fourth part of the header is a bit of encryption which specifies whether the frame utilizes wired equivalent privacy (WEP)-based encryption.
5. **MAC Address**: At last, the header has the MAC addresses of the sender and receiver in order to determine whether the frame is meant for them of the wireless devices within the range.

Payload:
After the frame header, the next key piece of the wireless data frame is the frame payload. For this frame, Wi-Fi data frames have data between 0 and 2047 bytes of payload data.

Frame Checksum sequence:
Finally, the third major part of the wireless data frame is the frame checksum sequence which is used to verify the integrity of the frame. In other words, the frame check sequence is used to make certain that the data frame was not modified or corrupted while transmitting between the sender and receiver. The checksum sequence for a data frame generally uses the cyclic redundancy check (CRC) value.

2 Vulnerabilities in Wireless Network

Although wireless networks have many advantages for both network users and network providers, it is important to realize that using wireless networks can potentially create many security vulnerabilities. This section explores the vulnerabilities in wireless network such as insecure physical location, rogue access points, lack of network monitoring, inadequate encryption and decryption standards, and unknown network boundary.

2.1 Vulnerability Definition

Vulnerability can be defined as a type of weakness in a wireless network itself, in a set of protocols, or whatever thing that leaves wireless network security exposed to a threat. Vulnerability is the internal weakness of wireless network systems. On the other hand, a threat is an external factor that could harm the wireless network system [6].

2.2   Insecure Physical Location

Wireless access points (WAPs) should not be positioned from where they are easily accessible. For this reason, they can be detached and tampered with copied configurations or altered configurations then returned.

2.3   Rogue Access Points

When an access point is installed on a wireless network without the apparent authorization from a local wireless network administrator, then it is known as rogue access points. These points are usually added by the employees of an organization or by a malicious attacker [7].

A rogue access point may also be effortlessly smuggled onto enterprise premises by a stranger. Moreover, a rogue access points cause grave security threat to a wired enterprise wireless network. In view of the fact that, it provides a wireless backdoor into the enterprise wireless network for outside users, evading every one of wired safety measures such as network access control and firewalls. A typical scenario of rogue access points is presented in Fig. 3. As shown in the figure even the firewall cannot protect from rogue access points since the firewall operates at traffic transfer point between the Internet and local area network. Firewalls are not able to monitor traffic through rogue access points.

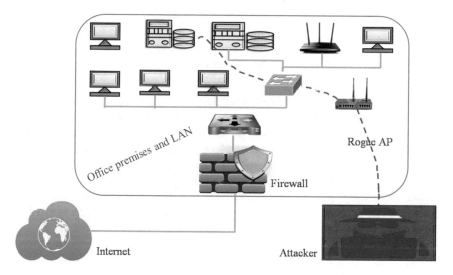

Fig. 3   Rogue access points

2.4 Physical Access Is Not Required

No physical boundary exists between a malicious attacker and the data that is transmitted over a wireless network. In other words, a wired network gives an extra layer of protection that a wireless network does not provide insofar as eavesdropping on a wired network needs physical access to network infrastructure components, for instance, network cables, routers, or switches.

Wireless access points may also lose signals because of office wall, doors, cabins, and other office building materials. These signals may also penetrate into the airspace of other office and could connect with their wireless network. This is known as accidental associations and could take place in densely populated areas where numerous end users or organizations use wireless technology [8].

2.5 Inadequate Encryption and Decryption Standards

Wired equivalent privacy also referred as WEP is an encryption algorithm for wireless computer networks [9].

Wired equivalent privacy (WEP) encryption standard has a number of weaknesses such as key size and the initialization vector (IV) is very small. As a result, some end users will not even enable it. This can prove to be damaging to the wireless local area network since weak encryption is far better than the no encryption. Furthermore, some users make use of the longer encryption key, but this is not going to make the local area network extra secure.

2.6 Unidentified Network Boundary

Another vulnerability, connected with wireless network systems is that they often create an unknown boundary for the wireless network. In a wired network, administrator and security personnel know precisely where the network boundary ends. This is not generally possible with a wireless network because clients often establish or drop connections to a wireless network on an ad hoc basis. Moreover, these clients can further complicate the network boundary by using network extenders or by enabling internet connection sharing for the other users through which computing devices are could connect to the wireless network by piggybacking on an existing client connection [10].

2.7   Insecure Wireless Network

The most substantial threats to information security emerge in the context of inse-cure wireless networks. Unencrypted wireless data frames can be easily sniffed and analyzed using free software by anyone within range of the wireless network.

Insecure wireless networks are in fact such appealing targets for networks and have become a widespread activity among malicious parties searching for and mark-ing unsecured wireless networks. The most common forms are "War Driving" and "Warchalking" [11, 12].

2.7.1   War Driving

War driving sometimes also referred as an access point mapping is a method used to take an advantage of wireless networks by the unauthorized use of insecure wireless local area network. It involves driving around a town neighborhood or city to search for open wireless local area network.

If the open wireless networks are found, they could be exploited for many pur-poses, from accessing organizational documents to simply browsing the Internet. A lot of wireless networks remains unlock to whatever happens since the individual user or organizations make use of these networks that do not consider taking an added security measures to be essential.

Global positioning system (GPS) technology is exploited for war driving, creating these networks with no trouble to spot. The war driver could locate wireless networks on a map and distribute it using the apps, for instance, Google maps or other GPS softwares.

2.7.2   Warchalking

It is a process of sketching of a particular set of codes to mark the streets and sidewalks in order that other well-informed parties can take advantage of the wireless networks. Warchalking is motivated by hobo symbols, given away by Matt Jones who draw the set of icons and made an open access document containing these symbols. Warchalking symbols of open node, closed node, and WEP node are shown in Fig. 4.

3   Attacks in Wireless Network

This section introduces the attacks in the wireless networks. First, the attack definition is presented, and then the classifications of attacks such as active attack, passive attacks, cryptographic, and noncryptographic attacks are explored.

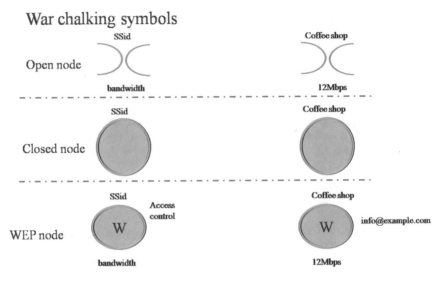

Fig. 4 Warchalking symbols

3.1 Attack Definition

Wireless network attacks can be defined as an attempt to harm, expose, modify, damage, embezzle, or gain unauthorized access to a wireless network asset. Internet Engineering Task Force (IETF) describes the term attack as an assault on the computer network security or planned attempt especially in method or technique to dodge security services and violate the security guiding principle of a system. An attack could be carried out by an insider known as inside attack or from outside the company known as outside attack [13].

In an inside attack, the attack is started by an attacker who is inside the security boundary, i.e., an insider. This is usually carried out by an employee who is authorized to access the system resources and, however, uses these resources in such a way that it is not permitted by the administrator who granted the authorization. In an outside attack, the attack is started by an attacker from outside the security boundary, by an unauthorized or illegal user of the system that is an outsider.

3.2 Attacks Classification in Wireless Networks

Attacks can be categorized into two main categories, according to the disruption in wireless communication, i.e., passive wireless attacks and active wireless attacks. The taxonomy of attacks in wireless networks is presented in Fig. 5.

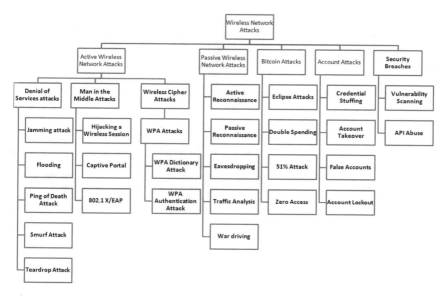

Fig. 5 Taxonomy of attacks in wireless networks

3.2.1 Passive Wireless Attacks

In passive wireless attack, a wireless network system is scanned and monitor for vulnerabilities and open ports. In this type of attack, an attacker acquires the data exchanged in the wireless network without disrupting the wireless communication. The main idea behind these attacks is exclusive to obtain the information about the target, and no data is changed or modified on the target systems [14]. Examples of passive wireless attacks include the following:

- **Active Reconnaissance**: In active reconnaissance, the passive attacker connects to the target system to collect the information about weaknesses and the vulnerabilities. An attacker examines systems for vulnerabilities without interruption, with the techniques like session capture.
- **Passive Reconnaissance**: In active reconnaissance, the passive attacker connects with the target system with methods such as port scanning.
- **Eavesdropping:** The literal meaning of eavesdrop is to secretly listen to private conversation over a confidential communication channel in a way that is not legally authorized. In wireless system, it is the process of collecting the information from a wireless network by snooping while the data is being transmitted. The information remains unaltered, but the privacy is compromised.
- **Traffic Analysis**: Traffic analysis is the method of capturing and monitoring wireless frames or messages with the purpose of driving the information for patterns in communication [15].

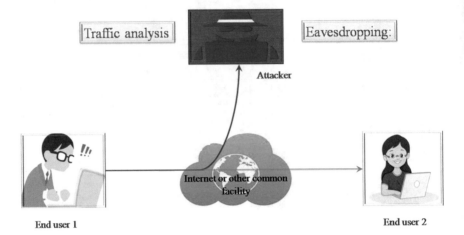

Fig. 6 Passive attacks

- **War driving:** In this type of passive attack, vulnerable Wi-Fi networks are scanned in close proximity places with a portable antenna. This attack is usually carried out from a moving motor vehicle, with GPS systems that the attackers use to mark areas with vulnerabilities on a Google map or other GPS software. Passive attacks are shown in Fig. 6.

3.2.2 Active Attacks

In active wireless attack, a wireless network system is attacked by an attacker attempting to penetrate into the wireless system. During this type of attack, the attacker damages data of the system as well as potentially modifies the data within the system. During this attack, attacker interrupts the normal functionality of the wireless network [16]. It is characterized by the following characteristics:

- Modification of information,
- Disruption of information, and
- Fabrication.

Active attacks are further classified as denial-of-service attacks, man-in-the-middle attacks, and cipherattacks.

3.3 Denial-of-Service Attacks

During denial of service, a mostly single computer is used by the attacker to exploit the vulnerabilities of wireless network. It is performed in numerous ways, for instance,

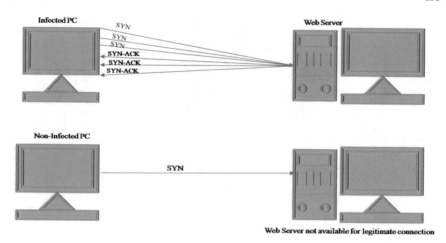

Fig. 7 DoS attacks

flooding a network by redundant traffic thereby preventing legitimate user request, twisting the connection information, for instance, resetting a TCP session, and blocking the access by disrupting the connection among the communicating systems [17]. A typical DoS attack is shown in Fig. 7.

Denial-of-service attack in wireless network can be done in different ways including jamming attack, flooding, ping of death attack, smurf attack, and teardrop attack.

3.3.1 Jamming Attack

In jamming attack, the attacker uses deliberate radio interference to damage the wireless network communications by making wireless communication medium busy. Jamming attack causes a transmitter to pull back when it senses busy medium, or it may corrupt the wireless frames received at receiver site. Jamming attacks typically target at the physical layer; however, attacks on other layers attacks are also possible [18].

This attack is generally carried out by producing the radio frequency noise in the frequency domain used by the wireless equipment. Devices such as laptop, mobile, and PC which operate with the similar frequency may be affected by the jamming attack. Typical jamming attack is shown in Fig. 8.

3.3.2 Flooding

In flooding denial-of-service attack, continuous and particular types of wireless frames are sent into the wireless network.

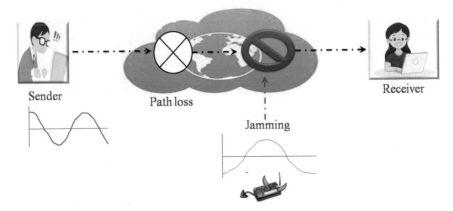

Fig. 8 DoS jamming attack

In flood attack, attacker floods victim with unbounded number of request but does not respond with acknowledgment packet when a packet is received from the server their by making half-open connection with the server. As the number of pending request increases, the server resource get consumed and fails to provide service to other genuine clients. Flooding in a wireless network can be of two types: authentication flooding and de-authentication flooding.

3.3.3 Ping of Death Attack

Ping is used to check the reachability of a client on a network. Ping transmits ICMP request packets and stays for ICMP echo reply. Ping of death attack is performed by sending packets of size larger as large as 65,535 bytes. A ping packet is of size 56 bytes if it is correctly formed. If a packet is of size larger than the allowed limit, packet gets divided into multiple packets [19]. Victim system crashes when it performs reassembling of these malformed packets. Ping of death attack is shown in Fig. 9.

3.3.4 Smurf Attack

Smurf attacks are those attacks caused because of misconfiguration of network devices. Source IP is spoofed to victim IP address by the attacker [20]. Smurf attack is shown in Fig. 10.

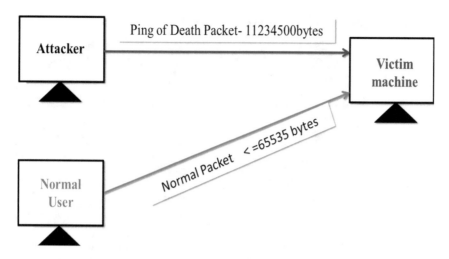

Fig. 9 Ping of death attack

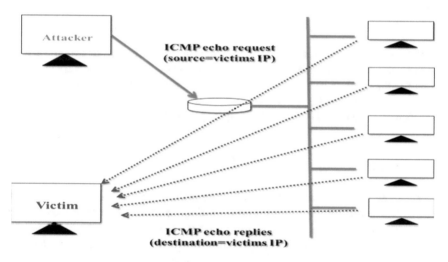

Fig. 10 Smurf attack

3.4 Man-in-the-Middle Attacks

After denial-of-device attack (DoS), the most serious attack to wireless network is the man-in-the-middle attack. This attack typically works in three steps. In the first step, an attacker waits for wireless access point (WAP) for a genuine user to initiate an association with the WAP. In the second step, as soon as an association is established, the attacker captures the MAC addresses of the genuine user and the wireless access point continuing the attack. After that, the attacker changes his MAC address to match the MAC address of the wireless access point and sends a disassociate request

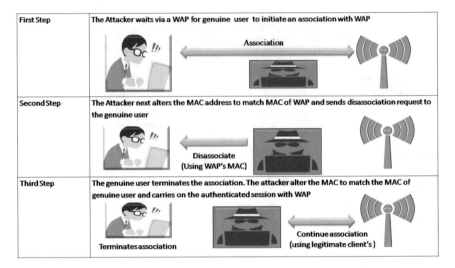

Fig. 11 Man-in-the-middle attack

to the legitimate user. Legitimate user terminates the association with the real wireless access point. Meanwhile, the attacker changes his MAC address once more this time adopting the legitimate users MAC address; the attacker is then able to continue the legitimate users authenticated session with the wireless access point [21]. Systematic description of man-in-the-middle is depicted in Fig. 11.

3.4.1   Captive Portal

A captive portal is generally a web page which is shown to new users before they are allowed further access to network resources. Captive portals are usually used login page which may require authentication in the form of username and password, or other valid credentials that both the provider and the user agree. Captive portals are used for a wide range of mobile, Wi-Fi, and home hotspots.

In captive portals, end users merely use a browser to login to the wireless network, a malicious party attacker only requires to make the similar login page, which should look indistinguishable from the real page, and capture credentials such as username and password as people attempt to login. The malicious party could even operate as proxy transferring the username and password onto the real authentication server. This kind of attack is also known as wireless phishing [22].

3.4.2   802.1x/Eap

The IEEE standard for port-based network access control is IEEE 802.1x. This standard is part of networking protocols of the IEEE 802.1. It offers an authentication method to computers wishing to connect wireless local area network.

Even as a correctly employed, WPA/WPA2 wireless network using 802.1x authentications is protected and not exposed to a man-in-the-middle attack, numerous clients are mistakenly configured, leaving them vulnerable to an attack. The weakness occurs from certificate used to authenticate the remote authentication dial-in user service (RADIUS) server. A lot of clients configure their computers in order that it does not discard certificates given by the RADIUS server [23].

3.5   Wireless Cipher Attacks

In wireless cipher attacks, the basic purpose is to decode ciphertext by breaking a cryptosystem and then obtain the plaintext from the ciphertext. To get the plaintext from the ciphertext, the attacker only requires getting the decryption key, since the encryption and decryption algorithms are already in the public domain.

Therefore, the attacker applies utmost effort to obtain the secret key of encryption and decryption algorithms. If the key is obtained by an attacker, then system is considered as a compromised and broken system. Different types of wireless cipher attacks are WEP attacks, WPA dictionary attack, and WPA/TKIP.

3.5.1   WEP Attacks

The threats of security associated with the wireless transmission of data are known since the inception of the age of wireless communications. In an effort to offer a level of protection to Wi-Fi networks, wired equivalent privacy (WEP) was released for the 802.11 protocols in the 1997 and was officially accepted in the year 1999 [24].

WEP was intended to give wireless communications with a level of security and privacy analogous to that provided by wired communications. However, numerous weaknesses have been identified in wired equivalent privacy.

- The very first weakness is discovered within 2 years after its formal acceptance. For instance, WEP makes use of a shared static key, i.e., the same key sequence can be used for multiple data frames belonging to the same message.
- Second, WEP uses a very short encryption key comprised of as few as 40 bits. A 40-bit key can be easily cracked with advanced parallel computers using brute force algorithms.
- Further, WEP implemented a weak encryption method based on the RC4 algorithm. In the web encryption model if a malicious party is able to guess the decrypted value of any single frame, then the key sequence used to encrypt that

frame can be recovered, as WEP reuses the same key sequence again and again. This means that other data frames can be hacked too.

- Fourth, WEP used an unencrypted integrity check value that was generated from a well-known algorithm. A malicious party could thus send modified data frames with appropriate check values that would appear to be legitimate.
- Lastly, WEP has no proper method for authentication and without authentication malicious parties could gain access to the network if they can get a correct SSID and MAC address.

3.5.2 WPA Attacks

With regards to all of the vulnerabilities that were recognized with wired equivalent privacy, the IEEE design and developed an improved way of protecting wireless data while they were in transmission between a client and a wireless access point.

To overcome these attacks, Wi-Fi protected access or WPA and later Wi-Fi protected access version 2 or WPA2 were developed as a replacement of wired equivalent privacy standards. Wi-Fi protected access is designed to improve wireless security by particularly addressing the problems that had been identified with the wired equivalent privacy standards [24].

- First, WPA2 uses a dynamic encryption key as different to the static encryption key which was used with WEP. This makes the compulsion for the encryption key to be changed for each data frame involved in a wireless message.
- Second, WPA2 includes real authentication in practice. Nearly, all WPA2-enabled wireless networks use password-based authentication but the standard also allows for additional authentication mechanisms, for instance, tokens certificates.
- Third, WPA2 makes use of strong encryption with a long encryption key. The WPA2 standard at the present supports AES encryption with a 256-bit key.
- Fourth, WPA2 makes use of enhanced cryptographic integrity protection for wireless data frames. Particularly, the standard supports 64-bit encrypted data integrity values.
- Fifth, WPA2 implements a lot improved process of initiating sessions which relies on a four-way handshaking operation.

Collectively, all of these characteristics formulate WPA2 wireless networks more safe and secure than WEP-protected wireless networks. In spite of the many security enhancements, they were incorporated into the Wi-Fi protected access standard. There are numerous known ways in which WPA2-protected wireless networks can be attacked. The two most common WPA attacks are WPA dictionary attack and WPA authentication attack.

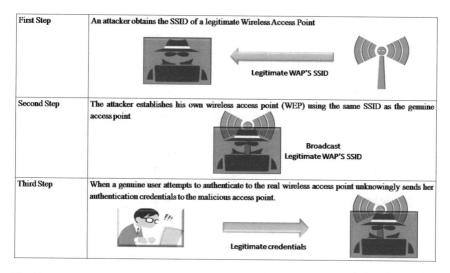

Fig. 12 WPA authentication attack

WPA Authentication Attack

The first way in which Wi-Fi protected access networks might be compromised is through an authentication attack. In authentication attack, SSID masquerading is used to gain the legitimate authentication credentials. Authentication attack in Wi-Fi protected access networks is three-step process. To initiate the attack, an attacker obtains the SSID of a legitimate wireless access point. In the second step, the attacker establishes his own wireless access point using the same SSID as the genuine access point, thus permitting the malicious party to masquerade as the genuine access point. In the third step, when a genuine user attempts to authenticate, the real wireless access point unknowingly sends her authentication credentials to the malicious access point. The attacker can then use the legitimate user's credentials to access the network using the real wireless access point [23]. Systematic description WPA authentication attack is shown in Fig. 12.

WPA Dictionary Attack

The dictionary attack has numerous forms; nearly in all of the forms, attackers compile a dictionary. In very basic method of this attack, the attacker compiles a dictionary of ciphers and equivalent plaintexts that has decoded over a span of time. Henceforth, as soon as an attacker obtains the cipher, then the dictionary can be referred to discover the equivalent plaintext.

WPA and WPA 2 allow for password-based authentication, so standard password cracking approaches such as a brute-force attack can be used in an attempt to gain access to the protected network if the password for the wireless network is short or easy to guess; then, a dictionary attack might be very effective [24].

3.5.3 WPA/TKIP

Temporal key integrity protocol (TKIP) was developed in the year 2003 and, among other improvements, incorporated an additional per packet hashing algorithm, known as message integrity check (MIC). It is likely to decrypt wireless frames which are protected via Wi-Fi protected access/temporal key integrity protocol (WPA/TKIP). The temporal key integrity protocol (TKIP) attack could give the plaintext data but does not depict the key. This attack lies on the attacker knowing nearly all of the bytes of the IPv4 range on the wireless network [23].

3.6 Cryptocurrency Attacks

A cryptocurrency or sometimes also written as crypto currency is a kind of digital currency designed to work as a medium of trade using cryptography to make safe the transactions and to prove the transfer of currency. For instance, the most common cryptocurrencies are bitcoin, litecoin, peercoin, and namecoin.

Bitcoin is the earliest decentralized digital cryptocurrency, since the bitcoin system works with no central depository or particular administrator [25].

The main reason behind the success of bitcoin is the decentralization of the system. As a replacement of using a central financial institute to control currency, bitcoin makes use of a decentralized network of computers which uses the computational proof-of-work to accomplish agreement on a distributed public ledger of transactions known as *blockchain*.

Blockchains keep data across a network of individual computers creating them not decentralized but distributed as well. Therefore, no single firm or individual possesses the system; however, one and all can use it and help out run it. It is difficult for any one person to take down the network or corrupt it. The individuals who use the system employ their computer to keep bundles of records submitted by others in a sequential chain. The blockchain uses cryptography to ensure that records cannot be counterfeited or changed by anyone else [26].

Many research studies have emphasized techniques to compromise one or numerous bitcoin nodes. In this section, large-scale bitcoin network-level attacks are explored.

3.6.1 Eclipse Attacks on Network of Bitcoin

In this attack, the attacker takes the charge over an access node to information in the bitcoin peer-to-peer network. With appropriate manipulation of the bitcoin peer-to-peer network, an attacker could eclipse a node in order that it is merely communicating with the malicious nodes [27].

To launch this attack, an attacker could maneuver the node in order that all its outgoing TCP connections are to the attacker IP address. This could be achieved in three basic steps: (1) Load the peer tables of node with the attacker IP address, (2) The node starts again and unable to find its existing outgoing connections, and (3) The compromised node makes fresh connections merely to the attacker IP address.

3.6.2 Double Spending Attack

As bitcoin is essentially a digital file, it is possible to duplicate than actual money. As a result, some users could manipulate their way to paying more than once with the same bitcoin. This is referred as double spending. A typical type of double spending attack is known as race attack which is generally exploited by agents and merchants who allow a zero confirmation transaction [28].

In this attack, the fraudulent user sends two transactions in quick succession, one to the trader and another to his own address. In a nutshell, it becomes a chase between two transactions, fifty percent chance of hitting and missing. The fraudulent user could get better his chance of hitting by adding up a bigger fee to his transaction; however, any trader will be able to distinguish that the double spend is happening, and a single confirmation is adequate to avoid it.

3.6.3 ZeroAccess Botnet Bitcoin Mining Attack

ZeroAccess is a type of botnet (Network of Bots) that attacks Windows operating systems. The bitcoin is spread by exploiting the ZeroAccess rootkit malware by numerous attack vectors. The most common attack vector is a type of social engineering, where an end user is convinced to run malicious code either by masquerade it as a genuine file, or attaching in hidden way as an extra payload in an executable file [29].

If a computer is infected with the ZeroAccess botnet, it would start either of the two key botnet tasks: (1) Bitcoin mining: machines involved in bitcoin mining make bitcoins by solving cryptographic problem for their wallet; and (2) click fraud, the computers used for click fraud imitate clicks on advertisements in website paid as pay per click basis.

3.7 Accounts-Linked Attacks (Semantic Attacks)

These types of attacks are performed on the user accounts and these attacks are also referred as semantic attacks. Fundamentally, these attacks do not target the application or any physical infrastructure rather they target the users of application [30]. They are classified into the following categories.

3.7.1 Credential Stuffing

The phrase "Credential Stuffing" consists of two words credential which means claim and proof. Usually, it is implemented by username and password. The second part is stuffing; here, the meaning of stuffing is large-scale stealing. So, the credential stuffing is taking over large number of credential, that is, the username and passwords.

3.7.2 Account Takeover

Unlike credential stuffing, where large numbers of credentials are stolen, in this attack, attacker targets the individual account.

3.7.3 False Accounts

In this attack, the attacker creates fake or false accounts at large scale in order to be able to form fake followers and fake likes as well as to be able to use those fake accounts for other types of purposes like money laundering.

3.7.4 Account Lockout

An account lockout attack is an attractive way for an attacker who does not have credentials for legitimate users on the website, and still be able to harm those users. In this case, they are taking an advantage of the fact that there is a security mechanism that has developed to be able to deal with brute-force attacks on websites which is the 24 h lockout [31].

Numerous web applications implement security mechanism where if the user tries to log into their account unsuccessfully four times, websites presume that the account is under attack and they lock out the account for 12 h or 24 h depending on what rule has been applied. Unluckily, an attacker could go through and attempt with different usernames that are associated with that websites or service and even without legitimate password they could make multiple password attempts and lock out thousands or even millions of users accounts. This creates a different type of problem for the service because now they have to deal with huge number of annoyed and irritated users who are calling them and asking the resetting of the user account passwords.

3.8 *Security Breaches*

A security breach is an event that results in illegal access of applications, network data, web services, and network devices by bypassing their fundamental security mechanisms. Generally, security violation is used for the term security breach. These types of attacks can be further classified as vulnerability scanning and API abuse.

3.8.1 Vulnerability Scanning

The vulnerability scanning is an automated process of constantly identifying security weakness of websites in a network in order to exploit, threaten, and attack the websites. Vulnerability scanning makes use of bots that search for security flaws and compare in a database with identified flaws [32].

3.8.2 API Abuse

The problem of API abuse is a growing one, as attackers are taking interest in the mobile application as a way of targeting APIs (Application Programming Interface) to steal sensitive information. An API Abuse is illegal or unauthorized access to the servers API through mobile applications or web applications. This may result in stealing of precious intellectual property application codes. Mobile application development is a vital area, where the applications are becoming a growing target for the attackers. It is important for both end users and application programmers to ensure that APIs are protected against attacks, and an important measure is ensuring that the application use to access the servers API is recognized and well-known [33].

4 Wireless Defense Mechanism

In this section, the key techniques behind the wireless security and defense mechanisms are discussed. The main purpose of security and defense mechanism is to detect, defend, and recover from the wireless network attacks. In the first part, goals of wireless security such as confidentiality, availability, integrity, access control, and non-repudiation are discussed.

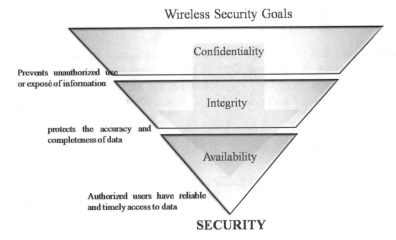

Fig. 13 Wireless security goals

4.1 Wireless Security Goals

Similar to information security and network security, the wireless security has three main goals: confidentiality, integrity, and availability [34]. This is usually referred as CIA model and its pictorial representation is shown in Fig. 13.

4.1.1 Confidentiality

Confidentiality guarantees that only authorized end users with sufficient privileges can view the information. To achieve the confidentiality in wireless networks, numerous encryption techniques are used such as Wi-Fi protected access.

4.1.2 Integrity

Integrity guarantees that the data stored on devices and data being in transit is correct, and no unauthorized persons or malicious user has altered the data. It is probably more critical than either confidentiality or availability measures.

As data frames in wireless networks are transmitted via the medium of air, they could be intercepted and altered with no trouble by the attackers. This means that integrity of data in wireless networks is additionally vulnerable to attacks. To protect integrity in wireless networks, error checking methods such as checksums and file hashing are used.

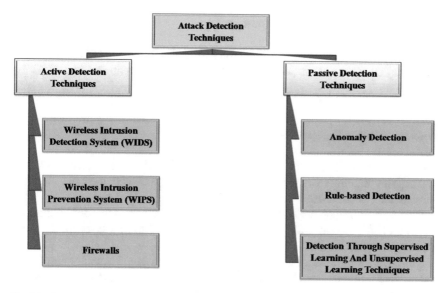

Fig. 14 Taxonomy attack detection techniques in wireless network

4.1.3 Availability

Availability is pretty straightforward; it means that wireless network resources are readily available to authorized users. However, a safe computer must provide limited access attempts by illegal users. It must allow immediate access to authorized users, for example, a banking client should be able to check their balance or withdraw their money in a timely manner.

Wireless networks are typically susceptible to DoS. Different from the wired networks, it necessitates the attacker to be physically coupled with the network in a way before the attacker could launch an attack.

4.2 Attack Detection Techniques

Depending on the methodology applied in detection of wireless attacks, they can be classified as active detection technique and passive detection technique. Taxonomy of attack detection techniques in wireless network is shown in Fig. 14.

4.2.1   Active Detection Techniques

Active detection techniques in wireless networks are further classified as wireless intrusion detection system (WIDS), wireless intrusion prevention system (WIPS), and firewalls.

Wireless Intrusion Detection System (WIDS)

An intrusion detection system (IDS) is any physical device or software application that actively monitors the wireless network for malicious activities and alerts the administrator when it detects an attack [35].

There are different types of intrusion detection system, each one monitoring incoming and outgoing traffic, notifies security administrators of abnormal activity, and generates reports.

Moreover, a number of intrusion detection system could actually respond to and prevent attempted attacks. The majority of intrusion detection system identifies threats using two common techniques, signature-based detection and anomaly-based detection. Similar to any antivirus software, intrusion detection system keeps a list of malware signatures. It compares incoming threats to this list and blocks any malicious request that is on the list. Intrusion detection system also examines the wireless network for any abnormalities. It identifies abnormalities by establishing a system baseline and looking for variations from that baseline.

In general, WIDS monitors a wireless local area network using a combination of hardware and software known as intrusion detection sensors. The sensor resides on the IEEE 802.11 network and monitors all wireless network traffic. The main problem while installing WIDS is to make a decision on the suitable place to locate the intrusion detection sensors. Typical examples of wireless intrusion detection system are AirMagnet, AirDefense, and Red-M.

Wireless Intrusion Prevention System (WIPS)

Wireless intrusion prevention system (WIPS) is generally a network device that examines the radio frequency spectrum for the existence of unauthorized access points or rogue access point, and be able to take countermeasures automatically.

The main idea of a WIPS is to check unauthorized network access to wireless local area networks. These systems are usually implemented as a cover to wireless LAN, even though they could be implemented standalone to inflict no wireless rules within business. A number of advanced wireless infrastructure has inbuilt WIPS capabilities [36].

Firewalls

A firewall is a security mechanism in wireless networks that constantly monitor and manage the wireless traffic both the incoming and outgoing based on predefined security policies. A firewall normally set up a boundary between a genuine, protected internal network and outside network. Firewalls could be made in hardware as well as software or sometimes a mixture of both. Firewalls are often used to check illegal users from gain an access to wireless network of some organization linked to the Internet, particularly the Intranet. Every incoming or outgoing message of intranet goes through the firewall, which monitors a message and stops those that do not qualify the security criteria.

Firewalls are generally categorized as network firewalls, host-based firewalls, and web application firewall. Network firewalls set up a boundary between two networks. Host-based firewalls give an extra layer of software on the system that monitors network incoming and outgoing of that single machine. A third category is specifically made for the web application is web application firewall (WAF).

4.2.2 Passive Detection Techniques

Numerous passive detection techniques using machine learning techniques, such as supervised learning, unsupervised learning, rule-based classification, and anomaly-based detection can be used effectively and efficiently to detect attacks on wireless networks [37].

Anomaly Detection

In anomaly detection methods, they try to distinguish attack traffic from genuine traffic based on a number of network traffic variance, i.e., anomalies, for instance, large volume traffic, high latency of network, traffic of unusual ports, and abnormal behavior of system that may specify the presence of attacks in the wireless network. The effective method for attack detection is to find out the attack traffic from genuine traffic.

Rule-Based Detection

Information of patterns and behavior of known attacks is used for attack detection and identification. For instance, SNORT is the most common IDS that monitors the traffic of network to identify the patterns of intrusions. Similar to all the intrusion detection system, Snort can be configured with signatures or patterns or a set of rules or to keep a record of traffic which is considered as suspicious. On the other hand, rule-based detection techniques mostly could be used for identification and detection of known wireless attacks. As a result, this technique is not helpful for detecting the unknown wireless attacks.

Detection Through Supervised Learning and Unsupervised Learning Techniques

Supervised learning and unsupervised learning techniques are based on machine learning algorithms such as classification including decision trees, random forest, K-nearest neighbor, and clustering techniques including flat and hierarchical clustering.

5 Conclusion

As described in the chapter, although there are tremendous advantages of using wireless networks, there are a lot of practical issues related to wireless security needs to be solved. Similar to any technology, several security issues confront wireless networks. In this chapter, security mechanism of (Detection methods and prevention methods) wireless networks from attacks are reviewed, and the primary vulnerabilities in wireless networks are also discussed. It is revealed that these attacks have a severe impact on wireless networks and the attacks may lead to the serious problem for wireless networks. Nearly, all central attacks in wireless networks along with the possible impacts and the available countermeasures have been described.

References

1. Pahlavan, K. (2011). Principles of wireless networks: A unified approach. John Wiley & Sons, Inc.
2. Welch, D., & Lathrop, S. (2003, June). Wireless security threat taxonomy. In Information Assurance Workshop, 2003. IEEE Systems, Man and Cybernetics Society (pp. 76–83). IEEE.
3. Nicopolitidis, P., Pomportsis, A. S., Papadimitriou, G. I., & Obaidat, M. S. (2003). Wireless networks. John Wiley & Sons, Inc.
4. Cali, F., Conti, M., & Gregori, E. (2000). IEEE 802.11 protocol: design and performance evaluation of an adaptive backoff mechanism. *IEEE journal on selected areas in communications*, *18*(9), 1774–1786.
5. Forouzan, B. A. (2002). *TCP/IP protocol suite*. McGraw-Hill, Inc.
6. Stallings, W., & Tahiliani, M. P. (2014). *Cryptography and network security: principles and practice* (Vol. 6). London: Pearson.
7. Shivaraj, G., Song, M., & Shetty, S. (2008, November). A hidden Markov model based approach to detect rogue access points. In *Military communications conference, 2008. MILCOM 2008. IEEE* (pp. 1–7). IEEE.
8. Gast, M. (2005). *802.11 wireless networks: the definitive guide*. "O'Reilly Media, Inc.".
9. Stubblefield, A., Ioannidis, J., & Rubin, A. D. (2004). A key recovery attack on the 802.11 b wired equivalent privacy protocol (WEP). *ACM transactions on information and system security (TISSEC)*, *7*(2), 319–332.
10. Khabbazian, M., Mercier, H., & Bhargava, V. K. (2009). Severity analysis and countermeasure for the wormhole attack in wireless ad hoc networks. *IEEE Transactions on Wireless Communications*, *8*(2), 736–745.
11. Berghel, H. (2004). Wireless infidelity I: War driving. *Communications of the ACM*, *47*(9), 21–26.

12. Lawrence, E., & Lawrence, J. (2004, April). Threats to the mobile enterprise: jurisprudence analysis of wardriving and warchalking. In *Information Technology: Coding and Computing, 2004. Proceedings. ITCC 2004. International Conference on* (Vol. 2, pp. 268–273). IEEE.

13. McNab, C. (2007). *Network security assessment: know your network*. "O'Reilly Media, Inc.".

14. Kanawat, S. D., & Parihar, P. S. (2011). Attacks in wireless networks. *International Journal of smart sensors and Ad Hoc Networks*, *1*(1), 113–6.

15. Kizza, J. M. (2009). *Guide to computer network security* (pp. 2007–2008). Springer.

16. Meng, K., Xiao, Y., & Vrbsky, S. V. (2009). Building a wireless capturing tool for WiFi. *Security and Communication Networks*, *2*(6), 654–668.

17. Bellardo, J., & Savage, S. (2003, August). 802.11 Denial-of-Service Attacks: Real Vulnerabilities and Practical Solutions. In *USENIX security symposium* (Vol. 12, pp. 2–2).

18. Sun, Y. Q., Wang, X. D., & Zhou, X. M. (2012). Jamming attacks in wireless network. *Ruanjian Xuebao/Journal of Software*, *23*(5), 1207–1221.

19. Richards, K. (1999). Network based intrusion detection: a review of technologies. *Computers & Security*, *18*(8), 671–682.

20. Kumar, S. (2007, July). Smurf-based distributed denial of service (ddos) attack amplification in internet. In *Internet Monitoring and Protection, 2007. ICIMP 2007. Second International Conference on* (pp. 25–25). IEEE.

21. Hwang, H., Jung, G., Sohn, K., & Park, S. (2008, January). A study on MITM (Man in the Middle) vulnerability in wireless network using 802.1 X and EAP. In *Information Science and Security, 2008. ICISS. International Conference on* (pp. 164–170). IEEE.

22. Hole, K. J., Dyrnes, E., & Thorsheim, P. (2005). Securing wi-fi networks. *Computer*, *38*(7), 28–34.

23. Lashkari, A. H., Danesh, M. M. S., & Samadi, B. (2009, August). A survey on wireless security protocols (WEP, WPA and WPA2/802.11 i). In *Computer Science and Information Technology, 2009. ICCSIT 2009. 2nd IEEE International Conference on* (pp. 48–52). IEEE.

24. Tews, E., & Beck, M. (2009, March). Practical attacks against WEP and WPA. In *Proceedings of the second ACM conference on Wireless network security* (pp. 79–86). ACM.

25. Zhang, R., & Preneel, B. (2017, January). On the Necessity of a Prescribed Block Validity Consensus: Analyzing Bitcoin Unlimited Mining Protocol. In *International Conference on emerging Networking EXperiments and Technologies-CoNEXT 2017*. ACM.

26. Nadal, M. (2017). Critical mining, Blockchain and Bitcoin in contemporary art.

27. Apostolaki, M., Zohar, A., & Vanbever, L. (2017, May). Hijacking Bitcoin: Routing attacks on cryptocurrencies. In Security and Privacy (SP), 2017 IEEE Symposium on (pp. 375–392). IEEE.

28. Lin, I. C., & Liao, T. C. (2017). A Survey of Blockchain Security Issues and Challenges. IJ Network Security, 19(5), 653–659.

29. Putman, C. G. J. (2017). Business Model of Botnets.

30. Heartfield, R., Loukas, G., & Gan, D. (2017). An eye for deception: A case study in utilising the Human-As-A-Security-Sensor paradigm to detect zero-day semantic social engineering attacks.

31. Berghel, H. (2017). A Quick Take on Windows Security Evolution. Computer, 50(5), 120–124.

32. Wang, Y., & Yang, J. (2017, March). Ethical Hacking and Network Defense: Choose Your Best Network Vulnerability Scanning Tool. In Advanced Information Networking and Applications Workshops (WAINA), 2017 31st International Conference on (pp. 110–113). IEEE.

33. Shirazi, H., Hadavi, M. A., & Hamishagi, V. S. Vulnerability Prevention in Software Development Process.

34. Padmavathi, D. G., & Shanmugapriya, M. (2009). A survey of attacks, security mechanisms and challenges in wireless sensor networks. *arXiv preprint* arXiv:0909.0576.

35. Pleskonjic, D. (2003, December). Wireless intrusion detection systems (WIDS). In *19th Annual Computer Security Applications Conference*.

36. Timofte, J. (2008). Wireless intrusion prevention systems. *Revista Informatica Economica*, *47*(3), 129–132.

37. Gjomemo, R., Malik, H., Sumb, N., Venkatakrishnan, V. N., & Ansari, R. (2014, March). Digital check forgery attacks on client check truncation systems. In *International conference on financial cryptography and data security* (pp. 3–20). Springer, Berlin, Heidelberg.

Spectrum Decision Mechanisms in Cognitive Radio Networks

Rafael Aguilar-Gonzalez and Victor Ramos

Abstract An open issue in cognitive radio networks (CRNs) is spectrum decision, which is the capability of a cognitive radio to efficiently choose a spectrum band to accomplish the quality of service (QoS) requirements of secondary users (SU) so as not to interfere primary users (PU). A complete mechanism for spectrum decision must take into account a detailed set of information parameters, ranging from spectrum occupancy statistics to the final spectrum allocation for an SU. Spectrum decision is a very important issue in CRNs; however, to date, there is still plenty of research work to do. One solution for such a process that has attracted a lot of attention is based on multiple attribute decision-making (MADM) mechanisms fed with actual information of spectrum occupancy. In this chapter, we provide a brief review of several techniques for spectrum decision in CRNs. We describe the main mechanisms that have been proposed by providing a comparative characterization among them, as well as an overview of the affordability of such mechanisms according to the demands for SUs. Finally, we discuss the impact on CRNs of emerging trends such as cloud CRN and Internet of Things (IoT) in cognitive radio.

1 Introduction

Wireless communications have gained a lot of significance among their users, technologies like Wi-Fi, WiMax, cellular telephony, and Bluetooth are some examples of the most used services. Due to the high demand for these technologies, there is a great need to perform efficient transmission mechanisms in order to achieve a good Quality of Service (QoS). However, the capacity of the current and forthcoming telecommunications services depends mainly on a finite resource, which is the electromagnetic spectrum. The fixed spectrum access (FSA) technique has been the most usual choice to allocate the available spectrum; nevertheless, this has caused

R. Aguilar-Gonzalez (✉) · V. Ramos (✉)
Universidad Autonónoma Metropolitana, Mexico City, Mexico
e-mail: r.aguilar@xanum.uam.mx

V. Ramos
e-mail: vicman@xanum.uam.mx

© Springer Nature Singapore Pte Ltd. 2018
K. V. Arya et al. (eds.), *Emerging Wireless Communication and Network Technologies*,
https://doi.org/10.1007/978-981-13-0396-8_14

spectrum scarcity [4]. According to [40], parts of the spectrum spaces considered by FSA are used between 15 and 80%, and there is work validating such unoccupied range with spectrum measurement campaigns [12, 26, 37]. Thus, the presence of several spectrum holes is clear; such holes may be temporarily occupied by other users, consequently allowing a better usage of the electromagnetic spectrum.

Dynamic spectrum access (DSA) is a technique to reduce spectrum scarcity. Such a mechanism adapts in real time the spectrum occupancy depending on the environment. According to the FSA policies, the licenses to transmit in a particular frequency are allocated to primary users (PU). However, in order to improve efficiency, using DSA whenever a PU reduces its transmission during a certain time period, another user called secondary user (SU) is allowed to use temporarily such frequency. SUs are able to occupy opportunistically the spectrum even if they have or have not allocated a transmission frequency [4, 23].

Cognitive radio (CR) is a technology able to execute DSA. CR has been proposed as an alternative during the last few years to achieve the necessary tasks in DSA [38]. A widely accepted definition for CR is the following: "A Cognitive Radio is a radio that can change its transmission parameters based on interaction with the environment in which it operates" [4]. A vital part of CR is software-defined radio (SDR), which is a reconfigurable system with the ability to modify parameters such as transmission frequency, transmission power, and modulation, just to mention a few [3]. Hence, it is possible to assess the spectrum holes considering CR as a serious candidate.

Communication among several CR devices is supported by a cognitive radio network (CRN). Basically, a CRN is composed at least of a primary and a secondary network. The primary network is where all PUs are distributed, while the secondary network is where all SUs or CR users are deployed [3]. A successful CRN achieves a good interaction among PUs and SUs, with a low user interference rate, providing quality of service and keeping seamless communications. A well-studied framework for CRNs may be found in [3]. Such a framework executes a cognitive cycle in four main functions, as it is shown in Fig. 1.

The definitions of these functions are the following. First, we have *spectrum sensing*, which determines the occupation state of the frequency spectrum [3, 4]. Then, *spectrum sharing* coordinates activities of PUs and SUs in order to minimize interferences among them [3, 4]. Next, *spectrum mobility* whose purpose in CRNs consists in carrying out the spectrum hole transition in a fast and correct fashion [3, 4]. Finally, *spectrum decision* where the selection of spectrum holes is achieved in order to fulfill SUs links requirements. An adequate spectrum allocation mechanism based on SU's needs allows to increase the spectral efficiency [3, 4]. Such a function is described in next section.

2   Spectrum Decision Overview

Spectrum decision is the ability to select a spectrum hole satisfying the quality requirements of SUs. The procedure is as follows: once spectrum holes are detected, it is necessary to decide which one of them is the best according to the quality

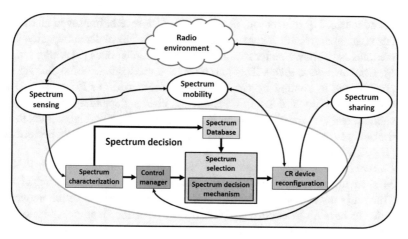

Fig. 1 Relation among the four functions of spectrum management and some of the main task of spectrum decision

parameters required by the SU without causing interference with the PU. Next, it is necessary to reconfigure transmission parameters in order to reach the new spectrum hole [31, 36]. We describe the three main tasks of spectrum decision and two recent reinforcement steps. The relation among the tasks of the spectrum decision and the other functions are shown in Fig. 1.

2.1 Spectrum Characterization

The knowledge of the characteristics of the available frequency bands is quite important. A robust characterization of spectrum holes increases the probability of making a right decision. We describe here the steps of spectrum characterization.

- Characterization of the radio-electric environment: Since the available spectrum holes are dynamically changing, it is necessary to estimate some parameters:
  - *Channel identification*: The main task is to identify the channel assigned to PUs. These channels are classified according to the traffic, deterministic, or stochastic. For the case of deterministic traffic, the process is expressed as an ON or OFF digital signal. However, for the case of stochastic traffic, it is necessary to compute some probabilities because of the intrinsic variation of time and space, as is the case of cellular networks. For both cases, it is possible to predict PUs behavior and apply the right spectrum decision method [19, 24].
  - *Channel capacity estimation*: With the knowledge of the spectrum hole size, we may be able to estimate if such space will fulfill or not the SU requirements. A traditional parameter estimation mechanism is the signal-to-noise ratio; however, it has been shown that the precision of the results with such a method is

not adequate. For this reason, new methods with several improved characteristics have emerged, for instance, they compute the normalized capacity. These methods have shown better results than the traditional ones [31, 50].

- *Channel switching delay*: This parameter is a consequence of switching from one channel to another by channel quality degradation or PU presence. This causes an additional delay in CRNs; it can change depending on the physical capacity and the type of algorithms implemented by the SU. It is essential to keep the delay as small as possible in order to keep an acceptable CRN performance [16, 22].
- *Channel interference estimation*: CRNs should coexist mainly with PUs and SUs, but also with several CRNs or primary networks around a particular zone. Thus, it is necessary to measure the interference generated in the network in order to be able to control it. With this information, it is possible to know the quality of the network and the available channels. The QoS and error rate depends straightly from the interference [15, 42, 57].

- PU activity modeling: Spectrum hole apparitions are not guaranteed, thus once an SU has selected an idle space, a PU could appear, reducing the provided quality of service. In order to reduce this effect, it is crucial to have a knowledge about the environment and predict PU behavior. By getting this information, it is possible to improve the network performance. Also, modeling PU activity may increase spectrum occupancy efficiency. Modeling depends on characterization and may be divided into the following areas:

  - *Poisson modeling based on PU activity*: This type of traffic has been well studied for some years. It is modeled as a two-state process; user apparition is an independent and identically distributed random variable assuming a Poisson process for arrivals. When a channel is occupied, it is represented as an ON state; consequently, an OFF state means an idle channel. Most of the applications assume this behavior. However, there is some evidence that it is not completely applicable; as a conclusion, Poisson modeling is not always recommended for practical applications [11].
  - *Modeling from real data*: The electromagnetic spectrum provides information for a better understanding of real PUs behavior. This information is measured through spectrum measurement campaigns. From the data spectrum, channels are classified as available or occupied and, according to the behavior, as stochastic or deterministic channels [6, 47]. Also, this information is useful for developing empirical distribution functions, among other possibilities [33, 34].

2.2 Spectrum Selection

After spectrum characterization of spectrum holes, it is essential to select them based on SU quality requirements. Due to constant changes in the available spaces, the network topologies change as well. In this section, centralized and distributed CRNs are described.

- *Centralized spectrum selection in CRN.* The IEEE 802.22 standard is one of the most common examples for spectrum selection, and it is charged with dynamic channel management [14, 49]. Spectrum availability depends on TV channels occupation. In this case, base stations and access points are considered; bandwidth is between 6 and 8 MHz. An important challenge is channel fragmentation of available channels, which varies depending on the number of TV channels [18].
- *Distributed spectrum selection in CRN.* This topology faces several challenges because of the several jumps caused by channel properties changes. To date, some of the main metrics in this subject are spectrum conjunction and design selection [29, 46].

2.3 CR Device Reconfiguration

Traditional communication devices are designed to operate on a particular frequency for transmission. However, in order to get a better spectrum occupancy, it is necessary to consider systems with the ability to adapt their parameters according to the environment. This process is executed after those of characterization and selection. Afterward, a parameter classification should be done when making a decision, such as the following:

- *Reconfigurable parameters*: Depending on SU quality requirements, some parameters should be reconfigured. For example, the schemes of modulations have to be adaptive to increase the data rate [39]. The transmission power control is implemented to reduce co-channel interference, maximize capacity, and reduce energy consumption [32]. The channel bandwidth size and the transmission frequency are some common parameters to be modified in CRNs [7, 56].
- *Energy efficiency*: The energy consumption is an open issue in all wireless communications systems. This parameter is highly important in CRNs due to the number of activities performed. The CRN energy consumption depends mainly on the network environment. Thus, several external variables should be considered in the CRN design. However, good energy savings are achieved by adjusting radio components characteristic [20].

2.4 Reinforcement Steps for Spectrum Decision

The following are additional steps often made in order to strengthen spectrum decision:

- *Control manager*: In a CRN, several SUs compete for a spectrum hole. However, due to the network capabilities, there is not enough space for all of them. In order to keep the QoS for current SUs and to avoid unnecessary operations, a control manager is needed. The control consists in allowing the access of SUs to the network. Some recent contributions consider this subject. Information coming from spectrum sharing and channel characterization provides a better panorama for this manager [9, 35].
- *Spectrum databases*: The recent administration of big data is suitable in CRNs. All the useful information generated in the management of the CRNs should be stored in order to improve the performance of all functions. If data from spectrum measurements, channels characterizations, PU apparitions, and other parameters are stored in a database, it could help to increase the certainty in spectrum decision or other functions as spectrum sensing. Recent work has shown databases as a trend in several wireless communications systems [17, 55].

3 An Overview of Spectrum Decision Frameworks

3.1 Definition of Spectrum Decision Frameworks

Several functions need to be executed in order to guarantee an efficient spectrum decision. However, they require to be coordinated because of the constant information exchange between adjacent functions and among internal decision functions. Spectrum decision frameworks work dynamically to provide an efficient decision-making process and to respond to all possible perturbations in the communication process. In this chapter, some representative spectrum decision frameworks are described. Also, a spectrum decision framework with an MADM mechanism is introduced.

3.2 Framework 1: A Spectrum Decision Framework for Cognitive Radio Networks

The proposal presented in [31] is one of the first frameworks focused on spectrum decision. Among its several features, this structure is characterized by a strong connection between spectrum sharing and spectrum sensing. The authors consider two types of applications, decision events and a capacity model. The main contribution of this work is the introduction of two decision schemes based on the application.

3.2.1 System Model

In this work, a secondary centralized network is considered, where SUs with SDR capabilities are deployed inside the coverage range of a base station which controls them. SUs perform spectrum sensing and send such information to the base station that characterizes spectrum activity. Uplink and downlink signals are split through a frequency division duplex (FDD) system. Authors consider multiple non-contiguous bands to solve the issue of communications break occurring when an SU has to move to another frequency band to transmit. This system model is deployed in the next framework.

3.2.2 Framework Characteristics

The network status is classified according to the capacity: underloaded (available space), overloaded (middle available), and outage state (not available space). PUs are modeled according to an exponential distribution, with busy and idle states. This behavior is considered for the cognitive radio capacity model proposal.

   The full process of the frameworks starts according to the next event. The structure of the framework is composed basically of three parts: resource manager, spectrum decision, and event detection. First, a decision is made when one of the next events occurs: A new CR user wants to transmit in the network, a PU appears in the spectrum band assigned, and the quality of a channel does not satisfy CR user requirements. Afterward, if a CR user wants to access the network, it should pass through the next block.

- *Resource Manager*: This is the first filter to access the network. This block is composed of two parts: admission control and decision control. The admission control part should decide if an SU is accepted or not for transmitting on the CRN. Here, the SU characteristics are measured; with this information, the network computes the network degradation caused by the SU. Afterward, based on the outage probability, an admission criteria unit decides if the new SU is accepted or not.
  Decision control is executed for SU or PU apparitions; there is a specific scheme for each case where the state of the network is considered. If the network may provide an affordable space for the SU, it is passed to the decision schemes explained in the next section. For the case of a PU apparition, contrary to the SU case, the occupancy is checked and SUs are redistributed in order to keep the provided service.
- *Event detection*: This block considers spectrum sensing and monitoring of quality. SUs are the only ones that can execute event detection; they also provide the spectrum sensing part. When an event as a PU apparition is detected, the CRN has

to reconfigure the spectrum assigned in order to maintain the QoS. This part is related to spectrum sharing and is the main source of information to the resource manager unit.

3.2.3 Spectrum Decision Mechanism

In this framework, the spectrum decision is applied according to the application type: real-time or best-effort. Both applications are described here:

- *Real-time applications*: These applications require a spectrum space with low latency and low jitter. Thus, it is strictly precise to have a stable frequency in order to guarantee the QoS offered. The same scheme is applied to single and multiple selections. In the case of multiple selections, users are sorted from highest to lowest loss rate, where users with high loss rate select the best capacities. In this method, there are three steps:

  - **Spectrum selection**: by using a linear integer optimization, SUs select spectrum bands, where the total number of SU's transceivers, SU's capacity, PU's activity, and currently available spectrum is considered. To satisfy real-time applications, available spectrum bands and the spectrum occupied by best-effort applications are considered in this selection.
  - **Resource allocation**: In this step, the variance minimization of the total capacity of the network takes place. Here, for each SU, the data loss rate and the capacity are considered. First, the CRN obtains the expected capacity, then the variance, and finally the minimization.
  - **QoS checkup**: This step is executed when the target loss rate is not achieved after the other two steps. In this case, there are two strategies to compensate this issue: aggressive and conservative approaches. The first one tries to find the best space for SU requirements. If it does not work, the conservative approach adapts and reduces the data rates.

- *Best-effort applications*: These applications also try to maximize the capacity of the network; however, here the PU activity and long-term channels are considered. This method works for simple or multiple selections. Basically, this proposal avoids the optimal maximum capacity method by considering a complex process and computes the decision again. Such a gain is defined as the sum of the expected capacity gain when a new SU joins the CRN, minus the expected capacity loss of other SUs over the available bandwidth.

3.3 Framework 2: A QoS-Aware Framework for Available Spectrum Characterization and Decision in Cognitive Radio Networks

The spectrum decision framework presented in [10] has a strong relation with spectrum sensing and spectrum mobility. Basically, this proposal considers two new parameters that help to increase network performance. The first one, called opportunity index, considers PU fluctuations. The second one is named request index and captures the SUs QoS requirements. The structure looks to provide good QoS to an SU while maintaining the overall fairness among all the SUs. An important contribution of this work is the inclusion of four types of traffic in the network.

3.3.1 Framework Characteristics

The decision framework is shaped by five modules: PU activity, QoS-aware, admission control, spectrum mobility, and spectrum decision. The former four are described next, while the latter is explained in the next subsection.

- PU activity module: all spectrum bands are modeled and analyzed separately. Here, the parameter called *opportunity index* is proposed. It computes the relation between available and occupied spectrum. This parameter shows what portion of the spectrum is available to be used by SUs.
- QoS-aware characterization module for CR users: Here, another parameter called *index request* is computed. It is a ratio between an instantaneous SU request and the maximum number of requests of all SUs. Authors classify SUs into four types with the following priorities: 1—constant bit rate, 2—video conference, 3—voice over IP, and 4—best-effort users.
- Admission control: this module stabilizes the QoS requests of new CR users. Before admitting a new user, an evaluation about its QoS requirements is done. If the network has the capacity of fulfilling these requirements, the CR user is accepted. The main target here is to not degrade the current QoS.
- Spectrum mobility: This module provides information about the CR users that are moving in order to send such information to the spectrum decision module. Basically, this information is about the current opportunities for CR users and so the spectrum mobility module decides if the user should move or stay in the same space.

3.3.2 Spectrum Decision Mechanism

This part collects information from all the other blocks; here, the proposal parameters help during the decision-making process. Three types of decision are considered: *perfect* decision, *smooth* decision, and *aggressive* decision. An algorithm makes the decision according to the SUs QoS requirements; this is repeated until all SUs receive a frequency space.

The algorithm works for a specific number of SUs in a determined number of spectrum bands. When the requested index is lower than the opportunity index, a perfect decision is made. For a smooth decision, the priority and request indexes have to be lower than the opportunity index. Finally, an aggressive decision is made when the opportunities are reduced. This last case means that there is a high demand for spectrum bands with low availability. The results of the algorithm are returned in terms of fairness and throughput.

3.4 Framework 3: A Belief-Based Decision-Making Framework for Spectrum Selection in Cognitive Radio Networks

The work in [43] considers a CRN architecture where a belief decision-making is a central part. The current work contributes to an interference model that helps to decide which spectrum band will fulfill the SU requirements. Also, the dynamic smart scenario characterization is considered and an eigenvalue-based metric is proposed. Finally, the framework is evaluated in different scenarios and compared with other frameworks of the literature.

3.4.1 System Model

The authors consider a centralized network with a set of radio links or SUs and a base station. The communications process can be possible among users or among users and a base station. Each SU has a determined data rate with a specific duration time. Also, the spectrum is organized in sets, each with a central frequency and a bandwidth size. The selection process avoids internal interferences; however, external transmissions may produce interference in the system considered.

3.4.2 Framework Characteristics

The framework is a centralized cognitive management entity composed of four blocks. One of them is in charge of control tasks between blocks, and another one requests spectrum measurements to the nodes of the network. The last two blocks are responsible for the main tasks, which are the following:

- Decision-making: This is the main block of the framework. Also, it presents a full interaction with other entities to process the information needed for selection.

Here, two other blocks have been placed: a spectrum selection decision-making unit that is explained in the next subsection and the observations strategy decision-making unit. This unit decides when the spectrum measurements need to be made by each user.

- Knowledge management: this block acquires and processes spectrum measurements from the environment. Also, it stores and provides important information for the other blocks of the framework. Knowledge management is formed by three parts: knowledge database, knowledge manager for processing, and knowledge manager for acquisition.

3.4.3 Spectrum Decision Mechanism

The spectrum selection decision-making unit operates with the following consideration. It allocates a spectrum hole each time that a new radio link session is started. This mechanism does not consider spectrum sharing, thus when there is no available space for a new user, it is blocked. The spectrum decision here is done considering the measured interference spectral density, where a discrete-time Markov process is used to model the interference evolution.

Based on the interference of each of the spectrum holes, each user will receive a reward that depends on the obtained performance. The reward is a metric that points out the convenience of a spectrum hole. Thus, the spectrum selection mechanism maximizes the vector of rewards considering the time of duration sessions among a group of spectrum holes. In order to predict the reward along the link duration, decision functions are generated. These functions consider spectrum measurements and the interference characterization in order to estimate the spectrum reward of the spectrum hole in a future instant of time. Also, these functions consider a belief vector where the conditional probabilities of spectrum holes are allocated. The decision functions have three modalities called *strategies*, each one according to the type of selection, duration time, and belief vector.

3.5 Framework 4: Reducing Spectrum Handoffs and Energy Switching Consumption of MADM-Based Decision in Cognitive Radio Networks

3.5.1 System Model

A centralized CRN coexists along a centralized primary network where users for both networks are considered in [1]. Each SU has two tasks: it performs spectrum measurements and sends its spectrum requirements through a connection profile.

The secondary base station processes this information and sends it to the spectrum decision framework. There, the main decision is made and the result is sent using the base station to the SU. The contributions of this work are the consideration of extracting information from spectrum measurements, reduction in energy consumption, the inclusion of MADM algorithms as decision mechanism, and the adaptation of this to CRNs.

3.5.2 Framework Characteristics

In this framework, spectrum decision takes information from spectrum sensing and shares information with spectrum mobility. This framework considers the next blocks: power spectrum measurements, user connection profile, power spectrum measurements processing, spectrum repository, spectrum decision with MADM algorithms, and a *Comparison Function*. A description of all of them is presented here. Also, Fig. 2 shows the interaction among the participants of this framework.

- Power spectrum measurements: In several frameworks, the PUs behavior is simulated with a particular probability distribution function (PDF). In order to get good results in implementation, it is preferable to consider real data information. In this regard, spectrum measurement campaigns have been used just for knowing spectrum occupancy. However, a recent trend uses spectrum samples to generate models of the behavior of PUs. The information coming from a spectrum measurement campaign is included as spectrum sensing. The power spectrum samples were taken during each second for 3 days in three observation points in several frequencies as it is mentioned in [1]. Each time a decision is requested, an amount of power spectrum samples is analyzed in order to know the current spectrum occupancy.
- Processing of power spectrum measurements: In this type of studies, most of the time the main target consists in knowing the power spectrum. Besides, by analyzing this information it is possible to get more data in order to characterize spectrum activity. Each time a request of a spectrum hole occurs, the spectrum measurements are averaged and analyzed as follows:

  - *Bandwidth size (bw)*: The importance of this parameter lies in the size of the bandwidth of the spectrum hole. The channel capacity is proportional to the bandwidth, thus larger is better. For any transmission, a larger bandwidth represents a higher data rate. An algorithm presented in [1] is used to find spectrum holes considering spectrum occupancy. Every time a decision is made, a set of spectrum holes appears where only spectrum holes from 5 to 8 MHz are considered. Each of them is characterized by a central frequency and a particular size.
  - *Stability of spectrum holes (st)*: This is one of the additional parameters that are hidden in spectrum measurements. A spectrum hole can be attractive in bandwidth size; however, this size can change frequently. This parameter measures

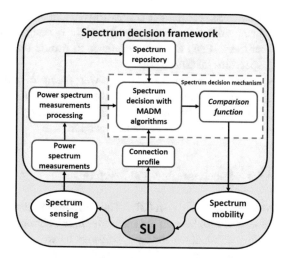

the changing rate of a spectrum hole each time a decision is made. The ratio
obtained provides information to the decision mechanism. A more stable spectrum hole is better because SUs will avoid frequently switching from one space
to another.

- *Power index (pi)*: Avoiding interferences between PUs and SUs is highly important to guarantee an efficient network functionality. This parameter shows the
  power spectrum ratio when a PU is present and absent. The target here is to
  quantify how much an SU transmission can be affected by a PU apparition. A
  higher power index means that there is not a big change in the power spectrum
  after a PU apparition.
- *Duty cycle (dc)*: This parameter provides the percentage of occupied spectrum
  band. After setting a threshold for each power spectrum sample, the result indicates if a frequency is occupied or available. In general with this parameter,
  it is possible to know the percentage of time a spectrum band is occupied [1].
  A lower value of this parameter is better, meaning that the spectrum band is
  available most of the time.
- *Interference temperature (it)*: A spectrum hole can be available; however, its
  neighbors can interfere with this band. This parameter computes the rate of
  interference caused by PUs adjacent to the spectrum hole. After knowing the
  location of the spectrum hole along with the bandwidth size, it is possible to
  compute this parameter.
- *Selected frequency (sf)*: A transmission with a specific power can reach a higher
  covering area if it is done in lower frequency bands rather than in higher ones [1]
  This parameter assigns high weights to spectrum holes in low frequencies and
  small weights in high frequencies.

- User connection profile: Each SU has different requirements, in this proposal their
  preferences are adjusted to each of the parameters mentioned before. For example,

for a video conference service, an SU might be interested in the largest and stable bandwidth size. Thus, a high weight is assigned to these parameters among all of them. Each SU has a connection profile that is sent at the same time as the spectrum measurements.

- Spectrum repository: After power spectrum measurements processing is performed, this module stores information each time that a decision is requested. The spectrum samples analyzed are utilized by the spectrum decision mechanism which is explained in the next section.

3.5.3   Spectrum Decision Mechanism

MADM algorithms in CRN. These algorithms have been considered in the solutions of several problems in the literature. In telecommunications, they have been applied mainly in handover for heterogeneous networks. Nevertheless, this decision tool is almost unexplored for CRN. In the first approach presented in [2], MADM algorithms exhibit a good performance according to the SU demand. However, in order to fulfill SU requirements, MADM algorithms execute a considerable number of frequency changes or spectrum handoffs. This phenomenon named the *ping-pong effect* can reduce drastically the performance of the CRN and consumes a reasonable amount of energy. Thus, in order to improve MADM algorithms, a function is proposed. In the next, the MADM algorithms considered in this work are presented and the decision function is explained.

A presentation of MADM algorithms. MADM algorithms are a branch in the multi-criteria decision-making (MCDM) field [48]. MADM provides a decision as long as the problem to be solved can be raised through alternatives, multiple attributes describing the alternatives in different units, and a set of weights that rank the importance of each attribute.

Among many MADM algorithms in the literature, we select three of the most common ones for testing them in CRNs. A decision matrix is set with the parameters mentioned previously. For this case, there are J parameters such as *bandwidth*, *stability*, and *power index* corresponding to **benefit parameters**, while *duty cycle*, *interference temperature*, and *selected frequency* are **cost parameters**. High values of the first three parameters are preferred, while for the others low values are desired. Each spectrum hole is an alternative, k; these parameters are called *attributes*. The MADM algorithms considered are as follows:

- *Simple additive weighting (SAW)*: This method is widely known among all MADM algorithms. SAW was presented for first time in [13]. SAW has been considered for the solution of diverse problems. For example, it was introduced for financial decision problems in [25], for engineering materials in [28], and in the pharmaceutical field in [27], among other applications.

  However, in the branch of electrical engineering, SAW has been used in wireless communications. This method is used for handover in heterogeneous networks in [52], where a particular number of candidate networks is considered.

The network attributes considered are price, bandwidth, SNR, time, seamlessness, and battery consumption. The services provided by the network are voice and file downloads. After a numerical demonstration, among several algorithms SAW proved affordable ranking results.

In this work, from the decision matrix that is composed by c_{kj} elements, SAW normalizes the parameters according to benefits or costs in the vector, at [1]. If at_{kj} is a benefit parameter, it is normalized with Eq. 1, where $c_j^+ = \max_{k \in K} c_{kj}$.

$$at_{kj} = \frac{c_{kj}}{c_j^+}. \tag{1}$$

If at_{kj} is a cost parameter, it is normalized according to Eq. 2, where $c_j^- = \min_{k \in K} c_{kj}$.

$$at_{kj} = \frac{c_j^-}{c_{kj}}. \tag{2}$$

The characteristic and final equation of SAW can be seen in Eq. 3, where the weight vector w is multiplied by the vector of normalized attributes, at. Finally, SAW selects an alternative that maximizes the operation and assigns the result to the hole selected vector $\mathbf{hs}^*_{SAW}$.

$$\mathbf{hs}^*_{SAW} = \operatorname*{argmax}_{k} \sum_{j=1}^{J} w_j \, at_{kj}. \tag{3}$$

- *Technique for Order Preference by Similarity to Ideal Solution (TOPSIS)*: This method is presented in [25]. Basically, its algorithm computes the ideal and negative-ideal solutions by calculating the shortest Euclidean distances between each alternative. The ideal alternative has the best values of each attribute, while the negative-ideal has the worst ones [48]. TOPSIS is a method that has been explored in different fields, for example, in supply chain management and logistics, business and marketing management, human resources management, and chemical engineering [8].

Similar to SAW in the handover decision, TOPSIS is also presented in [52]. After testing this method, the results point out that TOPSIS is sensitive to attributes with high scores and to users' preferences. TOPSIS is also explored and compared with more MADM algorithms in [45]. In that work, a 3GPP network is considered with the next classes of traffic: conversational, streaming, interactive, and background, with attributes as bandwidth, delay, jitter, and BER. The results shown that TOPSIS is a good method with a performance similar to SAW.

In order to make a decision, TOPSIS performs the next steps:

Step 1: Build the at_{kj} normalized decision matrix. This allows a fair comparison across the attributes, which is characterized by

$$at_{kj} = \frac{c_{kj}}{\sqrt{\sum_{k=1}^{K} c_{kj}^2}}. \tag{4}$$

Step 2: Construct the vl_{kj} weighted normalized decision matrix with $vl_{kj} = w_j * at_{Kj}$.

Step 3: Determination of alternative solutions. The ideal one with (5) and the negative-ideal with (6):

$$Al^+ = \{(\max_{k \in K} vl_{kj} | j \in BC), (\min_{k \in K} vl_{kj} | j \in BC')\}, \tag{5}$$

$$Al^- = \{(\min_{k \in K} vl_{kj} | j \in BC), (\max_{k \in K} vl_{kj} | j \in BC')\}, \tag{6}$$

where BC is the set of benefit parameters and BC' is the set of cost parameters.

Step 4: Compute the distance between the positive ideal alternatives with (7) and the negative-ideal alternatives with (8):

$$ss_k^+ = \sqrt{\sum_{j \in J} (vl_{kj} - vl_j^+)^2}, \tag{7}$$

$$ss_k^- = \sqrt{\sum_{j \in J} (vl_{kj} - vl_j^-)^2}. \tag{8}$$

Step 5: Calculate the relative closeness to ideal alternative.

$$cc_k^* = \frac{sc_k^-}{(ss_k^+ + ss_k^-)}. \tag{9}$$

The last equation of the process appears in (10), where TOPSIS maximizes the relative closeness to the ideal alternative, cc_k^*. The result is assigned to $\mathbf{hs}_{\mathrm{TOP}}^*$.

$$\mathbf{hs}_{\mathrm{TOP}}^* = \underset{k}{\mathrm{argmax}}\ cc_k^*. \tag{10}$$

- *VIKOR*: The VlseKriterijumska Optimizacija I Kompromisno Resenje (VIKOR) method is introduced in [41]. This mechanism was developed for the optimization of multi-criteria decision-making processes. It works as follows: after receiving initial weights, VIKOR obtains the preference stability compromise solution to determine the compromise ranking list, the compromise solutions, and weights stability intervals. VIKOR presents a multi-criteria index measuring the closeness to the ideal solution [48].

Similar to other MADM algorithms, VIKOR has been considered as a making-decision tool in several fields. For example, for operation research in manufacturing process in [5]. The performance of VIKOR under an empirical case and its applications is presented in [51]. Also, VIKOR has been applied to heterogeneous wireless networks for vertical handoff. In [44], VIKOR shows a performance ranging from satisfactory to excellent compared to other MADM algorithms in the selection of four different types of connections.

VIKOR executes four steps, where maximizations and minimizations are calculated [1].

Step 1: For each parameter $j = 1, 2, 3, \ldots, J$, select the best (11) and the worst (12) of the values given by

$$F_j^+ = \{(\max_{k \in K} x_{ij} | j \in J_b), (\min_{k \in K} c_{kj} | j \in J_c)\}, \tag{11}$$

$$F_j^- = \{(\min_{k \in K} c_{kj} | j \in J_b), (\max_{k \in K} c_{kj} | j \in J_c)\}, \tag{12}$$

where $J_b \subset BC$ is the set of benefit parameters and $J_c \subset BC$ is the set of cost parameters.

Step 2: Calculate the values of S_k y R_k for $k = 1, 2, 3, \ldots, K$ with

$$S_k = \sum_{j \in J} w_j \frac{(F_j^+ - c_{kj})}{(F_j^+ - F_j^-)}, \tag{13}$$

and

$$R_k = \max_{j \in J} \left[w_j \frac{(F_j^+ - c_{kj})}{(F_j^+ - F_j^-)} \right], \tag{14}$$

where w_j is the parameter weight importance j.

Step 3: Compute the values of Q_k for $k = 1, 2, 3, \ldots, K$ with

$$Q_k = \gamma \left(\frac{S_k - S^+}{S^- - S^+} \right) + (1 - \gamma) \left(\frac{R_k - R^+}{R^- - R^+} \right), \tag{15}$$

where

$$S^+ = \min_{k \in K} S_k, \quad S^- = \max_{k \in K} S_k, \qquad (16)$$

$$R^+ = \min_{k \in K} R_k, \quad R^- = \max_{k \in K} R_K, \qquad (17)$$

and the parameter γ with $0 \leq \gamma \leq 1$ is the strategic weight.

Step 4: The values of Q for all $k \in K$ are a list where all candidates are ranked in an increasing order.

Finally, this algorithm selects among all alternatives the closest to the ideal solution. It is defined with Eq. 18, where Q_k^* is the best solution according to SU's requirements.

$$\mathbf{hs}_{\text{VIK}}^* = \operatorname*{argmax}_k Q_k^*. \qquad (18)$$

Comparison function (CF). This proposed function establishes a trade-off between the energy consumed by channel switching and the goodness spectrum holes by the MADM algorithms [1]. As we mentioned before, a spectrum request occurs under different cases. Here, it is assumed that an SU requests a decision in order to improve the parameters of its current spectrum hole. Thus, when a spectrum decision is required, the spectrum decision framework obtains a spectrum hole $\mathbf{hs}_n$, expecting the spectrum hole $\mathbf{hs}_{n-1}$ has not disappeared, and n is the number of the spectrum decision requests. Under this consideration, CF calculates the energy consumption due to channel switching (ECCS) E_{CS_n} from the frequency of $f_{\mathbf{hs}_{n-1}}$ to the frequency $f_{\mathbf{hs}_n}$ with Eq. 19, where P_{CS} is the power dissipation and t_{CS} is the time delay for the channel switching [7].

$$E_{\text{CS}} = P_{\text{CS}}(t_{\text{CS}}|f_{\mathbf{hs}_{n-1}} - f_{\mathbf{hs}_n}|). \qquad (19)$$

Then, CF measures the benefits/costs provided by the new spectrum hole selected $\mathbf{hs}_n$ with respect to $\mathbf{hs}_{n-1}$. The process is done normalizing the differences of parameters of the same type, as can be seen in Eq. 20 for the case of stability of a benefit parameter. In Eq. 21, it is shown the process for selected frequency, which is a cost parameter.

$$\text{st}' = \frac{|\text{st}_{\mathbf{hs}_n} - \text{st}_{\mathbf{hs}_{n-1}}|}{\max(\text{st}) - \min(\text{st})}[0, 1]. \qquad (20)$$

$$\text{sf}' = \frac{\max(\text{sf}) - \min(\text{sf})}{|\text{sf}_{\mathbf{hs}_n} - \text{sf}_{\mathbf{hs}_{n-1}}|}[0, 1]. \qquad (21)$$

Next, an $\mathbf{r}_n$ vector stores the results of the six decision parameters. This vector is normalized and summed, and finally it gets a representative quantity of the benefit/cost parameters in r_n'. The ECCS also is normalized in E_{CS_n}'. By minimizing these values, the CF decides if the SU should remain in the current spectrum hole or move forward according to MADM, as can be shown in Eq. 22:

$$d_n = \min(E'_{CS_n}, r'_n). \tag{22}$$

The final spectrum hole, $\mathbf{fh}_n$, is obtained according to Eq. 23. If E'_{CS_n} is the minimum value, it means that ECCS is affordable and that the benefit/cost parameters of $\mathbf{hs}_n$ will provide a better performance than the current one. Thus, the CF recommends moving forward to $\mathbf{hs}_n$. The contrary case occurs when r'_n is the minimum, meaning that the benefits/costs of $\mathbf{hs}_n$ do not provide any gain and the ECCS is not affordable. For that case, CF suggests staying in $\mathbf{hs}_{n-1}$:

$$\mathbf{fh}_n = \begin{cases} \mathbf{hs}_n, & \text{if } d_n = E'_{CS_{n'}}, \\ \mathbf{hs}_{n-1}, & \text{if } d_n = r'_n. \end{cases} \tag{23}$$

3.5.4   Results

In the previous work shown in [1], the spectrum framework proposal is analyzed under two user connection profiles. In that first approach, the results exhibit considerable savings for spectrum handoffs and ECCS. The CF has shown a better performance for MADM algorithms in CRNs. Now, following the services mentioned in [10, 31], a real-time and best-effort applications are considered. The scenario, spectrum measurements, and methodology are the same as those presented in [1].

Real-time application connection profile. For this case, the SU connection profile looks for a spectrum hole with a constant data rate and low variance. In a subjective way and according to the spectrum parameters presented here, we assign high values to *stability* and *interference temperature*. The proportion considered is of 1/3 to each one of the mentioned parameters and the rest divided equally among the other parameters. Figure 3(left) shows the number of spectrum handoffs done by the MADM algorithms with and without the CF. It is possible to notice that the results considering the CF reduce for SAW and VIKOR almost the half of the spectrum handoffs. For the case of TOPSIS, the CF decreases the handoffs in more than a half respect to the algorithm alone. Figure 3(right) presents the ECCS in joules, where the CF still reduces the energy. In this figure, TOPSIS+CF obtains the lowest consumption among all the options. The purpose of the random case is considered to realize the importance of a decision mechanism.

Best-effort application connection profile. This user receives as much as the network can deliver; it has the lowest priority in the allocation order. Thus, the SU tries to maximize the spectrum hole capacity in order to get a good performance. We distribute weights in the same way as the real-time application. However, the considered parameters with high weights are *bandwidth size* and *duty cycle*. These two parameters are selected since they are highly related to the channel capacity. In Fig. 4(left), it is possible to appreciate again the reduction of spectrum handoffs for

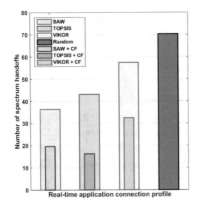

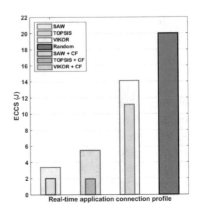

Fig. 3 Average performance of MADM algorithms, MADM algorithms + *CF* and a random case for a real-time application connection profile. Left: Number of spectrum handoffs. Right: Energy consumption due to channel switching

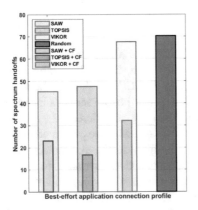

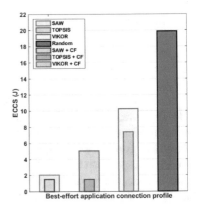

Fig. 4 Average performance of MADM algorithms, MADM algorithms + *CF* and a random case for a best-effort application connection profile. Left: Number of spectrum handoffs. Right: Energy consumption due to channel switching

all the MADM algorithms when they consider the CF. For this case, the number of spectrum handoff is lower than the half obtained with the MADM algorithms by their own. This reduction impacts directly the delay in channel switching. Figure 4(right) shows how the algorithm TOPSIS obtains the lowest ECCS among all combinations.

The results presented here demonstrate how the MADM algorithms work as a spectrum decision mechanism. Affordable outcomes can be obtained considering the proposal *Comparison Function*. There are still a lot of combinations of weight, requirements, and additional parameters from spectrum measurements; however, this approach offers another application for MADM algorithms.

4   Discussion

In this work, all the relevant elements of spectrum decision have been described along with a detailed explanation of how to achieve an efficient decision-making process. Besides, the importance of spectrum decision and how the collaboration with other functions allows to improve the CRN performance have been discussed. The main characteristic of spectrum decision frameworks lies on how they are organized; they are designed to coordinate efficiently the information needed for the decision-making process. Basically, spectrum decision consists of two parts: the framework and the decision mechanisms. The first one is needed to coordinate and organize the information moving around spectrum decision. The second one is basically the pure decision. There are different ways of making a decision. The different spectrum decision frameworks cited in this chapter point to several choices for making a decision. However, all of them depend on the scenario, user requirements, and the number of users, just to mention a few. To date, it would be quite difficult to state which is the most efficient spectrum decision mechanism. Nevertheless, the work presented here provides several ideas to decide which spectrum decision technique is the best according to the problem.

4.1   Open Issues

An important parameter that has been often ignored is computational complexity. Spectrum decision mechanisms can provide the best selection; however, in CRNs the time an algorithm spends on its execution is of key importance. Spectrum decision is just a part of the total CRN spectrum management, and thus a fast and efficient decision is needed. All spectrum decision frameworks should provide computational complexity information so as to consider them for the solution proposal.

Computational complexity is also related to energy consumption, which may represent a serious problem depending on the network topology and the assigned task to each entity. The user experience will be still better if this parameter is taken into account. Thus, for example, in mobile devices, battery life should be extended to the maximum. Also, new proposals have to follow the trend of green communications. Then, energy consumption should be considered imperatively in spectrum decision networks.

A couple of important open issues is the implementation of all the spectrum decision mechanisms and the choice of one of them according to the CRN requirements. The multiple solutions described in this chapter are deployed in simulated scenarios. Some of them consider real CRN characteristics as power spectrum, bandwidth sizes, services, and requirements. However, they have not been implemented in real time so as to measure their efficiency under realistic spectrum traffic patterns. Also, there is still work to do comparison of all the spectrum decision solutions. One single spectrum decision mechanism for all the possible variables in a CRN would not be the best choice. For each type of traffic or spectrum requirement, there should be a fair spectrum decision mechanism. This selection has to fulfill all the demands in the best way. Among the several solutions, one would be the best for a specific service. Thus, part of our future work is to look for decision mechanisms that are well suited for a particular traffic pattern.

4.2 Discussion of New Technologies in CRN

In the coming years, wireless communications in coming years require more available spectrum frequencies. However, it is almost sure that the spectrum scarcity problem will remain growing at the same rate than now or worst. In the last part of this work, we analyze some of the trend technologies and the incursion of CRN in them.

Internet of things (IoT) considering CRN [30]. IoT is a technology with tremendous possibilities of modifying and adapting the interactions among things. The most important goal of IoT is to account with smart sensors collaborating without any human intervention in order to get a new class of applications. Then, an object in which we are interested to retrieve information is equipped with sensors and a communication device. In most of the cases, this object will transmit information preferably in a wireless way. Thus, a set of objects will shape a network, considering that almost any object can be improved; the number of them will be huge. To fulfill with the offered, IoT networks need to take advantage of current networks solutions and more. As with other wireless technologies, they have to deal with spectrum scarcity among other issues. The combination of IoT with CRN may enhance the communications among these objects. CR-based IoT frameworks could reduce interference, provide information about network status, and dynamically access temporally unoccupied spectrum bands, among other benefits. However, some hardware and spectrum function issues need to be solved. It is highly likely that CRN should be adapted with more or fewer steps to work perfectly in IoT.

Long-Term Evolution (LTE) and Wireless Fidelity (Wi-Fi) coexistence. LTE is a technology that has provided good QoS in the last years. However, in order to reach modern QoS requirements, LTE has been complemented and improved several times. Because of its characteristics, LTE unlicensed (LTE-U) accesses dynamically the spectrum through several mechanisms, for example, listen-before-talk (LBT) which ensures that there are no users in the band before performing a transmission [21, 58]. Some CRN techniques are similar to LBT, and thus it is highly likely that some contributions of CRN can be tapped for LTE, especially for the case of the coexistence between LTE and Wi-Fi.

It is possible to improve some features of CRN; one of them is the storage. Taking advantage of the growth of cloud services, cognitive radio cloud networks (CRCN) emerge as a potential candidate [53, 54]. This combination will help CRNs to store huge amounts of information; for example, spectrum measurements among several benefits. Work in this regard is still in progress; however, the cloud is a tool that helps wireless communications technologies to improve performance.

5 Conclusions

Spectrum decision is a key factor for an affordable performance of a CRN. In this chapter, some spectrum decision frameworks were presented and applied to CRN. All of them provide different approaches and use several mathematical tools. However, the core idea is exactly the same: improving spectrum decision to satisfy SU requirements. The multiple applications of spectrum decision mechanisms presented here are difficult to compare. Part of our future work is to configure a specific scenario related to CRN standards with realistic services in order to decide which spectrum decision framework is the best for each user. Particularly, spectrum management should consider the different spectrum mechanisms and apply them according to the current scenario.

Here, MADM algorithms were presented as an option in the decision-making process for CRNs. However, the simple application of these methods does not provide considerable good results. They need to be adapted according to CRNs requirements. As we showed, considering a comparison function (CF) improves the performance of MADM algorithms. The case of TOPSIS is an example considering a CF with a reduction of more than a half in the number of spectrum handoffs and ECCS compared with the simple use of TOPSIS. The results show the need for new MADM algorithms adapted only to CRNs. Finally, it is worth to mention that CRNs should include MADM algorithms as an option in the spectrum decision process.

References

1. Aguilar-Gonzalez, R., Cardenas-Juarez, M., Pineda-Rico, U., Arce, A., Latva-aho, M., Stevens-Navarro, E.: Reducing spectrum handoffs and energy switching consumption of MADM-based decisions in cognitive radio networks. Mobile Information Systems **2016**, 1–14 (2016). https://doi.org/10.1155/2016/6157904
2. Aguilar-Gonzalez, R., Cardenas-Juarez, M., Pineda-Rico, U., Stevens-Navarro, E.: Performance of MADM algorithms with real spectrum measurements for spectrum decision in cognitive radio networks. In: 11th International Conference on Electrical Engineering, Computing Science and Automatic Control (CCE). IEEE (2014). https://doi.org/10.1109/iceee.2014.6978282
3. Akyildiz, I., Lee, W.Y., Vuran, M., Mohanty, S.: A survey on spectrum management in cognitive radio networks. IEEE Communications Magazine **46**(4), 40–48 (2008). https://doi.org/10.1109/mcom.2008.4481339
4. Akyildiz, I.F., Lee, W.Y., Vuran, M.C., Mohanty, S.: NeXt generation/dynamic spectrum access/cognitive radio wireless networks: A survey. Computer Networks **50**(13), 2127–2159 (2006). https://doi.org/10.1016/j.comnet.2006.05.001
5. Anvari, A., Zulkifli, N., Arghish, O.: Application of a modified vikor method for decision-making problems in lean tool selection. The international journal of advanced manufacturing technology **71**(5–8), 829–841 (2014). https://doi.org/10.1007/s00170-013-5520-x
6. Arienzo, L., Tarchi, D.: Statistical modeling of spectrum sensing energy in multi-hop cognitive radio networks. IEEE Signal Processing Letters **22**(3), 356–360 (2015). https://doi.org/10.1109/lsp.2014.2360234

7. Bayhan, S., Alagoz, F.: Scheduling in centralized cognitive radio networks for energy efficiency. IEEE Transactions on Vehicular Technology **62**(2), 582–595 (2013). https://doi.org/10.1109/tvt.2012.2225650

8. Behzadian, M., Otaghsara, S.K., Yazdani, M., Ignatius, J.: A state-of the-art survey of topsis applications. Expert Systems with Applications **39**(17), 13,051–13,069 (2012). https://doi.org/10.1016/j.eswa.2012.05.056

9. Bian, K., Park, J.M., Gao, B.: Cognitive Radio Networks. Springer International Publishing (2014). https://doi.org/10.1007/978-3-319-07329-3

10. Canberk, B., Akyildiz, I.F., Oktug, S.: A QoS-aware framework for available spectrum characterization and decision in cognitive radio networks. In: 21st Annual IEEE International Symposium on Personal, Indoor and Mobile Radio Communications (PIMRC), pp. 1533–1538. IEEE (2010). https://doi.org/10.1109/pimrc.2010.5671959

11. Castellanos-Lopez, S.L., Cruz-Perez, F.A., Hernandez-Valdez, G.: Performance of cognitive radio networks under ON/OFF and poisson primary arrival models. In: IEEE 22nd International Symposium on Personal, Indoor, and Mobile Radio Communications (PIMRC), pp. 609–613. IEEE (2011). https://doi.org/10.1109/pimrc.2011.6140034

12. Chen, D., Yang, J., Wu, J., Tang, H., Huang, M.: Spectrum occupancy analysis based on radio monitoring network. In: 1st IEEE International Conference on Communications in China (ICCC). IEEE (2012). https://doi.org/10.1109/iccchina.2012.6356981

13. Churchman, C.W., Ackoff, R.L.: An approximate measure of value. Journal of the Operations Research Society of America **2**(2), 172–187 (1954). https://doi.org/10.1287/opre.2.2.172

14. Cordeiro, C., Challapali, K., Birru, D., Shankar, S.: IEEE 802.22: the first worldwide wireless standard based on cognitive radios. In: First IEEE International Symposium on New Frontiers in Dynamic Spectrum Access Networks DySPAN, pp. 328–337. IEEE (2005). https://doi.org/10.1109/dyspan.2005.1542649

15. Das, D., Das, S.: Interference-aware power allocation in soft decision fusion (SDF) based cooperative spectrum sensing. In: Annual IEEE India Conference (INDICON). IEEE (2014). https://doi.org/10.1109/indicon.2014.7030618

16. Flores, A.B., Guerra, R.E., Knightly, E.W., Ecclesine, P., Pandey, S.: IEEE 802.11af: a standard for TV white space spectrum sharing. IEEE Communications Magazine **51**(10), 92–100 (2013). https://doi.org/10.1109/mcom.2013.6619571

17. Grissa, M., Hamdaoui, B., Yavuz, A.A.: Location privacy in cognitive radio networks: A survey. IEEE Communications Surveys & Tutorials pp. 1726–1760 (2017). https://doi.org/10.1109/comst.2017.2693965

18. Hasegawa, M., Hirai, H., Nagano, K., Harada, H., Aihara, K.: Optimization for centralized and decentralized cognitive radio networks. Proceedings of the IEEE **102**(4), 574–584 (2014). https://doi.org/10.1109/jproc.2014.2306255

19. Haykin, S.: Cognitive radio: brain-empowered wireless communications. IEEE Journal on Selected Areas in Communications **23**(2), 201–220 (2005). https://doi.org/10.1109/jsac.2004.839380

20. He, A., Srikanteswara, S., Reed, J.H., Chen, X., Tranter, W.H., Bae, K.K., Sajadieh, M.: Minimizing energy consumption using cognitive radio. In: IEEE International Performance, Computing and Communications Conference. IEEE (2008). https://doi.org/10.1109/pccc.2008.4745093

21. Hillery, W.J., Mangalvedhe, N., Bartlett, R., Huang, Z., Kovacs, I.Z.: A network performance study of LTE in unlicensed spectrum. In: IEEE Globecom Workshops (GC Wkshps). IEEE (2015). https://doi.org/10.1109/glocomw.2015.7414108

22. Hoan, T.N.K., Koo, I.: Partially observable markov decision process-based sensing scheduling for decentralised cognitive radio networks with the awareness of channel switching delay and imperfect sensing. IET Communications **10**(6), 651–660 (2016). https://doi.org/10.1049/iet-com.2014.1260

23. Hossian, E., Niyato, D., Han, Z.: Dynamic Spectrum Access and Management in Cognitive Radio. Cambridge University Press (2009)

24. Hoyhtya, M., Pollin, S., Mammela, A.: Improving the performance of cognitive radios through classification, learning, and predictive channel selection. Advances in Electronics and Telecommunications **2**(4), 28–38 (2011)
25. Hwang, C.L., Yoon, K.: Multiple attribute decision making, methods and applications. Tech. rep., Springer Verlag (1981)
26. Islam, M.H., Koh, C.L., Oh, S.W., Qing, X., Lai, Y.Y., Wang, C., Liang, Y.C., Toh, B.E., Chin, F., Tan, G.L., Toh, W.: Spectrum survey in singapore: Occupancy measurements and analyses. In: 3rd International Conference on Cognitive Radio Oriented Wireless Networks and Communications (CrownCom). IEEE (2008). https://doi.org/10.1109/crowncom.2008.4562457
27. Jaberidoost, M., Olfat, L., Hosseini, A., Kebriaeezadeh, A., Abdollahi, M., Alaeddini, M., Dinarvand, R.: Pharmaceutical supply chain risk assessment in iran using analytic hierarchy process (ahp) and simple additive weighting (saw) methods. Journal of the Operations Research Society of America **8**(9), 1–10 (2015). https://doi.org/10.1186/s40545-015-0029-3
28. Jahan, A., Edwards, K.L., Bahraminasab, M.: Multi-criteria Decision Analysis for Supporting the Selection of Engineering Materials in Product Design. Butterworth-Heinemann (2013). https://doi.org/10.1016/C2012-0-02834-7
29. Jha, S., Rashid, M., Bhargava, V., Despins, C.: Medium access control in distributed cognitive radio networks. IEEE Wireless Communications **18**(4), 41–51 (2011). https://doi.org/10.1109/mwc.2011.5999763
30. Khan, A.A., Rehmani, M.H., Rachedi, A.: Cognitive-radio-based internet of things: Applications, architectures, spectrum related functionalities, and future research directions. IEEE Wireless Communications **24**(3), 17–25 (2017). https://doi.org/10.1109/mwc.2017.1600404
31. Lee, W.Y., Akyldiz, I.F.: A spectrum decision framework for cognitive radio networks. IEEE Transactions on Mobile Computing **10**(2), 161–174 (2011). https://doi.org/10.1109/tmc.2010.147
32. Li, Y., Chen, Z., Gong, Y.: Optimal power allocation for coordinated transmission in cognitive radio networks. In: IEEE 81st Vehicular Technology Conference (VTC Spring). IEEE (2015). https://doi.org/10.1109/vtcspring.2015.7145983
33. Lopez-Benitez, M., Casadevall, F.: Discrete-time spectrum occupancy model based on markov chain and duty cycle models. In: IEEE International Symposium on Dynamic Spectrum Access Networks (DySPAN), pp. 90–99. IEEE (2011). https://doi.org/10.1109/dyspan.2011.5936273
34. Lopez-Benitez, M., Umbert, A., Casadevall, F.: Evaluation of spectrum occupancy in spain for cognitive radio applications. In: IEEE 69th Vehicular Technology Conference. IEEE (2009). https://doi.org/10.1109/vetecs.2009.5073544
35. Malon, K., Skokowski, P., Marszalek, P., Kelner, J.M., Lopatka, J.: Cognitive manager for hierarchical cluster networks based on multi-stage machine method. In: IEEE Military Communications Conference, pp. 428–433. IEEE (2014). https://doi.org/10.1109/milcom.2014.77
36. Masonta, M.T., Mzyece, M., Ntlatlapa, N.: Spectrum decision in cognitive radio networks: A survey. IEEE Communications Surveys & Tutorials **15**(3), 1088–1107 (2013). https://doi.org/10.1109/surv.2012.111412.00160
37. McHenry, M.A., McCloskey, D., Lane-Roberts, G.: Spectrum occupancy measurements, location 4 of 6: Republican national convention, new york city, new york, august 30, 2004 - september 3, 2004, revision 2. Tech. rep., Shared Spectrum Company Report (2005)
38. Mitola, J., Maguire, G.: Cognitive radio: making software radios more personal. IEEE Personal Communications **6**(4), 13–18 (1999). https://doi.org/10.1109/98.788210
39. Mokari, N., Saeedi, H., Navaie, K.: Channel coding increases the achievable rate of the cognitive networks. IEEE Communications Letters **17**(3), 495–498 (2013). https://doi.org/10.1109/lcomm.2013.012313.122437
40. Notice of proposed rulemaking and order, et docket no 03-222. Tech. rep., Federal Communications Commission (2003)
41. Opricovic, S.: Multicriteria optimization of civil engineering systems. Ph.D. thesis, Faculty of Civil Engineering, Belgrade (1998)
42. Ozcan, G., Gursoy, M.C., Gezici, S.: Error rate analysis of cognitive radio transmissions with imperfect channel sensing. In: IEEE 78th Vehicular Technology Conference (VTC Fall), pp. 1642–1655. IEEE (2013). https://doi.org/10.1109/vtcfall.2013.6692196

43. Perez-Romero, J., Raschella, A., Sallent, O., Umbert, A.: A belief-based decision-making framework for spectrum selection in cognitive radio networks. IEEE Transactions on Vehicular Technology **65**(10), 8283–8296 (2016). https://doi.org/10.1109/tvt.2015.2508646
44. Stevens-Navarro, E., Gallardo-Medina, R., Rico, U.P., Acosta-Elias, J.: Application of madm method vikor for vertical handoff in heterogeneous wireless networks. IEICE Transactions on Communications **E95B**(2), 599–602 (2012). https://doi.org/10.1587/transcom.E95.B.599
45. Stevens-Navarro, E., Wong, V.W.: Comparison between vertical handoff decision algorithms for heterogeneous wireless networks. In: IEEE Vehicular Technology Conference, (VTC-Spring), pp. 947–951. IEEE (2006). https://doi.org/10.1109/WCNC.2004.1311263
46. Su, W., Liao, Y.: A jury-based trust management mechanism in distributed cognitive radio networks. China Communications **12**(7), 119–126 (2015). https://doi.org/10.1109/cc.2015.7188530
47. Tadayon, N., Aissa, S.: Modeling and analysis framework for multi-interface multi-channel cognitive radio networks. IEEE Transactions on Wireless Communications **14**(2), 935–947 (2015). https://doi.org/10.1109/twc.2014.2362535
48. Tzeng, G.H., Huang, J.J.: Multiple Attribute Decision Making, Methods and Applications. CRC Press (2011). https://doi.org/10.1201/b11032
49. Wang, J., Song, M.S., Santhiveeran, S., Lim, K., Ko, G., Kim, K., Hwang, S.H., Ghosh, M., Gaddam, V., Challapali, K.: First cognitive radio networking standard for personal/portable devices in TV white spaces. In: IEEE Symposium on New Frontiers in Dynamic Spectrum (DySPAN). IEEE (2010). https://doi.org/10.1109/dyspan.2010.5457855
50. Wang, L.C., Wang, C.W., Adachi, F.: Load-balancing spectrum decision for cognitive radio networks. IEEE Journal on Selected Areas in Communications **29**(4), 757–769 (2011). https://doi.org/10.1109/jsac.2011.110408
51. Wei, J., Lin, X.: The multiple attribute decision-making vikor method and its application. In: IEEE International Conference on Wireless Communications, Networking and Mobile Computing (WiCom), pp. 1–4. IEEE (2008). https://doi.org/10.1109/WiCom.2008.2777
52. Wenhui, Z.: Handover decision using fuzzy madm in heterogeneous networks. In: IEEE Wireless Communications and Networking Conference, (WCNC), pp. 653–658. IEEE (2004). https://doi.org/10.1109/WCNC.2004.1311263
53. Wu, S.H., Chao, H.L., Jiang, C.T., Mo, S.R., Ko, C.H., Li, T.L., Liang, C.F., Cheng, C.C.: A conceptual model and prototype of cognitive radio cloud networks in TV white spaces. In: IEEE Wireless Communications and Networking Conference Workshops (WCNCW), pp. 425–430. IEEE (2012). https://doi.org/10.1109/wcncw.2012.6215536
54. Wu, S.H., Chao, H.L., Ko, C.H., Mo, S.R., Jiang, C.T., Li, T.L., Cheng, C.C., Liang, C.F.: A cloud model and concept prototype for cognitive radio networks. IEEE Wireless Communications **19**(4), 49–58 (2012). https://doi.org/10.1109/mwc.2012.6272423
55. Xiao, K., Mao, S., Tugnait, J.K.: MAQ: A multiple model predictive congestion control scheme for cognitive radio networks. IEEE Transactions on Wireless Communications **16**(4), 2614–2626 (2017). https://doi.org/10.1109/twc.2017.2669322
56. Yang, Z., Song, Y., Wang, D.: An optimal operating frequency selection scheme in spectrum handoff for cognitive radio networks. In: International Conference on Computing, Networking and Communications (ICNC), pp. 1066–1070. IEEE (2015). https://doi.org/10.1109/iccnc.2015.7069496
57. Yang, Y., Zhang, G.a., Ji, Y.c.: Impartial spectrum decision under interference temperature model in cognitive wireless mesh networks. In: 4th International Conference on Intelligent Networking and Collaborative Systems, pp. 566–570. IEEE (2012). https://doi.org/10.1109/incos.2012.15
58. Zhang, R., Wang, M., Cai, L.X., Zheng, Z., Shen, X., Xie, L.L.: LTE-unlicensed: the future of spectrum aggregation for cellular networks. IEEE Wireless Communications **22**(3), 150–159 (2015). https://doi.org/10.1109/mwc.2015.7143339

Vehicular Networks to Intelligent Transportation Systems

Felipe Cunha, Guilherme Maia, Heitor S. Ramos, Bruno Perreira,
Clayson Celes, André Campolina, Paulo Rettore, Daniel Guidoni,
Fernanda Sumika, Leandro Villas, Raquel Mini and Antonio Loureiro

Abstract Urban mobility is a current problem of modern society and large cities, which leads to economic and time losses, high fuel consumption, and high CO_2 emission. Some studies point out Intelligent Transportation Systems (ITS) as a solution to this problem. Hence, Vehicular Ad hoc Networks (VANETs) emerge as a component of ITS that provides cooperative communication among vehicles and the necessary infrastructure to improve the flow of vehicles in large cities. The primary goal of this chapter is to discuss ITS, present an overview of the area, its challenges, and opportunities. This chapter will introduce the main concepts involved in the ITS architecture, the role of vehicular networks to promote communication, and its integration with other computer networks. We will also show applications that leverage the existence of ITS, as well as challenges and opportunities related to VANETs such as data collection and fusion, characterization, prediction, security, and privacy.

1 Introduction

The disorderly growth of large urban centers has caused severe socioeconomic and structural problems for the population, which contributes to the increase of social inequalities and a significant stress on the structure of cities. In this way, services

F. Cunha (✉) · R. Mini
Department of Computer Science, Pontifical Catholic University of Minas Gerais,
Belo Horizonte, Brazil
e-mail: felipe@pucminas.br

G. Maia · B. Perreira · C. Celes · A. Campolina · P. Rettore · A. Loureiro
Federal University of Minas Gerais, Belo Horizonte, Brazil

H. S. Ramos
Federal University of Alagoas, Maccio, Brazil

D. Guidoni · F. Sumika
Federal University of São João del-Rei, São João del-Rei, Brazil

L. Villas
University of Campinas, Campinas, Brazil

© Springer Nature Singapore Pte Ltd. 2018
K. V. Arya et al. (eds.), *Emerging Wireless Communication and Network Technologies*,
https://doi.org/10.1007/978-981-13-0396-8_15

and resources must be provided in a way that tackles and minimizes these problems. Among them, it can be mentioned the incorrect occupation of the urban space that collaborates to generate diverse problems of mobility in big cities. In this context, public transport systems are an essential part of improving urban mobility. For example, in São Paulo—Brazil, 23% of the residents spend at least 2 hours commuting to their destination every day [1, 2].

Over the years, traffic-related problems have been increasing due to the number of vehicles in circulation and the vast concentration of people in the same region. According to studies conducted by IBM, the current quantity of automotive vehicles in the world currently exceeds 1 billion, and this number can double in the year 2020. With this, big cities are the most affected by this increase of vehicles, with the constant presence of traffic jams. For example, recent surveys show that São Paulo has an annulling loss of 20 billion, and this loss is related to 85% lost time in traffic; 13% increase in fuel consumption; And only 2% of the growth in the emission of polluting gases. Which also contributes to the increase in warming in these urban centers.

Aiming to solve the problem of mobility, some solutions are proposed, for e.g., the plate casters and incentives for the use of public transport. However, these solutions have not been very successful. In many scenarios, they affect the routine of the population and do not achieve engagement. On the other hand, intelligent solutions that make use of communication can contribute to greater success, improving traffic in these scenarios. These solutions can provide applications that enable the control and management of traffic, with services ranging from a more assertive control of the schedules and routes of public transport to the intelligent synchronization of traffic lights. These services make up the Intelligent Transportation Systems (ITS) [3].

ITS use data, communication, and processing to provide services and applications to solve various transportation problems. These systems, in addition to providing services to manage and improve security for people in transit, also can enable comfort services for drivers and passengers, such as access to social networks and video stream services while traveling. These applications rely on collaboration between the elements that integrate the system such as vehicles, sensors and other mobile devices. Each of these elements plays an important role, collaborating and sensing data that will be evaluated by the system. All this collaboration of elements is made possible by the communication between them. For this, elements such as antennas and control stations can intermediate this communication. In the context of the direct communication between the vehicles, vehicular networks arise, a type of network that has been exerting a significant influence on the scene of the ITS [4].

The services and applications provided by ITS have their characteristics and peculiarities, which differs to other traditional applications. They are services that generate and consume a varied amount of data, use different communication technologies with different bandwidths, reach, and latency. Besides, vehicles have high mobility, moreover, speed limits and directions determined by public roads become communication a challenging task in this scenario. For this reason, designing a service part of these systems becomes a major challenge. In this chapter, we discuss ITS and present an overview of the area, defining its central concepts, integration, the role of VANETS to provide communication, and the cooperation with other networks. Also, we describe challenges of the infrastructure to promote the services and the open issues about data and security.

The remainder of this chapter is organized as follows. Section 2 discusses the concept of ITS presenting all definitions, architecture, and integration with other networks. Section 3 presents features and challenges related to infrastructure and services in ITS. Section 4 discusses opportunities for the current research topics related to data and security in ITS. Finally, Sect. 5 presents the conclusion.

2 Intelligent Transportation Systems

Intelligent Transportation Systems (ITS) aims to improve transport safety and mobility, as well as to increase people's productivity and reducing the harmful effects of traffic. This improvement is achieved through the integration of communication technologies in vehicles and infrastructure.

ITS is not only proposed to improve vehicle traffic conditions but also intends to make the transportation safer, more sustainable, and efficient, avoiding the inconvenience caused by traffic congestion and the effects of climate problems on traffic. To this end, the focus is on improving the management of cities' resources and increasing the convenience of people using information and alert services. This improvement, therefore, helps to ease the flow in the city, reducing the time spent on congestion and consequently reducing fuel consumption, CO_2 emissions and monetary losses. In the following sections, the central concepts related to ITS will be presented.

2.1 Architecture

Considering the evolution of computing and communication technologies and the increasing demand for ITS services with different requirements, the necessity for standardization to define how devices and components can interact with each other arises. Among the proposed architectures, it is worth to mention the North American, the European, and the Japanese.

The National ITS Architecture, defined by the US Department of Transportation, describes how communication between its elements and subsystems occurs, with a precise definition of the role of each one of them. This architecture is divided into four classes: *Center*, *Fields*, *Vehicles*, and *Travelers*. *Center* defines the center of control and management of the whole system, in which the services are executed. *Field* encompasses all the infrastructure of the environment (RSU, monitoring sensors, cameras). *Vehicles*, which are vehicles and embedded sensors, and *Travelers* that are defined by the devices people use during the trip.

The Japanese architecture proposed by the *SmartWay* defines the communication among vehicles and among vehicles and all the intelligent infrastructure of the roads (sensors, RSU, traffic lights) and uses as the standard of the DSRC [2], along with the proposed ARIB standard (similar to the WAVE protocol). The European architecture (ITS ISO CALM) has very similar characteristics to other architectures such as the adoption of RSUs and the DSRC [2] for communication. However, this architecture has the greatest difference in the utilization of the CALM communication protocol, which provides a communication interface between the transmission technologies such as 3G/4G, Wi-Fi, infra-red, and others.

Both Japanese and European architectures have disadvantages compared to North American architecture because they lack the flexibility to use new communication technologies and new paradigms of computing such as cloud and fog computing. Hence, one can observe a requirement to design architectures that allow the easy integration of new technologies, since they can cooperate for the development and improvement of services offered by ITS.

2.2 Vehicular Ad Hoc Networks

Vehicular networks are a type of emerging network that has attracted the interest of many research groups. These networks are made up of vehicles with processing capacity and wireless communication, traveling on streets and highways, sending and receiving information from other vehicles. They differ from traditional networks in many ways. The first of these is the nature of the nodes that form them, such as automobiles, trucks, buses, etc., which have wireless communication interfaces, and equipment attached to the roads. Also, these nodes have high mobility, and their trajectory follows the limits and direction defined by public roads [5–7].

Vehicles participating in the network are equipped with an onboard system with computing capability, communication interfaces, sensors, and user interfaces. The system supports a range of applications to improve transport security and also provide services to users. A network infrastructure along roadsides and streets called the Roadside Unit (RSU) is also part of VANETs and facilitates the communication of network nodes to the Internet. Also, passenger handhelds and the vehicle system can connect to the Internet through the RSU infrastructure. A management system can be adapted to control and authenticate the entrance of vehicles in the network, mainly in the aspect of computer security, such as the distribution of cryptographic keys, authentication servers, etc. The system can also provide services and manage node mobility during network exchanges.

Due to high node mobility, vehicular networks allow nodes to exchange information along with their trajectory without the need of any infrastructure, in an ad hoc fashion. Hence, vehicular networks can be considered as a type of Mobile Ad hoc Network (MANETs). However, there is a possibility of nodes communicating with the infrastructure of the highways, allowing an infrastructural communication [8–10]. In this way, considering these peculiar characteristics, the communication between the vehicles can be classified in three ways (as illustrated in Fig. 1).

(a)

(b)

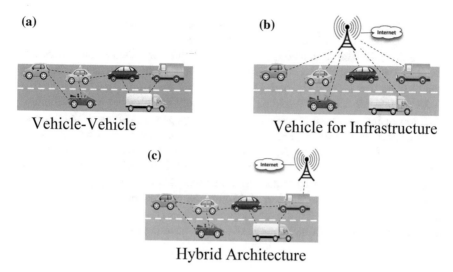

Vehicle-Vehicle   Vehicle for Infrastructure

(c)

Hybrid Architecture

Fig. 1 Types of communication in vehicular networks

- Vehicle-to-Vehicle (V2V): It allows a direct communication of vehicles without relying on fixed infrastructure support. In this type of communication, vehicles can exchange data of the conditions of the highway, detect the presence of other vehicles, and even information about vehicles in unsafe movement.
- Infrastructure-to-Vehicle (V2I): It allows a vehicle to communicate with the road infrastructure. In this way, the vehicle can receive from the road infrastructure information about obstacles and the presence of pedestrians, road conditions data, advertisements, and safety information.
- Hybrid Architecture: It combines V2V and V2I solutions. In this case, a vehicle can communicate with the road infrastructure in a single or multiple hops according to its location about the point of connection with the infrastructure for different purposes.

Currently, car manufacturers already put into circulation cars with onboard computers, wireless communication devices, sensors, and navigation systems. These resources enable the establishment of the vehicular networks. An example of the application of these features is vehicles that have sensors to collect weather conditions, vehicle status, road conditions, and even road speed limit. In this scenario, vehicles can interact with the infrastructure of the highways, obtaining information of traffic which generates improvements in the conditions for the driver to make decisions about the traffic.

An interaction between vehicles can prevent the occurrence of collisions on public roads. Traffic surveys show that in Brazil an average of 110,000 traffic accidents occurs per year, around 300 per day. Also, 6 thousand people die, and another 68 thousand are injured, generating to the government an expense about US$7 billion [11]. The primary cause of these accidents was the lack of attention of the drivers, followed by drivers who do not obey the safety distance and speed limit [12]. Studies show that about 60% of accidents can be avoided if the driver is warned a second before the collision. In this context, the use of vehicular networks can provide the reduction of these accidents rates, through the vehicle–vehicle interaction the drivers can be alerted of hazards on roads [13].

In vehicle networks, information should usually be delivered within vehicles in a region of interest, taking into account the geographical position of the node and the relevance of the information to the node. One challenge in this context is how to distribute information to vehicles efficiently, considering the dynamics and mobility of vehicles on the network and even the urgency of delivering information to avoid a collision. For this, an important tool to be studied is the routing protocol, which must be efficient, reliable, support multi-hop communication, and delay intolerant. Moreover, it is important that the vehicle receives the warning of the possible obstacle, even if they are not in the same range of communication [14].

2.3 Integration with Other Networks

Wireless technologies are becoming ubiquitous. It provides network access to a variety of standards, such as IEEE 802.11, 3G/4G, LTE and Bluetooth, which can be used to equip sensor networks, unmanned vehicle networks, and vehicular networks. Hence, cellular networks (4G/LTE) may provide long-distance communication and Internet access for vehicles, and, in short distance, DSRC (Dedicated short-range communications) ad hoc should be more suitable. Hence, ITSs must provide services to drivers and passengers at any time and place. And the success and availability of this service will depend on the integration of different technologies and networks.

In [15], the authors present a performance analysis of the two communication patterns in vehicular networks for different scenarios, densities, and speeds of vehicles. It can be observed that the DSRC scores good results in scattered networks. But because of its communication radius limitations, its support for vehicle mobility is limited. On the other hand, the LTE standard presented a good performance regarding scalability, reliability, and mobility support. However, it presents some challenges in dealing with the delay constraints in some applications.

Regarding data collection, ITSs should make use of an integration with sensor networks (WSN) and unmanned vehicle networks (FANET). Sensory data can be combined with other data collected by the vehicles to, for example, infer the positioning of a network node (vehicle, RSU, mobile user device), provide vehicle density in roads, point out the presence of points of floods and obstacles, etc. Taking into account the unmanned vehicles, they can be applied in special occasions like accidents or floods, to aid in the collection and dissemination of data. In such cases, they would assist in the diffusion of alert messages by establishing communication links in places where RSU infrastructure is in operation or unavailable.

Considering other aspects of data transmission technology, these standards can also be used to establish communication between ITSs and all intelligent traffic infrastructure. For instance, reprogramming traffic lights, reading data from cameras and sensors installed on public roads, communication with radars, etc., all such devices must be able to communicate with traffic monitoring centers to provide data that can assist with the management of all traffic.

3 Infrastructure and Services

In this section, we present the main current research topics related to infrastructure and services of traffic prediction and mobility in ITS. We listed the key features and opportunities of this topic.

3.1 ITS Infrastructure

The dynamic scenario of a transportation system has as main characteristic the high mobility of its components in an urban environment. Although people and goods' mobility are present for many years, it has never reached such a high scale as nowadays. Therefore, problems faced along those years, such as accidents, congestions, and dangerous situations have also worsened with the mobility increasing.

With the technology advancement, the communication evolved from radio, signs, and alerts from own drivers to onboard computers, sensors, smartphones, and other devices that receive real time notifications through wireless communication. New technologies enabled a more dynamic and instant communication.

ITSs have a flexible hybrid architecture, allowing the operation within Internet connectivity, either by infrastructure or taking full autonomy of the system, in an ad hoc manner. This architecture has benefits such as scalability and delay reduction, but it faces some challenges to perform efficiently and guarantees quality and safety, besides representing an additional cost that is not always feasible.

Many devices compose such architecture, including sensors, OBUs (*onboard units*), RSUs (*roadside units*), GPS (global positioning system), intelligent traffic lights, access points, portable devices (smartphones, tablets, laptops), satellites, specialized servers, and the Internet. To allow communication among those components, diverse technologies can be adopted, such as Wi-Fi, WiMAX, LTE, GSM, 3G, 4G, satellite, and Bluetooth, among others.

One of the biggest challenges consists of designing new communication solutions in this heterogeneous set of available technologies. Since an intelligent system operates in a collaborative manner, it is necessary to define standards to enable the integration of all components. Moreover, due to the high mobility, the infrastructure deployment becomes an issue (for instance, consider access points or RSUs location), besides delay and fault tolerance, inherent in such systems.

The components of ITS can be equipped with multiple types of wireless transceivers and can communicate over more than one wireless data channel.

The IEEE 802.11p protocol, a variant of Wi-Fi technology, provides the allocation of bandwidth for specific V2V and V2I communication. Communication can take place in short range, enabling V2V and V2I communication, through GPS and DSRC radios or long range, mainly for V2I and I2I, using cellular data transceivers, GSM-based, GPRS, UMTS.

The work [16] highlights the importance and the role played by the Internet infrastructure in the context of vehicular networks. Being ubiquitous and available in various urban environments, the wired Internet infrastructure can provide support to a variety of applications. For instance, the downloading of advertisements and entertainment or the storage of data gathered and sent by the vehicles. Also, content that is already in the possession of some vehicle may also be shared by opportunistic P2P connections between vehicles and other devices. The authors conclude that a big trend for the Future Internet is the interaction between wireless P2P communication side by side with a support infrastructure for the adequate provisioning of applications and services. Among them, we have navigation safety, navigation efficiency, entertainment, vehicle monitoring, urban sensing, participatory sensing, and emergencies.

In the following, we highlight some works on integrating infrastructure and ad hoc networks to show how ITS can become complete and efficient by using a hybrid architecture.

The problem of RSUs deployment for V2I communication through IEEE 802.11p is studied in [17]. The main goal consists of analyzing urban features' impacts, along with a suitable RSU deployment and communication configurations to guarantee a successful V2I communication. Results presented for a large set of experiments conducted in the city of Bologna show that the quality of V2I communication through IEEE 802.11p is strongly affected by street layout, terrain elevation, trees and vegetation, traffic density, and presence of heavy vehicles. Thus, it is necessary to take such factors into consideration in the proper deployment of RSUs and radio configuration. The authors propose guidelines to be followed for an efficient deployment in the design of vehicular networks.

In [18], the I2V data delivery problem is investigated. It consists of accurately estimating the destination position, considering the temporal and spatial encounter of the packet and destination vehicle. The proposed solution, named Trajectory-based Statistical Forwarding (TSF), uses a packet delay distribution and a vehicle delay distribution to select a target point aiming to minimize packet delivery delay while satisfying the packet delivery probability requested by the user. They consider the installation of RSUs as infrastructure, vehicles equipped with OBUs and DSRC communication, GPS present in both vehicles and stationary nodes and knowledge of the trajectory of the vehicle, which is shared on the Internet periodically through access points.

Infrastructure on the design of ITS is explored in many works of the literature. The employment of RSUs can be found in [19, 20]. The integration of VANETs and cloud computing is treated in [21–23]. Security strategies are studied in [24, 25].

3.2 Traffic Prediction

Traffic congestion impacts not only congested roads but also nearby streets and highways which are alternative paths to drivers avoiding it. A solution to avoid these situations is tracing more efficient routes, which depend on updated traffic information. Since obtaining real-time traffic state of all roads in a city is a hard task, alternative ways of sensing such aspect were developed.

Lippi et al. [26] presented a comparison between multiple short-term traffic prediction strategies. Predicting traffic state in shorter windows of time is an easier task and more effective, given that routes will be traced taking into consideration more recent information. Tostes et al. [27] and Abadi et al. [28] developed predictors of traffic levels in urban areas based on statistical tools. Regression models are especially useful in predicting short-term traffic because they capture the typical behavior of congestion levels and adapt in face of unusual situations, like accidents and roadblocks.

Recent communication technology advances have enabled vehicles to communicate among themselves and with a city's underlying infrastructure. Being able to communicate, vehicles can share sensor data which contains reflexes of traffic state. Wan et al. [29] and Pan et al. [30] present methods to aggregate sensor data from multiple individuals and extract traffic information to trace less crowded routes in a city.

3.3 Mobility and Traffic

Urban problems, especially regarding mobility and traffic, are one of the main research challenges related to the quality of life of people in the cities. In this sense, several efforts have been made to reduce congestion, provide safe means of locomotion, reduce environmental pollution, reduce noise pollution, among other objectives. ITSs can play a key role regarding technological solutions to achieve those objectives.

Understand the dynamics of cities is a fundamental aspect to provide mobility and traffic solutions. Thanks to the popularization of devices with the capacity for sensing and the evolution of ITSs, an enormous amount of data has been generated and made available to analyze the behavior of the entities (e.g., vehicles, people, and objects) in the cities, thus facilitating the understanding of human mobility and the behavior of traffic through the days. Many cities around the world provide several open datasets that can be freely used for the study of mobility, for example, Rio de Janeiro,[1] London,[2] and New York.[3]

---

[1] http://data.rio/.

[2] http://data.london.gov.uk/.

[3] http://opendata.cityofnewyork.us/.

Data sources such as social networks and applications (*Waze*[4] and *Bing Maps*[5]) also are a powerful form of data collection for the study of mobility and traffic. For example, [31] analyzed social network data (*Instagram*[6] and *Foursquare*[7]), [32] used Bing Maps data to analyze and predict congestion points in Chicago in the United States. These studies show how the discipline of data analysis can be interesting to facilitate the understanding of the dynamics of cities. For example, to identify the main routes used by the population, collect information on the demand for private and public vehicles, find out the causes of congestion, etc. Also, there are several opportunities related to the use of heterogeneous data sources, large-volume data manipulation and processing, and techniques for summarizing and understanding the semantics of the data.

In addition to the analysis to understand the mobility of the population in the cities, another critical perspective is the offer of services that optimize resources and efficiently use the means of transport, considering the particularities of each city such as territorial and population size, road topology, culture, and other aspects. In that sense, the remainder of this section focuses on exposing mobility and transit solutions, highlighting the key opportunities and challenges associated with them, as is illustrated in Fig. 2.

Shared mobility: In this case, new transport solutions allow users to use systems of shared means of transportation such as cars and bicycles for a particular time. Generally, in these systems, vehicles are available at stations and users can use them for a fee. In this context, several research challenges are related. Yang et al. [33] proposed a predictive method for balancing bicycles in stations based on the study of mobility data in Hangzhou in China. Similarly, some efforts have focused on studying vehicle sharing [34, 35].

Carpooling: A common occurrence in several cities is the action of drivers offering car rides to lower travel costs by considering their mobility routines. The digital media leveraged this behavior, as people started to organize themselves in social networks and message groups to plan the rides like the service provided by BlaBlaCar.[8] In this sense, one of the main challenges for this type of system is the creation of recommendation services that explore the infrastructure of ITS such as VANET and data generated by vehicles and people.

Integrated systems and multimodal transport: This refers to integrating the various modes of transport to provide the mobility of people. For example, an integrated system between bus lines, subways, bicycles, or shared cars. Therefore, several challenges must be considered when designing multimodal transport systems, such as real-time information manipulation, multicriteria analysis, making route recommendations, and user preferences.

---

[4]http://www.waze.com/.

[5]http://www.bing.com/maps/.

[6]http://www.instagram.com/.

[7]http://www.foursquare.com/.

[8]http://www.blablacar.com.br/.

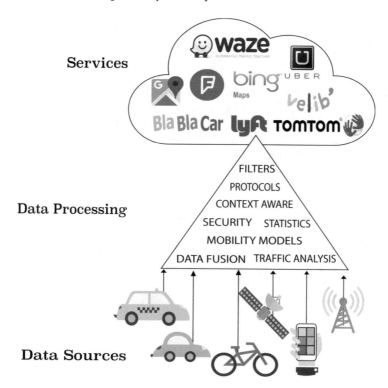

Fig. 2 Data flow to promote the services in ITS

Mobile applications: The popularization of smartphones has leveraged the development of mobile applications that provide services for both mobilities (e.g., Uber[9] and Lyft[10]) to get traffic information (e.g., Waze[11]). In this sense, new initiatives exploring supportive technologies (e.g., mobile and ubiquitous computing, things based on location systems) to ITS are highly recommended in the current scenario.

Traffic control: Monitoring and controlling traffic flow of vehicles (traffic) is an important topic in ITSs. Tian et al. [36] have reviewed the literature on studies that use cameras to monitor and assist the traffic in urban areas. They proposed a taxonomy of methods for detecting, tracking, and recognizing vehicles. Another topic related to the problem of vehicle traffic is the control of intersections, especially at peak times, to improve the flow of vehicles and safety of drivers and pedestrians. In this case, the challenge is to manage traffic lights and intersections for the synchronization of traffic between lanes as discussed in [37, 38].

Detection and management of traffic incidents: Detection and mitigation of traffic incidents is one of the leading research opportunities in the context of ITS,

---

[9]https://www.uber.com.

[10]https://www.lyft.com/.

[11]https://www.waze.com.

since it is possible to explore the large volume of data generated by vehicles or made available by users through mobile applications and social networks. Pan et al. [39] proposed a system for detecting incidents (e.g., accidents, sporting events) and suggested routes using vehicle location data and information shared by social networks. In this topic, there are some open challenges such as spatially determining the impact of an incident, time duration, and semantics.

In summary, technological solutions in mobility and traffic seek people to spend less time in traffic safely using the various types of transport, prioritizing the conscious consumption of energy resources and reducing the environmental impact.

4 Data and Security in ITS

Data and security become important research topics in ITS due to a series of restricts and challenges to deal with personal data and its peculiarities. Thus, we discuss aspects of data collection, quality, and security issues in ITS, highlighting some research opportunities in the following.

4.1 Data Collection and Quality

Nowadays, modern vehicles have high-technology embedded systems that aim to improve driving safety, performance, and fuel consumption. To achieve these goals, manufacturers have invested both in the quantity and quality of sensors that vehicles have [40]. Currently, a vehicle collects information from hundreds of sensors that are connected to the Engine Control Unit (ECU) through an internal wired sensor network [3] and the Output data is accessible via an Onboard Diagnostic (OBD) interface.

The control system of modern vehicles relies on data collected from embedded sensors. These systems allow to control vehicle's stability and contribute to safer driving. Sensor data is available through the OBD interface, which has been introduced for regulatory and maintenance issues but has been exploited for various other purposes due to the information it provides.

Some of the data collected from vehicle's sensors do not represent relevant information for drivers since most of this data is used by the ECU and does not have a clear meaning for the driver (e.g. oxygen and fuel pressure sensors). Besides, sensors that indicate meaningful information to the driver are displayed by indicators in vehicles such as rotation per minute, speed, and temperature of the engine.

Thus, the challenge is to extract useful information from vehicle's sensors to correlate them with internal and external variables, enabling personalized services for drivers and a transportation system. To better illustrate this subject, data were collected from Bluetooth adapters connected to the OBD interface and *smartphones*.

Fig. 3 Data collection schematic using the OBD interface and smartphone

Table 1 Used protocols of the OBD interface

Protocols	Bitrate (kbit/s)
SAE J1850 PWM	41.6
SAE J1850 VPW	10.4
ISO 9141-2	10.4
ISO 14230 KWP 2000	10.4
ISO 15765 CAN	250 or 500

The OBD-II interface was introduced to standardize the physical connector, protocols, and message formats. The system is used to monitor and regulate gas emission, and it is present in all produced cars in Europe and the United States since 1996. The OBD interface also assists maintenance services, tracking the origin of mechanical problems [41]. By enabling the storage of engine failure codes, this information provides a history of problems and possible associated sources. Figure 3 illustrates the collection process: the acquired data from the sensors through the OBD interface are transferred to a smartphone with the Android operating system where they are processed and registered.

Table 1 shows five protocols allowed with the OBD interface: These protocols use the same OBD connector, but the pins have different functions except those that provide battery power. The collected data from sensors are available through OBD parameters IDs (PIDs). Table 2 shows some of the information available considering *smartphone*, vehicle, and data from virtual sensors (values are generated from physical sensor data and mathematical processing and data fusion). There are also hundreds of other sensors that can be accessed through PIDs, some of which are defined by OBD standards and others by vehicle manufacturers.

It is important to note that data from physical sensors are inherently subject to errors caused by some reasons, including the accuracy of the sensor itself, and even operation failures of the vehicle and sensor [42]. Therefore, the first step of the processing and analysis of virtual sensor data is its verification to ensure it is in correlation to the measured data. Among the observed issues at this stage we can highlight discrepant or outliers' data, conflicting information from two or more sensors, incomplete or ambiguous data. Once the data are verified, it is possible to apply

Table 2 Data collected from ECU and mobile phone

Data collected				
Mobile phone		Vehicle		Virtual sensor
Device time	Trip distance	Engine load	Engine RPM	Acceleration
GPS location	Fuel remaining	Fuel flow	Speed	Reaction time
GPS speed	Ambient air temp	Engine coolant temp	CO_2 average	Air drag force
GPS precision	Barometer	Adapter voltage	CO_2 instant	KPL instant
GPS bearing	GPS altitude	Fuel level	Pedal	Speed/RPM relation
Gravity		Intake air temp		Gear

data fusion, which aims to obtain new values with a more significant meaning than the individual data.

4.2 Security

This section presents data security and privacy as a major issue when developing ITSs. ITS services data may contain personal information, enabling people and vehicles tracking. Because data can be transmitted through multiple hops and administrative domains, malicious entities can capture this data. By implementing security in ITSs, one can mitigate those issues and avoid high degradation of ITSs services factors such as response time, network overload, and desirable quality of service. The overview presented in the following was based on [43–45], which introduced guidelines and good practices for Internet security, cyber security for general propose ITS and intelligent public transport, respectively.

One can define the security in ITS into three aspects: *objectives, threats, and services. Security objectives* are the goals to keep the ITS as safe as possible. *Security threats* may affect the security objectives causing degradation of ITS services. Finally, *security services* are aimed to counter the threats and to drive the system towards the security objectives. Each security component is discussed in more details in the following section.

4.2.1 Security Objectives

One can divide security objectives for ITSs into four major categories[12]: confidentiality, availability, integrity, and peer authentication.

---

[12]Note that different authors can consider others security objectives [43–45].

Confidentiality: It aims to keep user and system data free of unauthorized malicious entities, processes, or systems. Hence, the ITSs may selectively grant access only for entities, processes or system with right permissions.

Availability: This objective intends to keep ITSs information available when it is needed. ITS available is accessible at all times and has ways to overcome threats (such as natural disasters, accidental, or intentional ones) to its proper functioning.

Integrity: The basic goal here is to ensure that ITS data sent and received was the same. In another word, the data over the communication channel should maintain its meaning, completeness, and consistency.

Peer authentication: It is aimed to make sure that the one in the endpoint of the communication is the one intended to be. In [43], the authors highlight that without peer authentication, it is hard to reach confidentiality and integrity.

4.2.2 Security Threats in ITS

This section concerns security threats in ITS. Here, threats are everything that potentially can cause problems to proper ITS functioning. One can loosely divide ITS security threats according to its consequences to a system (such as unauthorized usage, denial of service, manipulation). Also, one can classify ITS security threats concerning its origins: *natural disasters, accidental, or intentional ones*. These understanding promote a basis to create security services to counter or mitigate threats. The remaining section discusses mainly threats and consequences to the system, which are pertinent to a wide range of ITSs and applications.

Unauthorized usage: ITS must not be freely accessible to most of the public. Indeed, only authorized users with worth permission level have to receive access to a given ITS function. Although several ITS are available for public users, some sub-services are intended to specific users. Suppose, for instance, ITS RESTFUL service like Google Maps[13] or HERE,[14] the servers will serve data to its ordinary users, but they restrict the ability to modify data from the servers or even insert/remove data to specific users. Thus, if the regular users perform service data modification, it would be an unauthorized usage, and then action would be taken. Unauthorized usage arises from accidental or intentional origins ranging from misconfiguration to malicious attackers.

Denial of Service (DoS): One can classify an attack as DoS when actions are taken to block access or interrupt the proper ITS functioning. DoS arises from intentional, accidental, or natural events. However, usually, the cause of DoS is an intentional insertion of malicious codes into the system or by executing inappropriate actions. Critical hazards can emerge when DoS attacks occur in ITS. For instance, if a system of safe driving detection suffers a DoS, then accidents can happen.

Manipulation: The practice of altering data and other information from systems to produce unauthorized effects is namely known as manipulation. Natural, accidental,

---

[13]https://developers.google.com/maps/.
[14]https://developer.here.com/.

or intentional events can cause manipulation issues. Within ITS, consider a sign system controlling speed limits in roadways. If system manipulations are made in the signs to display incorrect or inappropriate information, several traffic issues might happen such as poor system performance, traffic jams or even accidents.

Replay: In such threat, an attacker records a sequence of data messages and sends them back to the intended receiver.[15] Thus, an attacker replays valid data under invalid circumstances to promote unauthorized effects to the system. In the ITS context, consider a toll system being used by Alice to perform a payment. Bob (an attacker) could capture the messages of the Alice payment and replays it, although he cannot read the messages, he causes twice transactions.

Message insertion, deletion, or modification: Several threats come from insertion, deletion, or modification of spurious messages. On the message insertion case, an attacker forges a message and inserts it into the system to promote malicious effects, for example, in a DoS, the attacker can open a bulk of TCP connections with the victim to drain memory resources and deny the service quickly. On message deletion cases, messages are dropped from the system. An example is the black hole attack [46], where a misconfigured router has zero cost to any destination, and then all traffic loads are forwarded to this router. Consequently, the router does not support the burden and fails. Finally, on the modification case, the attacker removes a message from the system, modifies it, and then reinserts the modified message again into the ITS network. Consider a ITS fast food service, where a user does an order to the service. An attacker wants to attack the order and receives the food. The attacker does not know the victim credit card number. Thus, the attacker waits for the victim to perform an order, then he intercepts the messages order, modifies them by replacing some properties (such as address, goods, and order description) and put the messages again to the system.

Repudiation: When users deny that they performed actions or transaction in the system, one can say a repudiation has occurred. Hence, repudiation attacks are hard to prove without an auditing. Repudiation arises from accidental or intentional events. Such threat usually affects the system integrity and peer authentication. In the ITS context, repudiation, usually, occurs in electronic transactions, for instance, suppose an order service facility. Using an automatic payment scheme without proper security audition, the user could deny ordering a service and refuse to pay.

4.2.3 Security Services

Developing security services is a natural step derived from the identification of security objectives and threats categories. Security services are protections usually employed to enhance confidentiality, availability, integrity, and peer authentication. In the following, it is listed some useful security services for ITS[16]:

---

[15]The attacker does not need to know the message content to replay messages.

[16]In [44], the reader can find a more exhaustive list of ITS security services.

- **Authentication service**: It aims to verify entities identity and ensure that the ones in the endpoints or even in the middle of the communication channel are those who are supposed to be. Usually, the own entity performs its identification to the system. Such services, enhance the system confidentiality, integrity, and peer authentication objectives.
- **Integrity services**: These services support integrity analyses over information flowing through the system, aiming to minimize manipulation threats. Error detection and correction, cryptographic checksums, digital signatures are basic integrity services that provide some confidence that the data has not been modified during the communication.
- **Access control services**: It aims to provide specific permissions/limits for system entities (such as users, managers, process) to access system resources, according to entity rule in the system. Usually after an entity authentication action, then it is applied some system access permission level to the entity. Access control helps to mitigate unauthorized usage, DoS, manipulation and furthers.

5 Conclusion

This chapter discusses the main concepts related to intelligent transportation systems. Issues related to existing architectures, communication network standards, and integration of systems with different types of communication were pointed out and discussed, showing the demand for the standardization and integration of these systems.

Additionally, it is presented the main types of ITS applications, to show the works found in the literature that already uses these concepts, giving some directions of new works. Finally, it points out the main topics of current research and the challenges that are found in ITS with the purpose of guiding future research in the area. We believe that there are new challenges that can arise as these systems evolve and new users join.

References

1. M. Cintra, "A crise do trânsito em São Paulo e seus custos," *GV-executivo*, vol. 12, no. 2, pp. 58–61, 2013.
2. "Intelligent Transport Systems—Communications Access for Land Mobiles ({CALM})—Architecture," ISO, Geneva, Switzerland, Apr. 2010.
3. F. Qu, F. Y. Wang, and L. Yang, "Intelligent transportation spaces: Vehicles, traffic, communications, and beyond," *IEEE Commun. Mag.*, vol. 48, no. 11, pp. 136–142, Nov. 2010.
4. G. Karagiannis *et al.*, "Vehicular Networking: A Survey and Tutorial on Requirements, Architectures, Challenges, Standards and Solutions," *Commun. Surv. Tutorials, IEEE*, vol. 13, no. 4, pp. 584–616, 2011.
5. M. Faezipour, M. Nourani, A. Saeed, and S. Addepalli, "Progress and Challenges in Intelligent Vehicle Area Networks," *Commun. ACM*, vol. 55, no. 2, pp. 90–100, 2012.

6. A. Boukerche *et al.*, "A new solution for the time-space localization problem in wireless sensor network using UAV," in *DIVANet 2013 - Proceedings of the 3rd ACM International Symposium on Design and Analysis of Intelligent Vehicular Networks and Applications, Co-located with ACM MSWiM 2013*, 2013, pp. 153–160.

7. A. Boukerche, H. A. B. F. Oliveira, E. F. Nakamura, and A. A. F. Loureiro, "Vehicular Ad Hoc Networks: A New Challenge for Localization-Based Systems," *Comput. Commun.*, vol. 31, no. 12, pp. 2838–2849, Jul. 2008.

8. R. S. Alves *et al.*, "Redes veiculares: Princípios, aplicações e desafios," in *Minicursos do Simpósio Brasileiro de Redes de Computadores*, 2009, pp. 199–254.

9. H. Hartenstein and K. P. Laberteaux, "A tutorial survey on vehicular ad hoc networks," *Commun. Mag. IEEE*, vol. 46, no. 6, pp. 164–171, 2008.

10. S. Yousefi, M. Mousavi, and M. Fathy, "Vehicular ad hoc networks (VANETs): challenges and perspectives," ... *Proceedings, 2006 6th ...*, pp. 761–766, 2006.

11. I. -, "Instituto Brasileiro de Pesquisas Econômicas." May-2012.

12. CESVI, "Centro de Experimentação e Segurança Viária." May-2012.

13. X. Yang, J. Liu, F. Zhao, and N. Vaidya, "A Vehicle-to-Vehicle Communication Protocol for Cooperative Collision Warning," in *First Annual International Conference on Mobile and Ubiquitous Systems: Networking and Services (MobiQuitous'04)*, 2004, pp. 114–123.

14. F. Li and Y. Wang, "Routing in vehicular ad hoc networks: A survey," *Veh. Technol. Mag. IEEE*, vol. 2, no. 2, pp. 12–22, Jun. 2007.

15. Z. Hameed Mir and F. Filali, "LTE and IEEE 802.11p for vehicular networking: a performance evaluation," *EURASIP J. Wirel. Commun. Netw.*, vol. 2014, no. 1, p. 89, 2014.

16. M. Gerla and L. Kleinrock, "Vehicular networks and the future of the mobile internet," *Comput. Networks*, vol. 55, no. 2, pp. 457–469, 2011.

17. J. Gozálvez, M. Sepulcre, and R. Bauza, "IEEE 802.11 p vehicle to infrastructure communications in urban environments," *IEEE Commun. Mag.*, vol. 50, no. 5, 2012.

18. J. Jeong, S. Guo, Y. Gu, T. He, and D. H. C. Du, "TSF: Trajectory-based statistical forwarding for infrastructure-to-vehicle data delivery in vehicular networks," in *Distributed Computing Systems (ICDCS), 2010 IEEE 30th International Conference On*, 2010, pp. 557–566.

19. Y. Peng, Z. Abichar, and J. M. Chang, "Roadside-aided routing (RAR) in vehicular networks," in *Communications, 2006. ICC'06. IEEE International Conference on*, 2006, vol. 8, pp. 3602–3607.

20. O. Trullols, M. Fiore, C. Casetti, C. F. Chiasserini, and J. M. B. Ordinas, "Planning roadside infrastructure for information dissemination in intelligent transportation systems," *Comput. Commun.*, vol. 33, no. 4, pp. 432–442, 2010.

21. S. Olariu, I. Khalil, and M. Abuelela, "Taking VANET to the clouds," *Int. J. Pervasive Comput. Commun.*, vol. 7, no. 1, pp. 7–21, 2011.

22. R. Hussain, J. Son, H. Eun, S. Kim, and H. Oh, "Rethinking vehicular communications: Merging VANET with cloud computing," in *Cloud Computing Technology and Science (CloudCom), 2012 IEEE 4th International Conference on*, 2012, pp. 606–609.

23. W. He, G. Yan, and L. Da Xu, "Developing vehicular data cloud services in the IoT environment," *IEEE Trans. Ind. Informatics*, vol. 10, no. 2, pp. 1587–1595, 2014.

24. K. Plößl and H. Federrath, "A privacy aware and efficient security infrastructure for vehicular ad hoc networks," *Comput. Stand. Interfaces*, vol. 30, no. 6, pp. 390–397, 2008.

25. A. Studer, E. Shi, F. Bai, and A. Perrig, "TACKing together efficient authentication, revocation, and privacy in VANETs," in *Sensor, Mesh and Ad Hoc Communications and Networks, 2009. SECON'09. 6th Annual IEEE Communications Society Conference on*, 2009, pp. 1–9.

26. M. Lippi, M. Bertini, and P. Frasconi, "Short-term traffic flow forecasting: An experimental comparison of time-series analysis and supervised learning," *Intell. Transp. Syst. IEEE Trans.*, vol. 14, no. 2, pp. 871–882, 2013.

27. A. I. J. Tostes, F. de L. P. Duarte-Figueiredo, R. Assunção, J. Salles, and A. A. F. Loureiro, "From Data to Knowledge: City-wide Traffic Flows Analysis and Prediction Using Bing Maps," in *Proceedings of the 2Nd ACM SIGKDD International Workshop on Urban Computing*, 2013, p. 12:1–12:8.

28. A. Abadi, T. Rajabioun, and P. A. Ioannou, "Traffic Flow Prediction for Road Transportation Networks With Limited Traffic Data," *IEEE Trans. Intell. Transp. Syst.*, vol. 16, no. 2, pp. 653–662, Apr. 2015.

29. J. Wan, J. Liu, Z. Shao, A. V Vasilakos, M. Imran, and K. Zhou, "Mobile crowd sensing for traffic prediction in internet of vehicles," *Sensors*, vol. 16, no. 1, p. 88, 2016.

30. B. Pan, Y. Zheng, D. Wilkie, and C. Shahabi, "Crowd Sensing of Traffic Anomalies Based on Human Mobility and Social Media," in *Proceedings of the 21st ACM SIGSPATIAL International Conference on Advances in Geographic Information Systems*, 2013, pp. 344–353.

31. T. H. Silva, P. O. S. V. De Melo, J. M. Almeida, and A. A. F. Loureiro, "Large-scale study of city dynamics and urban social behavior using participatory sensing," *IEEE Wirel. Commun.*, vol. 21, no. 1, pp. 42–51, 2014.

32. A. I. J. Tostes, F. de LP Duarte-Figueiredo, R. Assunção, J. Salles, and A. A. F. Loureiro, "From data to knowledge: city-wide traffic flows analysis and prediction using bing maps," in *Proceedings of the 2nd ACM SIGKDD International Workshop on Urban Computing*, 2013, p. 12.

33. Z. Yang, J. Hu, Y. Shu, P. Cheng, J. Chen, and T. Moscibroda, "Mobility Modeling and Prediction in Bike-Sharing Systems," in *Proceedings of the 14th Annual International Conference on Mobile Systems, Applications, and Services*, 2016, pp. 165–178.

34. R. Nair, E. Miller-Hooks, R. C. Hampshire, and A. Bušić, "Large-scale vehicle sharing systems: analysis of V{é}lib'," *Int. J. Sustain. Transp.*, vol. 7, no. 1, pp. 85–106, 2013.

35. C. Boldrini, R. Bruno, and M. Conti, "Characterising demand and usage patterns in a large station-based car sharing system," in *Computer Communications Workshops (INFOCOM WKSHPS), 2016 IEEE Conference on*, 2016, pp. 572–577.

36. B. Tian *et al.*, "Hierarchical and networked vehicle surveillance in its: A survey," *IEEE Trans. Intell. Transp. Syst.*, vol. 18, no. 1, pp. 25–48, 2017.

37. Z. Ye and M. Xu, "Decision Model for Resolving Conflicting Transit Signal Priority Requests," *IEEE Trans. Intell. Transp. Syst.*, vol. 18, no. 1, pp. 59–68, 2017.

38. M. S. Shirazi and B. T. Morris, "Looking at Intersections: A Survey of Intersection Monitoring, Behavior and Safety Analysis of Recent Studies," *IEEE Trans. Intell. Transp. Syst.*, vol. 18, no. 1, pp. 4–24, 2017.

39. B. Pan, Y. Zheng, D. Wilkie, and C. Shahabi, "Crowd sensing of traffic anomalies based on human mobility and social media," in *Proceedings of the 21st ACM SIGSPATIAL International Conference on Advances in Geographic Information Systems*, 2013, pp. 344–353.

40. W. J. Fleming, "Overview of Automotive Sensors," *IEEE Sens. J.*, vol. 1, no. 4, pp. 296–308, 2001.

41. J. Lin, S. Chen, Y Shih, and S. Chen, "A study on remote on-line diagnostic system for vehicles by integrating the technology of OBD, GPS, and 3G," *World Acad. Sci. Eng. Technol.*, vol. 32, no. 8, pp. 435–441, 2009.

42. P. H. Rettore, B. P. S. André, Campolina, L. A. Villas, and A. A.F. Loureiro, "Towards Intra-Vehicular Sensor Data Fusion," in *Advanced perception, Machine learning and Data sets (AMD'16) as part of the 2016 IEEE 19th International Conference on Intelligent Transportation Systems (ITSC 2016)*, 2016.

43. E. Rescorla and B. Korver, "Guidelines for writing RFC text on security considerations," 2003.

44. K. Biesecker, E. Foreman, K. Jones, and B. Staples, "Intelligent Transportation Systems (ITS) Information Security Analysis," 1997.

45. C. Levy-Bencheton and E. Darra, "Cyber security and resilience of intelligent public transport: good practices and recommendations," 2015.

46. J. R. Vacca, "Front Matter," in *Computer and Information Security Handbook (Third Edition)*, Third Edit., Boston: Morgan Kaufmann, 2017, p. iii.

State Estimation and Anomaly Detection in Wireless Sensor Networks

Aditi Chatterjee and Kiranmoy Das

Abstract Reliable co-operative wireless sensor networks are now used in a variety of disciplines including but not limited to geoscience, medical science and security management. Sensor nodes are low-powered microelectronic devices with limited communication and sensing range. In a cluster-based network, sensor nodes are grouped into a number of clusters based on their physical locations. Sometimes, an anomalous node is cleverly placed into the network by intruders for reducing power and efficiency of the network. These nodes collect and send the confidential information to outsiders. Additionally, an anomalous node can destroy the network by gradually damaging inter-node dependence within a cluster. We review some existing and recent statistical models for locating such anomalous node and recovering efficiency of the network. Then, we propose a novel dynamic linear mixed model for the simultaneous state estimation and anomaly detection. Our proposed model can efficiently locate an anomalous node in a relatively short time and the power of the model is evaluated by numerical studies. The proposed approach will be very useful in medical science, military surveillance and environmental studies.

1 Introduction

In the past 10 years, we have witnessed a revolution in the advancement and wide applications of wireless sensor networks (WSNs). WSNs are being applied in military surveillance [42], in health science [22], in battlefield [25], in environment monitoring [40], in geology [35] and in criminology [12]. WSNs are mainly task-oriented networks with the objective of obtaining and transmitting information of

A. Chatterjee
Department of Electronics & Telecommunication, Jadavpur University,
Kolkata 700032, India
e-mail: eice.aditi@gmail.com

K. Das (✉)
Interdisciplinary Statistical Research Unit (R. A. Fisher Bhavan),
Indian Statistical Institute, 203 B. T. Road, Kolkata 700108, India
e-mail: kiranmoy.das@gmail.com; kmd@isical.ac.in

© Springer Nature Singapore Pte Ltd. 2018
K. V. Arya et al. (eds.), *Emerging Wireless Communication and Network Technologies*,
https://doi.org/10.1007/978-981-13-0396-8_16

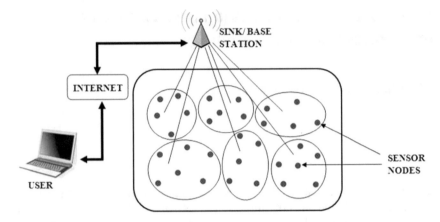

Fig. 1 A general wireless sensor network

interest [8]. Typically WSNs consist of ten to thousands of sensor nodes which are low powered, and have limited processing and storage capacity. Sensor nodes sense and obtain relevant information (e.g. temperature, humidity, event occurrence, etc.) and then pass it to sink. The sink is also known as base station, and it is a high-powered device linked to databases via satellite links. Relaying of data occurs via wireless multi-hop routing. Figure 1 shows a typical WSN for information collection and data processing.

WSNs can be used either for discrete times, or for continuous time monitoring. For a discrete-time WSN, the time points $t_1, t_2, \ldots, t_k$, are integers, and for a continuous-time WSN, these are real numbers. The quantity being measured is typically continuous, e.g. room temperature, humidity, etc. However, sensor nodes consume more energy for collecting, storing and relaying continuous measurements over time. Since energy conservation is always desirable [1], sometimes sensor nodes are made in such a way that they collect continuous measurements, and then based on some pre-fixed threshold they store only binary responses (e.g. normal/abnormal). The nodes transmit these binary outcomes to base station dynamically. Sometimes sensor nodes collect measurements at each time point, but store and transmit only "important" information (e.g. event based, or query based) to base station [20]. These protocols are energy efficient.

Reliable data delivery is the fundamental goal of WSNs. However, reliability of data delivered can be affected by several factors including but not limited to node failure, presence of anomalous node(s), lack of proper communication, defective or disturbed communication mechanism, etc. Such challenges have motivated the researchers in the network designing, reliability checking, state estimation and anomaly detection. The issues and challenges related to network designs for WSNs are discussed in Hac [14], Fischione et al. [10], Lazaropoulos [16], Gupta and Sikka [11] and the references therein. A number of recent articles propose reliability analysis of wireless sensor networks under different model settings. Examples include

Zhu et al. [43, 44], Mahmood et al. [19], Silva et al. [39], Dâmaso et al. [7]. Because of the wide applications, it has been extremely important to develop energy- efficient reliable WSNs in the recent years.

Most of the applications of WSNs, however, depend on some important aspects, namely, (i) time synchronization, (ii) node localization, (iii) state estimation of sensor nodes. The reliability of a network is affected if sensor-specific clocks are not well synchronized. There is a rich literature on time synchronization, and powerful statistical models have been proposed for this purpose. Examples include Noh et al. [23, 24], Kaur and Kaur [15], Chatterjee and Venkateswaran [2] and references therein. These authors proposed statistical models for estimating clock-offset and clock-skew parameters and used those estimated values for time synchronization. There is also a vast literature on node localization. By "node localization", we mean to determine the location of a sensor node. In many applications, it is extremely important to determine the location of a sensor node based on infrastructure [13]. The nodes send signals to base station, and based on those signals the system locates the position of the sensor node [38]. A real example of node localization system is GPS. GPS determines location of a sensor node by receiving signals from its nearest satellites. The literature for state estimation is equally vast, and probabilistic models have been proposed for efficient and robust state estimation of both discrete-time and continuous-time WSNs under different conditions.

Innovations-based state estimation for WSNs has been proposed by Quevedo et al. [26]. Liang et al. [17] proposed distributed state estimation of a discrete-time WSN. An energy-efficient state estimation method has been proposed by Quevedo et al. [27]. Mo et al. [21] proposed a powerful state estimation for WSN under false data injections. A constrained state estimation for individual localisation was proposed by Feng et al. [9]. Rana and Li [30] proposed a microgrid state estimation method for WSN. Chhade [5] developed a data fusion and collaborative state estimation method, and more recently Liu and Wang [18] developed robust state estimation with data-driven communication for WSNs. It is worth mentioning that most of the state estimation approaches proposed in the literature are based on Kalman filtering [31]. Chatterjee et al. [4] propose a regression-based statistical model for state estimation of a discrete-time WSN and estimate model parameters using the maximum likelihood estimation (MLE) method. These authors also propose an alternative Bayesian model for state estimation, where model parameters are estimated by Markov Chain Mote Carlo (MCMC). Chatterjee et al. [3] developed a Bayesian approach of the simultaneous state estimation for a cluster-based WSN.

Proper communication within a network can be ascertained by allowing "information exchange" among sensor nodes which are spatially close to each other. The model proposed by Chatterjee et al. [4] consider such spatial dependence by incorporating a neighbourhood effect in statistical model. Essentially, they assume that current state value of a sensor node will be affected by the average state value of its nearest neighbours at the previous time point. The temporal correlation is also captured in their model by considering the effect of the state value for the previous time point of a sensor node on its current state value. However, their model did not consider any sensor-specific effects, which could explain variations among the sensor nodes (in their state values) with similar other parameters. In this chapter, we are proposing a better version of their model which can explain such sensor-specific variations over time.

The reliability of a wireless network is also affected by the presence of one or more anomalous node. Anomalous nodes are the nodes which behave abnormally compared to other nodes in the network. Sometimes a sensor node stops working, and hence will be an anomalous node for the next time points. An external intruder can on purpose make a normal sensor node anomalous for information collection. Such nodes will collect information from other nodes and relay to the intruder but will not communicate with other nodes properly in the network. An early detection of such nodes is extremely important for obvious reasons, and there are some papers in the literature addressing this issue. Rajasegarar et al. [28, 29], Wang et al. [41], Roy et al. [33] developed methods for anomaly detection in WSNs. Traditionally, an anomalous node is detected by its behaviour compared to its neighbours. However, this method is quite subjective, and hence might not be reliable. Here, we will review model-based method proposed by Chatterjee et al. [4], where they used the estimated state values for anomaly detection. For a cluster-based WSN, these authors locate an anomalous node by successive splitting and merging of the clusters. This method is discussed in detail in Sect. 2.5.

The rest of the chapter is organized as the follows. In Sect. 2, we review the continuous state estimation methods for a discrete-time WSN. We review Kalman filter models and then propose a linear mixed model similar to Chatterjee et al. [4]. The motivation and different model components and estimation method are discussed in different subsections of Sect. 2. Numerical results for assessing the effectiveness of the proposed model are shown in Sect. 3. Finally, in Sect. 4, we give some concluding remarks, and discuss possible future research directions.

2 Statistical Models for State Estimation

Here, we will review some useful methods for state estimation of WSNs, and then propose a new linear mixed model for the same purpose. First, we will give a precise definition of the "State" of a dynamic system, and then discuss estimation procedure.

For any dynamic system, "state" refers to the smallest vector that fully summarizes the past of the system. In the context of WSNs, we note that for a particular sensor node we have the measurements say, $x_1, x_2, \ldots, x_t$, till the time point t. Based on these t measurements, we predict the measurement for the time point $t + 1$, and that predicted value denoted by $\hat{x}_{t+1}$ will be called "state" of the sensor node at time $t + 1$. For estimating the state of a dynamic system traditionally state-space models are used in the literature. State-space models assume that "true" state of the system at time t, which we denote by x_t is latent or unobserved, and can be modelled as the following:

$$x_t = f_t(x_{t-1}, v_{t-1}), \tag{1}$$

where v_{t-1} is random noise, and f_t is a time-dependent mathematical function (possibly nonlinear) describing the evaluation of states. Since "true" state is latent, we assume a measurement process in which true state is obtained from noisy measurements z_t as the following:

$$z_t = h_t(x_t, n_t), \tag{2}$$

where the time-dependent function h_t defines measurement process, and n_t denotes random noise. The particle filter-based estimation is carried out in two steps. In the "prediction step", we predict x_t based on $z_1, z_2, \ldots, z_{t-1}$ (denoted by $z_{1:t-1}$) as the following:

$$P(x_t|z_{1:t-1}) = \int P(x_t|x_{t-1}) \times P(x_{t-1}|z_{1:t-1})dx_{t-1}.$$

Note that here $P(A|B)$ denotes the conditional probability of event A given event B. Next, in the "update step" we update x_t based on $z_{1:t}$ as the following:

$$P(x_t|z_{1:t}) \propto P(z_t|x_t) \times P(x_t|z_{1:t-1}).$$

Note that the term "filter" is used because "true" state is obtained from noisy measurements through filtering. Under linearity of f_t and h_t, and under Gaussian distribution for noise terms v_t and n_t, for each time point t, estimation becomes simpler and is known as Kalman filter. Thus, the state-space model for Kalman filter can be written as the follows:

$$x_t = F_t x_{t-1} + v_{t-1}, \quad z_t = H_t x_t + n_t.$$

Here, F_t and H_t are process and measurement matrices; and noise terms v_{t-1} and n_t are Gaussian with zero means and unknown but fixed variances.

In this chapter, we will propose an alternative powerful linear mixed model for state estimation of sensor nodes in a cluster network. A cluster network is shown in Fig. 2, where the entire network is divided into a number of clusters, and each cluster has a cluster head. The nodes within a cluster are spatially close to each other. We consider the following approach for one particular cluster, and the model has to be repeated for each cluster separately.

Consider a particular cluster with N sensor nodes forming a discrete-time wireless sensor network. For each node, state values for t different time points $(1, 2, \ldots, t)$ are communicated to the cluster head. Let (θ_i, δ_i) be the coordinates of the ith sensor, $i = 1, 2, \ldots N$. The Euclidean distance between sensor i and sensor j is denoted by D_{ij}.

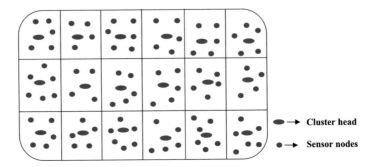

Fig. 2 Cluster formation in a general wireless sensor network

2.1 Linear Mixed Model

The state value of the ith sensor at time t, which we denote as $X_i(t)$, is modelled as the follows:

$$X_i(t) = f(t) + \alpha X_i(t-1) + \beta Z_i(t-1) + \gamma_{0i} + \gamma_{1i}t + \varepsilon_i(t), \qquad (3)$$

where the smooth function (twice differentiable) $f(\cdot)$ is the general effect of time on current state value and Z_i is a covariate which measures the average state values at time $t-1$ of r sensors which are closest (in terms of the Euclidean distance) to sensor i. Note that r can either be pre-fixed or can be estimated from given data. The regression coefficients α and β respectively measure the effect of the immediate predecessor state value and the effect of the nearest neighbours on current state value. The sensor-specific random intercepts and random slopes are denoted by γ_{0i} and γ_{1i} respectively. This explains random variation of state values between different sensors within a cluster at a given time point t. Note that here for the sake of simplicity, we assume linear random effects. However, one can easily consider higher order random effects and choose the optimal order using information criteria like AIC and/or BIC discussed later. The random errors $\varepsilon_i(t)$ are assumed to follow a Gaussian distribution with mean = 0 and unknown variance = σ_ε^2.

The motivation of the above linear mixed model is the following. We note that state value of a particular sensor is updated over time. The general effect of time explains how much of this change is due to the effect of time. Also, state value at time t will somehow depend on the state value at time $t-1$, and this effect will be reflected in the estimate of α. A statistically significant value of α will indicate that the effect of the immediate predecessor state value is significant for estimating current state value. Also, current state value of a particular sensor might be affected by the state values of its nearest neighbours at previous time points. This effect will be reflected in the estimate of β. Here, for the sake of simplicity, we assume a Markovian setup, i.e. we assume that given all the state values upto time t, current state value $X_i(t)$ depends only on the state value at time $t-1$. The random effects of time, as mentioned earlier, explains the difference (if any) between state values from different sensor nodes at each time point. The residual errors $\varepsilon_i(t)$ are assumed to be independent with subject-specific random effects vector $\gamma_i = [\gamma_{0i}, \gamma_{1i}]^T$.

2.2 Modelling the General Effect of Time

The smooth function (twice differentiable) $f(\cdot)$ can be modelled in various parametric or non-parametric approaches, e.g. polynomial functions, wavelets, B-splines, penalized splines, orthogonal Legendre polynomials, etc. Here, we will use polynomial function of time for simplicity of illustration, i.e. we consider $f(t) = a_0 + a_1 t + a_2 t^2 + \cdots + a_p t^p$. When state values of different sensors are measured at different time points and/or are expected to differ a lot from one sensor to other, then a polynomial function of time might not be a good choice for modelling general effect of time. However, in our current setup, we assume that state values are measured for all sensors at the same discrete-time points and those values do not differ too much from one sensor to other over time. So, the above formulation of $f(\cdot)$ will be good enough for modelling the general effect of time in our setting.

The optimal order (p) of the polynomial function can be obtained from information criteria like AIC, BIC given as follows:
$\text{AIC} = -2\log L + 2P^*$ and $\text{BIC} = -2\log L + \log(N)P^*$, where L denotes the joint likelihood function discussed later, P^* denotes total number of model parameters that need to be estimated. We fit the model given in Eq. (3) for $p = 1, 2, \ldots$ and choose the optimal order p for which the smallest AIC and BIC values are obtained. For large values of N, BIC provides more consistent model than AIC, and hence BIC is more preferable in the model selection literature.

2.3 Modelling the Random Effects of Time

The sensor-specific random effects $\gamma_i = (\gamma_{0i}, \gamma_{1i})^T$ are assumed to be independent with the residual error vectors. Traditionally, random effects are modelled with multivariate Gaussian distributions. In this paper, we consider the traditional approach of modelling sensor-specific random effects and assume that γ_is follow bivariate Gaussian distribution with mean vector $= (0, 0)^T$ and covariance matrix $= \Sigma_i = \begin{bmatrix} \sigma_0^2 & \rho\sigma_0\sigma_1 \\ \rho\sigma_0\sigma_1 & \sigma_1^2 \end{bmatrix}$, where ρ denotes the correlation between γ_{0i} and γ_{1i}, σ_0 and σ_1 respectively denote the standard deviation of γ_{0i} and γ_{1i}.

2.4   Joint Likelihood Function and Parameter Estimation

Note that because of the Markovian nature of our model in Eq. (3), the conditional distribution of $X_i(t)$ given all the previous time points can be expressed as $l(X_i(t)|1, \ldots, t-1) = l(X_i(t)|t-1)$. Here, by "given time $t-1$" we essentially mean "given the state values of all sensors at time $t-1$". Thus, the conditional distribution of $X_i(t)$ for given all previous time points (state values) and random effects is given by $X_i(t)|t-1, \gamma_{0i}, \gamma_{1i} \sim \text{Gaussian}(f(t) + \alpha X_i(t-1) + \beta Z_i(t-1) + \gamma_{0i} + \gamma_{1i}t, \sigma_\varepsilon^2)$. Thus, we have,

$$l(X_i(t)|t-1, \gamma_{0i}, \gamma_{1i}) = \frac{1}{\sqrt{2\pi\sigma_\varepsilon^2}} \exp\left[-\frac{(X_i(t) - f(t) - \alpha X_i(t-1) - \beta Z_i(t-1) - \gamma_{0i} - \gamma_{1i}t)^2}{2\sigma_\varepsilon^2}\right].$$

$$(4)$$

The marginal distribution of $X_i(t)$ given $t-1$ can be obtained by integrating out the random effects as follows:

$$l(X_i(t)|t-1) = \iint l(X_i(t)|t-1, \gamma_{0i}, \gamma_{1i})\pi(\gamma_{0i}, \gamma_{1i})\mathrm{d}\gamma_{0i}\mathrm{d}\gamma_{1i}, \qquad (5)$$

where $\pi(\gamma_{0i}, \gamma_{1i})$ denotes the density of the bivariate Gaussian distribution with mean vector $= (0, 0)^T$ and covariance matrix $= \Sigma_i$ as specified earlier.

Note that given all the previous time points, state values of different sensors at a fixed time point are independently distributed. We will exploit this conditional independence property to formulate joint likelihood function. However, we note that state values of different sensors at a fixed time point are not marginally independent.

The joint likelihood function of all N sensors for t time points can be expressed as

$$L = \prod_{i=1}^{N}[l(X_i(1))l(X_i(2)|1)\ldots l(X_i(t)|t-1)]. \qquad (6)$$

Model parameters are to be estimated by maximizing the above joint likelihood function. Note that in our setting, the maximum likelihood estimates of the model parameters cannot be written in explicit forms. Softwares like R, SAS (also MAT-LAB) can solve this optimization problem and provide the parameter estimates. An iterative algorithm is used to maximize the likelihood function and solutions are obtained when convergence criterion is satisfied. In other words, if θ denotes the set of all model parameters, and θ^k is the estimated parameter set at kth iteration, then we stop when $||\theta^k - \theta^{k-1}|| < \varepsilon$, for some pre-specified value of ε.

2.5   Detection of Anomalous Node

The above model is capable of detecting one or more anomalous node in a cluster-based sensor network, as illustrated in Chatterjee et al. [4]. We review the method from their work as the following. Note that an anomalous node is a node with abnormal behaviour compared to other nodes in the network. Typically an anomalous node can be a node which stops working after a certain time, and hence do not provide any information to the network. Also, it can be a selfish node which collects information from the network but does not share its own information to other nodes. Alternatively, an intruder makes a normal node anomalous in order to disturb the network communication. In all these cases, it is extremely important to detect such node and remove it from the network. We propose a model-based detection approach as the following.

Once the model parameters for the linear mixed model are estimated, one can compute average state value of each cluster. Suppose, there are K sensor nodes in a certain cluster, then the average state value of that cluster at time t will be $B(t) = \frac{1}{c} \sum_{i=1}^{c} \hat{X}_i(t)$, where $\hat{X}_i(t) = \hat{f}(t) + \hat{\alpha} X_i(t-1) + \hat{\beta} Z_i(t-1)$. Note that "hat" indicates the estimate of the respective parameter or function obtained from maximum likelihood estimation discussed earlier. We compute absolute difference between the average state values at two consecutive time points, i.e. $|B(t) - B(t-1)|$, and check if the differences are consistently below a pre-fixed threshold (η), i.e. if $|B(t) - B(t-1)| < \eta$, for all t, then we conclude that the cluster under consideration is less dynamic and hence possibly does not contain any anomalous node. However, for the other case, when we observe $|B(t) - B(t-1)| > \eta$, for many consecutive time points, we conclude the existence of an anomalous node in this cluster. Note that the choice of the threshold value depends on the desired accuracy level and application. We divide a dynamic cluster into two "sub-clusters" and repeat this process for each sub-cluster to detect the more dynamic sub-cluster. The process is continued until we obtain the target node and state trajectory of this node will be significantly different from the trajectories of the other nodes. We detect this node as an "anomalous node". The investigation method is shown in Fig. 3.

We note that sometimes it is difficult to fix a single η as a threshold for such anomaly detection. In such cases, one can consider a function $\eta(t)$ with known form, and use that function to capture dynamicity of the anomalous node. Form of the

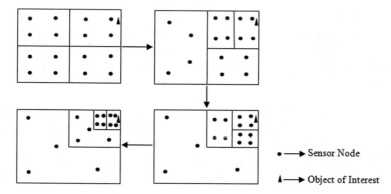

Fig. 3 Detection of an anomalous node in a cluster-based wireless sensor network

function $\eta(t)$ can be known form past experience or suggested by some experts. In patient monitoring or military surveillance such dynamic thresholds work better than a constant threshold.

3 Numerical Studies

We will perform two simulation studies for investigating the operating characteristics of the linear mixed model proposed in Eq. (3). The results of these simulation studies show the importance and usefulness of the proposed methodology for estimating the states of a discrete-time WSN.

3.1 Simulation 1

We consider a single cluster containing six sensor nodes and one cluster head. The position of the sensors are given by their respective coordinates as (2, 4), (5, 6), (3, 2), (4, 3), (3, 7), and (4, 6). For each sensor node, state values are measured at 5 discrete time points. For each sensor, the state value at time t (for $t = 1, 2, \ldots, 5$) is simulated using the following model:

$$X_i(t) = a_0 + a_1 t + a_2 t^2 + \alpha X_i(t-1) + \beta Z_i(t-1) + \gamma_{0i} + \gamma_{1i} t + \varepsilon_i(t), \quad (7)$$

where $(a_0, a_1, a_2) = (2, -0.6, 1.4)$, $\alpha = 2.5$, $\beta = 1.5$. The vector of random effects $\gamma_i = [\gamma_{0i}, \gamma_{1i}]^T$ are simulated from a bivariate Gaussian distribution with mean vector $= [0, 0]^T$ and variance–covariance matrix $= \begin{bmatrix} 2.25 & 1.8 \\ 1.8 & 4 \end{bmatrix}$. The random residual errors $\varepsilon_i(t)$ are simulated from Gaussian distribution (mean $= 0$, variance $= 1.44$). The Z_i values are obtained considering 3 nearest neighbours ($r = 3$) of each sensor.

We simulate 100 datasets (replications) from the above equation. For each dataset, we estimate the model parameters using the maximum likelihood approach discussed in Sect. 2.4. On the basis of 100 simulated datasets, we obtain the final estimates of the parameters by averaging the estimates from each dataset. The standard errors (standard deviation/$\sqrt{100}$) of the model parameters are also estimated based on 100 replicates.

Table 1 provides the model parameter estimates and the respective standard errors on the basis of 100 simulations. Note that true values of these parameters are also shown in this table. We observe that the estimated values are very close to the true parameter values with reasonable (not very high) standard error values. We also show the approximate 95% confidence intervals of the model parameters in Table 1.

Table 1 Parameter estimates, estimated standard errors, and 95% confidence intervals of the model parameters for Simulation 1

Model parameter	True value	Estimate	Standard error	C.I.
a_0	2	1.78	0.23	(1.33, 2.23)
a_1	−0.6	−0.63	0.14	(−0.90, −0.35)
a_2	1.4	1.32	0.38	(0.57, 2.06)
α	2.5	2.56	0.18	(2.20, 2.91)
β	1.5	1.46	0.26	(0.95, 1.97)
σ_ε	1.2	1.14	0.32	(0.51, 1.77)

Table 2 AIC-BIC table for selecting the optimal order of the polynomial function in general effect of time in Simulation 1

Order	AIC	BIC
1 (linear)	7.64	9.18
2 (quadratic)	**3.72**	**4.68**
3 (cubic)	6.93	8.71

Note that these approximate confidence intervals are computed using the following formula:

Lower confidence limit = parameter estimate − 1.96 × standard error,
Upper confidence limit = parameter estimate + 1.96 × standard error.

The confidence interval provides a reliable range of the estimated parameter values. A model or an estimation procedure will be preferable if it estimates the parameters with smaller width of the confidence intervals. We note that the confidence intervals for the parameters as shown in Table 1 are not wide and hence the proposed method does a decent job in parameter estimation.

To select the optimal order of the general effect of time, we calculate the AIC and BIC values for a linear, quadratic and cubic function of time for simulated dataset. Table 2 provides the calculated AIC and BIC values for three different choices of the polynomial function. We notice that the smallest value for AIC (and BIC) is obtained for the quadratic function, which verifies that the proposed method picked up correct order of the polynomial function. In practice, the recommendation is, one should consider upto third or fourth degree and select the best one among them.

To obtain the optimum number of the nearest neighbours to be considered for each sensor, we again consider AIC and BIC. We fit the model using $r = 1, 2, 3$ and 4. Table 3 shows AIC and BIC values for different choices of r. We notice that $r = 3$ provides the smallest AIC and BIC value. Thus, it is verified that the correct number of nearest neighbours is picked up through AIC and BIC.

This simulation study investigates the practical usefulness of the proposed approach of estimating the state values. We notice that model parameters are esti-

Table 3 AIC-BIC table for selecting the optimal number of nearest neighbours for each sensor in Simulation 1

Number of nearest neighbours	AIC	BIC
1	8.52	10.63
2	7.29	9.58
3	**4.42**	**5.89**
4	5.76	6.94

mated quite accurately (as shown in the estimates) and precisely (as shown in the standard errors of the estimates). The correct order of the polynomial function used to model the general effect of time is picked up and the effect of the nearest neighbours is also considered quite appropriately. For a general WSN, the proposed approach can be used separately for each cluster. In that case, different models (models with different parameter values) have to be used for different clusters.

3.2 Simulation 2

In this simulation study, we investigate the practical usefulness of the mixed model proposed in Eq. (3). Here, we consider the same 6 sensors (same coordinates as given in Simulation 1) in a single cluster network. We simulate the state values of each sensor at 10 different time points ($t = 10$) using the following model:

$$X_i(t) = a_0 + a_1 t + a_2 t^2 + \alpha X_i(t-1) + \beta Z_i(t-1) + \gamma_{0i} + \gamma_{1i} t + \varepsilon_i(t), \quad (8)$$

where $(a_0, a_1, a_2) = (1.4, 2.1, 0.8), \alpha = 2.7, \beta = 3.45$. The vector of random effects $\gamma_i = [\gamma_{0i}, \gamma_{1i}]^T$ are simulated from a mixture Gaussian distribution given by the following density:

$g = 0.6g_1 + 0.4g_2$, where g_1 is the density of a bivariate Gaussian distribution with mean vector $= [0, 0]^T$ and variance–covariance matrix $= \begin{bmatrix} 4 & 2.7 \\ 2.7 & 9 \end{bmatrix}$ and g_2 is the density of a bivariate Gaussian distribution with mean vector $= [0, 0]^T$ and variance–covariance matrix $= \begin{bmatrix} 1 & 0.56 \\ 0.56 & 2.56 \end{bmatrix}$.

The residual errors $\varepsilon_i(t)$ are simulated from Gaussian(mean $= 0$, variance $= 1$). Z_i values are obtained considering 5 nearest neighbours ($r = 5$) of each sensor.

First we compare two competing models. The first one is the model given in Eq. (3) without the random intercept (γ_{0i}) and the random slope (γ_{1i}). We will refer this as Model I. This model is essentially the one proposed in Chatterjee et al. [4].

Table 4 Estimated bias and 95% confidence intervals of the model parameters for Model I and Model II in Simulation 2

	Parameter	Bias estimate	Width of C.I.
Model I	a_0	0.63	1.05
	a_1	0.57	0.94
	a_2	0.71	1.14
	α	0.68	1.01
	β	0.52	0.88
Model II	a_0	0.42	0.68
	a_1	0.38	0.61
	a_2	0.32	0.73
	α	0.29	0.70
	β	0.24	0.51

The second one is the model given in Eq. (3), where $\gamma_i = [\gamma_{0i}, \gamma_{1i}]^T$ follows bivariate Gaussian distribution with mean vector $= (0, 0)^T$ and covariance matrix $= \Sigma_i = \begin{bmatrix} \sigma_0^2 & \rho\sigma_0\sigma_1 \\ \rho\sigma_0\sigma_1 & \sigma_1^2 \end{bmatrix}$. We will refer this as Model II.

We simulate 100 replications of the dataset similar to the previous simulation study. Then, we fit both Model I and Model II on the simulated datasets. Model parameters for both the models are estimated using the maximum likelihood estimation method discussed earlier. Then, we further simulate 400 datasets from the above equation. The fitted Model I and fitted Model II are now tested on newly simulated datasets. For each dataset, the mean squared errors (MSE) are calculated for both the fitted models using the following formula:

$MSE = \frac{1}{60} \sum_{i=1}^{6} \sum_{t=1}^{10} \left(X_i(t) - \hat{X}_i(t) \right)^2$, where $\hat{X}_i(t)$ is the estimated state value from the respective model. The average mean squared error (AMSE) value is calculated for Model I and Model II by averaging the MSE values for 400 newly simulated datasets. For Model I, the AMSE value is calculated as 11.36 with standard error 4.58, where for Model II it is 6.83 with standard error 2.49. This result indicates the better predictive power of Model II compared to Model I.

Table 4 shows the estimate of the bias and the width of the 95% confidence intervals for the mean parameters of Model I and Model II. Note that these estimates are on the basis of 100 simulations originally used to fit the models. For each dataset, bias of a parameter estimate is estimated as bias = true parameter value – estimated parameter value. Final estimate of bias is obtained by averaging the bias estimates over 100 simulated datasets. Clearly, smaller bias is desirable for better inference. Width of the confidence intervals is simply the difference between the upper confidence limit and lower confidence limit. We notice that Model II provides parameter estimates with smaller bias and shorter confidence intervals than Model I. This illustrates the superiority of Model II over Model I and hence a mixed model is appropriate for state estimation of a discrete-time WSN.

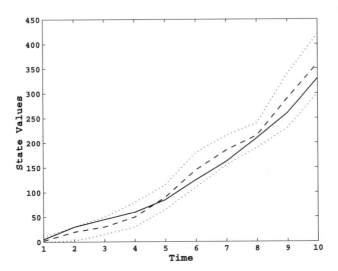

Fig. 4 Estimated mean curve and the 95% point-wise confidence interval for Model I in the simulation study 2. The solid curve is the true curve, broken curve is the fitted one and the dotted curves are the confidence intervals

Figure 4 shows true mean curve (solid) and estimated mean curve (broken) of the state values for sensor 1 at 10 time points using Model I. We also show 95% point-wise confidence interval (dotted) for the mean curve. In Fig. 5, we show the same true curve (solid), estimated curve (broken) and the point-wise confidence intervals

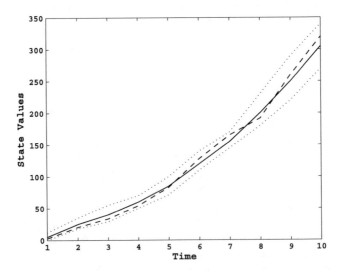

Fig. 5 Estimated mean curve and the 95% point-wise confidence interval for Model II in the simulation study 2. The solid curve is the true curve, broken curve is the fitted one and the dotted curves are the confidence intervals

(dotted) using Model II. As we observe in these figures, Model II provides a much better estimate of the state values over different time points than Model I. Similar results are obtained for all other sensors (graphs not shown for other sensors). Thus, the usefulness and importance of considering the subject-specific random effects of time is assessed and justified.

4 Conclusions and Future Work

The methodological developments for state estimation, event detection, node localization, etc., have been enriched in the recent years because of wide applications of WSNs under various complex environments and constraints. It is impossible to make an exhaustive list of applications of WSNs. In this chapter, we have discussed the basic structure of a WSN, and some basic state estimation algorithm. The numerical studies show the usefulness of the statistical model we discuss here, and the estimation accuracy and precisions.

An important aspect of state estimation, which we have not discussed here, is modelling incomplete data. Sometimes in an energy-constrained environment it is desirable to keep some of the sensor nodes in sleep mode for some time points, after which, these sensors become active but some other nodes are kept in the sleep mode. In a cluster-based network, at each time point at least one node from each cluster are kept in the active mode for the cluster to be active. This creates an incomplete dataset for state estimation, time synchronization, node localization, etc. Our proposed model considers the effects of the nearest neighbours, and hence can be used for estimating the missing state values for sensors kept in sleep mode. A statistical treatment for estimating the missing measurements can be found in Rubin [34]. However, we note that if the missingness is ignorable (e.g. missing at random), one can simply add a data-augmentation algorithm to the proposed estimation approach. However, for non-ignorable missingness (missing not at random) more complex algorithm will be needed.

Our current modelling and estimation approach is based on the assumption that sensor nodes, and base station are static. However, in many applications there are mobile sink and/or mobile sensor nodes. Mobile nodes are effective for covering larger geographical area, and hence in military surveillance, environment monitoring mobile sensors are used [6]. Recently, Shi et al. [36, 37] proposed methods for modelling the WSNs with mobile sink. We also note that even for a network with static sink and static sensor nodes, there might be a mobile anomalous node collecting information from different parts of the network. Extension of our proposed model for mobile nodes can be an interesting future research.

In recent years Bayesian statistics is playing a key role in various disciplines. The popularity of Bayesian statistics is due to the specification of "prior" information available from the previous study or provided by some domain experts. Such prior data are combined with observed measurements from the designed experiment, and then "posterior" inference is made. Posterior inference are powerful since prior and the experimental data are used together for such inference. In the Bayesian framework, model parameters are estimated by Markov Chain Monte Carlo (MCMC). Detail discussion on Bayesian modelling and the MCMC estimation method can be found in Robert and Casella [32]. In the context of state estimation of a cluster-based WSN, Bayesian modelling has been proposed by Chatterjee et al. [3, 4]. However, more in-depth applications of Bayesian statistics can be done in various problems related to WSNs.

References

1. Anastasi G. et al.: Energy conservation in wireless sensor networks: A survey. Ad hoc networks. **7(3)**, 537–568 (2009)
2. Chatterjee A., Venkateswaran P.: An efficient statistical approach for time synchronization in wireless sensor networks. International Journal of Communication Systems. **29(4)**, 722–733 (2016)
3. Chatterjee A., Venkateswaran P., Das K.: Simultaneous State Estimation of Cluster-Based Wireless Sensor Networks. IEEE Transactions on Wireless Communications, **15(12)**, 7985–7995 (2016)
4. Chatterjee A., Venkateswaran P., Mukherjee D.: A unified approach of simultaneous state estimation and anomalous node detection in distributed wireless sensor networks. International Journal of Communication Systems. **30(9)**, (2017)
5. Chhade H.: Data fusion and collaborative state estimation in wireless sensor networks. Doctoral dissertation, Universit de Technologie de Compigne (2015)
6. Chen X., Yu P.: Research on hierarchical mobile wireless sensor network architecture with mobile sensor nodes. Biomedical Engineering and Informatics (BMEI), 3rd IEEE Conference. **7**, (2010)
7. Dâmaso A., Rosa N., Maciel P.: Reliability of wireless sensor networks. Sensors **14(9)**, 15760–15785 (2014)
8. Davis A., Chang H.: A Survey of wireless sensor network architectures. International Journal of Computer Science and Engineering Survey. **3(6)**, 1 (2012)
9. Feng X. et al.: Constrained State Estimation for Individual Localization in Wireless Body Sensor Networks. Sensors. **14(11)**, 21195–21212 (2014)
10. Fischione C. et al.: Design principles of wireless sensor networks protocols for control applications. Wireless Networking Based Control. Springer New York. 203–238 (2011)
11. Gupta K., Sikka V.: Design Issues and Challenges in Wireless Sensor Networks. International Journal of Computer Applications. **112**, (2015)
12. Gong Y. et al.: Study on Security Issues in Wireless Sensor Network. International journal of Security and Its Applications. **10**, 295–302 (2016)
13. Gopakumar, A.: Node Localization in Wireless Sensor Networks by Artificial Immune System. Advances in Computing and Communications (ICACC). Fifth International Conference (2015)
14. Hac A.: Wireless sensor network designs. John Wiley & Sons Ltd. (2003)
15. Kaur B., Kaur A.: A Survey of Time Synchronization Protocols for Wireless Sensor Networks. International Journal of Computer Science and Mobile Computing. **2**, 100–106 (2013)
16. Lazaropoulos A.G.: Wireless sensor network design for transmission line monitoring, metering, and controlling: introducing broadband over power lines-enhanced network model (BPLeNM). ISRN Power Engineering **2014**, (2014)
17. Liang J. et al.: Distributed state estimation in sensor networks with randomly occurring nonlinearities subject to time delays. ACM Transactions on Sensor Networks (TOSN), **9(1)**, 4 (2012). State Estimation and Anomaly Detection in WSN 17

18. Liu H., Wang D.: Robust state estimation for wireless sensor networks with data-driven communication. International Journal of Robust and Nonlinear Control. (2017) https://doi.org/10.1002/rnc.3819
19. Mahmood M., Winston K.: Event reliability in wireless sensor networks. Intelligent Sensors, Sensor Networks and Information Processing (ISSNIP). Seventh International Conference IEEE (2011)
20. Michalik, M.: Base station for Wireless sensor network. Diss. Masarykova univerzita, Fakulta informatiky (2013).
21. Mo Y. et al.: False data injection attacks against state estimation in wireless sensor networks. Decision and Control (CDC). 49th IEEE Conference. IEEE (2010)
22. Nagarajan M., Karthikeyan S.: Health Condition Observation with High Level Efficiency in Wireless Sensor Network. Wireless Communication $1(5)$, 183–189 (2009)
23. Noh K., et al.: Novel clock phase offset and skew estimation using two-way timing message exchanges for wireless sensor networks. IEEE transactions on communications. $55(4)$, 766–777 (2007)
24. Noh K., et al.: A new approach for time synchronization in wireless sensor networks: Pairwise broadcast synchronization. IEEE Transactions on Wireless Communications, $7(9)$, 3318–3322 (2008)
25. Qian H., Sun P., Rong Y.: Design proposal of self-powered WSN node for battle field surveillance. Energy Procedia 16, 753–757 (2012)
26. Quevedo D. et al.: Predictive power control of wireless sensor networks for closed loop control. Nonlinear Model Predictive Control, 215–224 (2009)
27. Quevedo D. et al.: Energy efficient state estimation with wireless sensors through the use of predictive power control and coding. IEEE Transactions on Signal Processing. $58(9)$, 4811–4823 (2010)
28. Rajasegarar S., et al.: Anomaly detection in wireless sensor networks. IEEE Wireless Communications. $15(4)$, (2008)
29. Rajasegarar S. et al.: Quarter sphere based distributed anomaly detection in wireless sensor networks. Communications ICC'07. IEEE International Conference 3864–3869 (2007)
30. Rana M., Li L.: An overview of distributed microgrid state estimation and control for smart grids. Sensors $15(2)$, 4302–4325 (2015)
31. Ribeiro A. et al.: Kalman filtering in wireless sensor networks. IEEE Control Systems. $30(2)$, 66–86 (2010)
32. Robert C., Casella G.: Monte Carlo statistical methods. Springer (1999)
33. Roy A., Kar P., Misra S., Obaidat M.S.: D3: distributed approach for the detection of dumb nodes in wireless sensor networks. International Journal of Communication Systems. $30(1)$, (2015)
34. Rubin D.: Inference and missing data. Biometrika, $63(3)$, 581–592 (1976)
35. Shi-yong J. et al.: Geological disaster monitoring system based on WSN and GSM dual network integration technology. Communication Technology (ICCT), 14th International Conference. IEEE (2012)
36. Shi L. et al.: An efficient data-driven routing protocol for wireless sensor networks with mobile sinks. International Journal of Communication Systems. $26(10)$, 1341–1355 (2013)
37. Shi L. et al.: An efficient distributed routing protocol for wireless sensor networks with mobile sinks. International Journal of Communication Systems. $28(11)$, 1789–1804 (2015)
38. Singh P., Tripathi B., Singh N.: Node Localization in wireless sensor Networks. International Journal of Computer Science and Information Technologies. $2(6)$, 2568–2572 (2011)
39. Silva I., Guedes L., Portugal P., Vasques F.: Reliability and availability evaluation of wireless sensor networks for industrial applications. Sensors, $12(1)$, 806–838 (2012)
40. Valverde J. et al.: Wireless sensor network for environmental monitoring: application in a coffee factory. International Journal of Distributed Sensor Networks 8.1: 638067 (2011)
41. Wang X., Fu M., Zhang H.: Target Tracking in Wireless Sensor Networks Based on the Combination of KF and MLE Using Distance Measurements. IEEE Transactions on mobile computing. 11, 567–576 (2012)

42. Winkler M. et al.: Wireless sensor networks for military purposes. Autonomous Sensor Networks. Springer Berlin Heidelberg. 365–394 (2012)
43. Zhu J., Tan L., Xi H., Zhang Z.: Reliability analysis of wireless sensor networks using Markovian model. Journal of Applied Mathematics. (2012)
44. Zhu X., Lu Y., Han J., Shi L.: Transmission reliability evaluation for wireless sensor networks. International Journal of Distributed Sensor Networks, **12(2)**, 1346079 (2016)

Experimental Wireless Network Deployment of Software-Defined and Virtualized Networking in 5G Environments

Flávio Meneses, Carlos Guimarães, Daniel Corujo and Rui L. Aguiar

Abstract 5G research has been looking for flexible, dynamic and low-latency network architectures. In this regard, Software-Defined Networking (SDN) and Network Functions Virtualization (NFV) have been key enablers for the next generation networks. Although SDN was originally designed for wired networks, such as data-centre networks, its popularity in the research community has extended its boundaries, also becoming considered for wireless networks. With its flexibility and programmability, Software-Defined Wireless Networks (SDWN) have been suggested for mobile networks evolution. This chapter provides an overview of how virtualization of network entities (such as points of attachment and flow-based mobility management entities) can contribute to a new level of abstraction in heterogeneous wireless access environments. To assess the enhancement potential of such mechanisms, a framework case study is presented and evaluated, showcasing flow-based mobility scenarios in multi-technology environments.

1 Introduction

Upcoming 5G communication systems have been driven by evolutions in customers, technology and operator contexts. As such, consumers expect to get access to content from just one click away, independently of where and how they are connected, with smartphones considered the main connectivity device [1]. As of late, mobile data has been increasing along with the number of connected devices. In 2016, smartphones registered 81% of the total mobile traffic, with mobile video representing 60% [2].

The development of 5G architectures has been leveraging Software-Defined Networking (SDN), Network Function Virtualization (NFV) and cloud technologies, to extend the physical equipment capabilities to better support applications

F. Meneses (✉) · C. Guimarães · D. Corujo · R. L. Aguiar
Instituto de Telecomunicações and Universidade de Aveiro, Aveiro, Portugal
e-mail: flaviomeneses@av.it.pt; flaviosmeneses@gmail.com

F. Meneses · C. Guimarães · D. Corujo · R. L. Aguiar
Campus Universitário de Santiago, 3810-193 Santiago, Portugal

© Springer Nature Singapore Pte Ltd. 2018
K. V. Arya et al. (eds.), *Emerging Wireless Communication and Network Technologies*,
https://doi.org/10.1007/978-981-13-0396-8_17

requirements. The adoption of such technologies will break otherwise vertical systems into chainable and configurable building blocks, allowing greater flexibility.

SDN decouples control and data planes, centralizing management decisions in a Controller. Decisions are further transmitted to forwarding devices through a southbound API (with OpenFlow [3] as the de facto protocol). This adds greater network abstraction, allowing the deployment of a virtualized network independently of the underlying transport and protocols. In this context, NFV allows the deployment of network functions as software instances running on servers through virtualization, replacing hardware middleboxes [4]. It allows virtualization, in the cloud, of IP network functions (e.g. load balancers, firewalls) and Long-Term Evolution (LTE)/Evolved Packet Core (EPC) network functions (e.g. mobility management entities).

The adoption of SDN and NFV not only improves CAPEX and OPEX, but also provides new possibilities by bringing virtualization to network operations. For example, in bare-metal scenarios, operators overprovision hardware resources to face peak-hours utilization. Virtualization allows a more dynamic approach, with resources scaling as needed.

As such, operators not only can virtualize middleboxes but also edge and access entities such as Base Stations (BSs) and Access Points (APs), allowing them to scale dynamically. Virtualization capabilities can also be extended into mobile devices, complementing their limited capabilities with virtualized counterparts at the data centre, allowing for high-quality video content generation, cloud gaming, supporting smart scenarios [1] (e.g. smart-cities, smart-agriculture, smart-transportation) and others.

Approaches such as SoftRAN [5] and Cloud-RAN [6], evolve from the application of SDN in cellular networks, exploring SDN and NFV to centralize the control plane for Radio Access Networks (RANs), abstracting BSs belonging to a local area (SD-RAN), and providing different network control algorithms. Finally, with the proliferation of cloud-based services, applications became portable across multiple devices (which are able to exploit multiple wireless technologies simultaneously) capable of accessing services while on the move, increasing the complexity for full inter-operation of such technologies, in terms of wireless mobility management and performance. Additionally, industries are investing in machine-type communication and IoT to improve and automate their processes [7]. Billions of smart devices will use embedded communications and integrated sensors to act on the local environment and use remote triggers based on intelligent logic [1]. However, these devices differ in terms of requirements and communication capabilities. This allows the creation of new services for vertical industries (e.g. health and automotive) with different network requirements (e.g. latency and performance), forcing the development of a reliable network capable of adapting to each use case. In this context, '*Network slicing offers an effective way to meet all of the diverse use case requirements and exploit the benefits of a common network infrastructure, enabling operators to establish different capabilities, deployments, and architectural flavors for each use case*' [8]. Network Slices are seen as a tool for delivering differentiated services, as they provide optimized connectivity for each use case, application and user. In order to run the network in such way, researchers are laying efforts into SDN and NFV technologies [1, 2].

Under this setting, it becomes critical to study and validate the utilization of these mechanisms in different scenarios. This chapter provides a case study of key enabling 5G technologies in wireless environments, focusing on SDN and NFV, towards the exploitation and realization of mobile communication situations. First, existing IP-based mobility aspects are compared with the advances proposed for 5G (Sect. 2). Then, two key approaches are discussed and evaluated, namely virtualization of network endpoints (Sect. 3) and mobile nodes (Sect. 4), followed by an analysis of deployment possibilities (Sect. 5). These solutions are combined for flow optimization in 5G scenarios (Sect. 6), concluding in Sect. 7.

2　Mobility Management in 5G Networks

According to [9], by 2021 Wi-Fi and mobile will account for over 63% of all IP traffic. Among the reasons for this increase is the proliferation of mobile devices exploiting different wireless technologies (e.g., cellular and Wi-Fi) and more powerful capabilities (e.g., cameras). As such, multi-technology mobility becomes a key requirement to users.

Device mobility can be seen as changing Points of Attachment (PoA) (i.e. network end-point towards user devices, namely BSs or APs) along with associated procedures. Mobile IP (MIP) [10] was introduced by the Internet Engineering Task Force (IETF), to address mobility by mapping, in the home agent, a global address with the device's current IP. Thus, packets destined to the mobile node (MN) are received by the home agent and tunnelled towards the MN. However, even considering its IPv6 variant (MIPv6), IP-based mobility was not deployed Internet wide [11] due to handover delay, signalling overhead and software stack modifications. In alternative, network-based mobility management protocols such as Proxy Mobile IPv6 (PMIPv6) [12] and Distributed Mobility Management (DMM) [13] were proposed, shifting mobility signalling to network nodes.

PMIPv6 still inherited issues such as non-optimal path communication between the MN and the correspondent node, and tunnelling. Nonetheless, PMIPv6 was deployed as a network-based mobility management protocol, handling the signalling on behalf of the MN, improving the localized routing handover delay, signalling cost and local mobility anchor utilization [11]. Despite an evolution, PMIPv6 can be simultaneously deployed with MIPv6 [14], with the latter used globally and the former locally.

Conversely, DMM was designed to be flexible and distributed [15]. Evolving from IETF and 3GPP efforts, it complemented or replaced existing functions [13]. However, the inherent triangular routing and tunneling between mobile access routers are still an issue [15]. Solutions such as [16–18], integrate DMM with SDN flow-based capabilities, resulting in a framework providing mobility and Quality of Service (QoS) for the end-user, in terms of latency and throughput.

With SDN and NFV standing as base enabler technologies for 5G, studies propose their integration within the LTE network [19]. In this way, the IP-based network is maintained, while SDN improves its flexibility. Also, integrating the SDN controller with the Mobility Management Entity (MME), allows the former to be aware of mobility requirements. The adoption of SDN allows different grades of performance and complexity for core network services, unifying mobility and routing management.

2.1 Software-Defined Networking and Network Function Virtualization as Key Enablers

5G is addressing the development of a converged architectural solution supporting optimized device mobility in heterogeneous wireless environments. Mobility management procedures are deployed to allow users to exploit those features with the best Quality of Experience (QoE). Mobility management studies for 5G networks already started, with initial SDN-based DMM architectures [16, 17, 19, 20].

Despite that the initial deployment aim of SDN-based protocols and operations targeted wire-based technologies (i.e., Ethernet and fibre), the flow-based traffic reconfiguration capabilities were deemed of interest for wireless environments. Works, such as [21], applied OpenFlow capabilities to APs, further extended in [22], allowing SDN controllers to remotely configure flows traversing them. This established the foundation for using SDN and OpenFlow as mobility management mechanisms, with [23, 24] supporting an enhanced controller capable of managing handovers, using IEEE 802.21 and OpenFlow. DMM principles were also integrated with SDN mechanisms in [16, 17].

OpenFlow-based mobility support procedures were also applied into mobile devices [25]. Experimental deployments were presented in [26, 27] focusing on IEEE 802.11 technologies, and [28, 29] validating SDN mobility management capabilities in heterogeneous environments. This was progressed in [30] by deploying OpenFlow in end-to-end source mobility.

Works such as [31–33] have provided mobility management enhancements enabling the creation of virtual APs (vAP). In this context, CloudMAC [34] allows Wi-Fi MAC frames to be processed in a vAP instantiated in a data centre. Similarly, Odin [35] virtualizes a managing agent to control APs. These were enhanced in [31] with vAPs able to "follow" the MN, placing mobility procedures entirely inside the virtualized network of interconnected vAPs. As such, each MN is associated with its own vAP when it connects to the network, the latter moving along with its client, with the MN unaware of the physical handover. In [32], the authors take advantage of APs operating on multiple channels for seamless handovers. In [33] SDN-based procedures were proposed for creating, managing and migrating vAPs, without modifying MNs. Finally, in [36] the MN was also virtualized, allowing its virtual representation to enhance the capabilities of its physical counterpart and

Table 1 Mobility management frameworks with "key 5G technologies"

	Advantages	Disadvantages	Mobility impact
SDN-based DMM [16–20]	Overcome triangular routing issues	Does not support cross-technology seamless source handover	Per-flow mobility Simpler data-plane
802.21 with OpenFlow [23, 24]	Mobile offloading	Changes in the UE Requires 802.21	Inter-technology handover
meSDN [27]	TDMA scheduling	802.11 only	N/A
Forging client mobility [30]	Allows seamless mobile offloading	Changes in the transmitter and receiver	Inter-technology handover
Extended OpenFlow control path [28, 29]	Mobile offloading without additional mobility protocols	Changes in the UE	Inter-technology handover
Extended OpenFlow control path and MN virtualization [36]	Mobile offloading without additional mobility protocols Allows to virtualize the receiver in different devices	Changes in the UE	Inter-technology handover

supporting media-independent network control processes. Table 1 presents an overview of the state of the art in mobility management proposals for 5G networks.

2.2 The Role of Network Slicing

Slicing allows differentiating traffic traversing a shared network infrastructure, providing communication isolation. Presently, this is achieved through various mechanisms, such as VLANs or firewalls, adding configuration complexity and special devices in specific network locations. FlowVisor [37] leveraged SDN-centralized control to enable applications to control networks without interfering with each other, albeit requiring complex hypervisor software at the managing plane.

Network slicing represents a key enabler for 5G architectures [38]. Requirements in such environments are more demanding than today, envisioning an explosion of connected devices, massive traffic and data rates with reduced latency [1]. As such, the network needs to be built in a flexible and cost-efficient way, where speed and scale can be configured on-demand in logical slices to meet the demands of each use case [39]. Within initial 5G architectural proposals mostly based on SDN and NFV enablement, several key issues have surfaced, considering both the application of network slicing, as well as the transition towards novel 5G environments [40].

The level of network programmability provided by SDN allows network slices to be optimally configured using the same physical and logical network infrastructure [39]. These features were exploited in [41] with slicing strategies used for better user experience according to network requirements. On the other hand, NFV's flexibility allows network functions to be executed in a location-independent way, no longer bound to a specific network node, as well as chained together as a logical network slice [39].

Works such as [41] simulate control mechanisms that dynamically allocate network resources to different slices, albeit disregarding mobility issues. The authors in [42] proposed a "5G Network Slice Broker" to facilitate on-demand resource allocation and admission control, using traffic monitoring and forecasting. Similarly, [43] introduced hierarchical slices, allowing operators to customize E2E networks as a service. Despite enabling operators to build network slices for vertical industries more agilely, these approaches disregard mobility management issues, such as the role of the MME and its slice management interaction.

Studies show that despite the increase of users/devices, 5G solutions should not assume mobility support for all. Instead, on-demand mobility should be supported, ranging from very high mobility, such as trains/airplanes, to low mobility or stationary devices such as smart-metres [1]. In this context, [44] proposes to extend SDN to wireless networks, virtualizing the RAN and Core Network (CN), as a vAP and a Network Slice, respectively. Simulation results demonstrate the efficiency of the proposed scheme, but experimental results are still lacking. There is also the need to test a slice-based 5G architecture, in order to consider that multiple network slices can be present at the radio access and the core, simultaneously.

3   Points of Attachment Virtualization

This section addresses the PoA virtualization, presenting different approaches and comparing them with legacy solutions. 5G mobility communication systems need to go beyond mobile network technologies, thus providing mobility management and control procedures in a technology-agnostic way. In this line, the vPoA represents the virtualization of PoA management and control services (e.g. authentication, association, Dynamic Host Configuration Protocol (DHCP), etc.) in the cloud, exploiting resources therein to regard PoAs as virtualizable network functions that can be dynamically instantiated according to utilization needs. As such, the vPoA can be seen as working on behalf of the PoA, able to fully offload its operating mechanisms (Fig. 1b) or only selected services and applications (Fig. 1c). Figure 1 compares the different deployments, evaluated in Sect. 3.2.

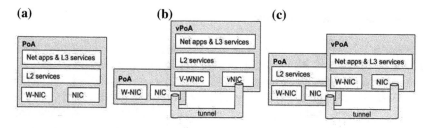

Fig. 1 PoA conceptual architecture for: **a** without virtualization; **b** L2 and L3 virtualization; and **c** L3 virtualization

3.1 Overview of Existing Works

NFV has been being considered as a flexible mechanism in mobile CN and RAN, exploring high-performance data centres [45]. Cloud-based RAN architectures (or Cloud-RAN), exploit network virtualization and centralized coordination techniques. CellSDN [46], SoftRAN [5], Cloud-RAN [6] and SoftAir [47] apply SDN allowing PoAs to be abstracted and connected to virtual representations of themselves in the data centre, providing optimized, flexible and scalable control.

Nonetheless, the proposed solutions have yet to become fully implemented, whereas existing experimental deployments (i.e. such as CAPWAP [48], Odin [35] and CloudMAC [34]), focus only in Wi-Fi. In CAPWAP [48], frame generation is split between the controller and APs, providing centralized configuration, authentication and association. Odin [35] uses different virtualized agents at each AP, with a unique BSSID for each connected MN. Finally, CloudMAC [34] performs IEEE 802.11 MAC processing in the vAP.

3.2 Virtualization Approaches and Their Signalling

PoA virtualization brings a trade-off between delay and both hardware and wireless mobility simplicity. By virtualizing the PoA in the cloud (i.e. vPoA), the physical PoA becomes a bare-metal device, but creates a longer control and management path between the attached device and the vPoA. Different approaches can be considered for PoA virtualization: (i) virtualize all its L2 and L3 management procedures; or (ii) virtualize only L3 mechanisms (e.g. DHCP). In this way, for an IEEE 802.11 AP, in (i) L2 management and control frames are offloaded to the vPoA, with wireless authentication and association procedures performed there. Otherwise, in (ii), only L3 mechanisms are offloaded to the vPoA. Figure 2 compares the signalling of both approaches with the legacy approach (i.e. a regular Wi-Fi AP network), summarized in Table 2.

3.2.1 Mobile Node Attachment

In a regular Wi-Fi (Fig. 1a) network, where virtualization is not considered, the AP deals with all wireless management and control frames (e.g. authentication, association and probes), while at the same time manages L3 connections and services. Nowadays, AP configuration is simple, but in large-scale environments, complexity increases substantially. Works such as [23, 24], apply SDN to wireless traffic control, facilitating dynamic reconfiguration. Notwithstanding, it relies on complex procedures and dedicated protocols, hindering heterogeneity.

Figure 2 illustrates exchanged messages for L2 (represented as the Wi-Fi open system authentication) and L3 attachment (represented as the DHCP). Table 2 presents the theoretical delay for each Wi-Fi virtualization approach. Discarding processing time, it is expected that the "*L2 + L3 virtualization*" approach has a higher delay for both L2 and L3 attachment, since besides the Wi-Fi round trip time (RTT), an Ethernet (eth) RTT is added. As such, "*L3 virtualization*" has a lower delay, since the

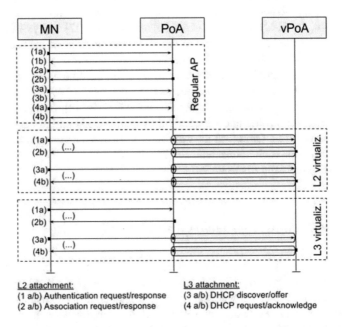

L2 attachment:
(1 a/b) Authentication request/response
(2 a/b) Association request/response

L3 attachment:
(3 a/b) DHCP discover/offer
(4 a/b) DHCP request/acknowledge

Fig. 2 Layer 2 (L2) and Layer 3 (L3) attachment signalling for different AP deployment approaches

Table 2 Delay calculation for different AP deployment approaches

AP implementation approaches	Delay calculation for L2 and L3 attachments
Without virtualization	4x wifi_rtt + processing
L2 + L3 virtualization	2x (2x wifi_rtt + 2x eth_rtt) + processing
L3 virtualization	2x (2x wifi_rtt + eth_rtt) + processing

"*eth*" delay is inputted only for DHCP messages. A regular AP deployment should have the lowest L2 and L3 attachments delay (evaluated in Sect. 3.3).

3.2.2　Mobility Enhancement Enabled by the Point of Attachment Virtualization

Under a L2 perspective, virtualizing the PoA in the cloud enables traffic management to be offloaded to the vPoA, decoupling the intelligence from the hardware and deploying it in any point of the network. This facilitates handover management both for L3 and L2, since the attachment is processed in the virtual instance instead of the physical node. As such, a vPoA can be transmitting/receiving management frames from different physical nodes, and a handover between two physical PoAs does not necessarily mean an L2 or L3 handover, since they may be connected to the same vPoA. CloudMAC [34] and Odin [35] identified that AP virtualization into a data centre improves intra-technology mobility at a cost in bandwidth and attachment delay. In contrast, physical PoAs become simpler, while vPoAs had increased computational power. This was further exploited in [36] with the proposal of a cloud-based mobility management procedure for L2 and L3 inter-technology handovers, which became seamless for end-nodes, supported by virtualization mechanisms. Table 3 compares the presented PoA deployments.

Despite Mobile-Edge Computing (MEC) scenarios where this approach is well suited [49], wireless networks may experience performance issues due to the delay between the physical AP and the corresponding vPoA. At the cost of losing seamless L2 handovers, this is overcome by keeping the L2 attachment in the physical equipment and virtualizing only L3 services (e.g. DHCP) (as in Table 2). In this case, the AP handles L2 connectivity, allowing MN attachment and remaining L2 communications to be managed therein. Otherwise, L3 services are migrated to the cloud. Considering a handover, where a MN switches the physical AP, a new L2

Table 3 Comparison of the different AP deployment approaches

	Advantages	Disadvantages	Mobility impact
Regular AP	Simple implementation	Manual reconfiguration	Complex mobility management Dedicated mobility protocols
L2 vPoA	Simpler network equipment	L2 + L3 attachment delay Lost performance due to the RTT delay	Seamless L2 + L3 handover
L3 vPoA	Simpler network equipment Reconfiguration on the fly	L3 attachment delay	Seamless L3 handover

connection is required. However, the MN may keep the L3 connection, facilitating handover management (more details in Sect. 4).

3.3 Virtual Wi-Fi Access Point Evaluation

This section provides an experimental overview of different vAP strategies. The MN's network (re)attachment delay, L2 and L3 handover and the MN signalling impact are studied. Physical nodes (i.e. MN and PoAs) were deployed in the AMazING wireless testbed [50], with the corresponding virtual instances (i.e. vMN and vPoAs) deployed in an in-house data centre, using a VM with 1 vCPU and 2 GB RAM. AMazING nodes were equipped with a 64-bit support AMD APU CPU with 1 GHz dual-core and 2 GB DRAM. In both types of network nodes, Ubuntu 64-bit 14.04 was used.

Figure 3 compares a regular AP with different vAP deployments in terms of attachment delay and bandwidth, related with Fig. 2. As such, "L2 attachment" corresponds to messages 1a–2b (i.e. Wi-Fi open system authentication), while "L3 attachment" corresponds to messages 3a–4b (i.e. DHCP). As a trade-off for hardware simplicity and L2 handover management, the "L2 virtualization" approach has the worst result, with a delay increase of 50% for a L2 attachment and DHCP IP address. As shown in Table 2 from the previous section, this is mainly due to the increase of RTT and overhead caused by increased forwarding of wireless management frames. This is reflected in the bandwidth, which decreased by 14%. By offloading L3 services only, not only the bandwidth value was re-established, but also the DHCP delay was improved, while maintaining the L2 attachment delay. This may be due to forwarding overhead decrease (since Wi-Fi management frames are processed in the AP), allied with the increased computational power of the vAP.

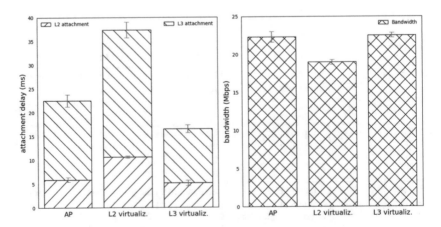

Fig. 3 Attachment delay and bandwidth for the deployed AP

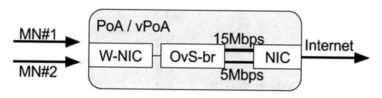

Fig. 4 QoS scenario deployment

AP behaviour was also evaluated in regard to QoS. The scenario, illustrated in Fig. 4, was deployed using Open vSwitch [51] and OpenFlow. Since APs have about 22 Mbps of upstream bandwidth, two queues with different bandwidths (15 and 5 Mbps) were created, mimicking different user classes. QoS performance was assessed for different traffic types (i.e. UDP, TCP and RTP). PoA deployments "without virtualization" and "L3 virtualization" have the queues implemented in the PoA, whereas "L2 virtualization" has it in the vPoA, evidencing the need of prioritizing wireless control and management frames.

In Fig. 5a, two TCP streams were sent via the AP, showing that QoS was accomplished both in the Wi-Fi and Internet link. Despite that MN#1 has all bandwidth (about 22 Mbps) available until 15 s of evaluation time, it only uses the predefined bandwidth (i.e. 15 Mbps). Moreover, when MN#2 started transmitting, it only occupied the pre-established bandwidth. In contrast, in Fig. 5b, two UDP streams were considered and, despite that QoS was accomplished for the traffic sent towards the Internet, in the wireless link both streams were competing for resources, with the first decreasing its bandwidth to 10 Mbps. From the figure, it is seen that MN#1 transmits at full bandwidth (22 Mbps), disregarding QoS. Such QoS shaping is only performed on the Internet link. When MN#2 begins transmitting at 15 s, it also tried to use full bandwidth. This resulted in a QoS downgrade for MN#1, while the QoS for MN#2 was achieved.

The TCP and UDP impact was also studied over a RTP video stream. Figure 5c shows that since the TCP stream adapts its traffic, the video stream QoS was not affected. Otherwise, Fig. 5d shows that, since UDP does not realize adaptations, QoS was downgraded for others. Since RTP is usually transported over UDP, if the upstreaming video outperforms the pre-establish bandwidth, the traffic shaping will be verified only on the Ethernet, allowing the video stream to compete for the wireless bandwidth. These behaviours provide indicators for future SDN-enabled PoAs deployments, in order to allow SDN protocols (such as OpenFlow) to (re)configure wireless QoS on the fly.

4 Bringing User Context to the Cloud

This section addresses end-virtualization. This enables information about the MN's link to be collected in the CN by its virtual instantiation (i.e. vMN), enabling the

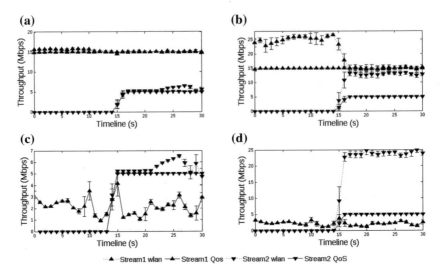

Fig. 5 QoS evaluation for: **a** two TCP stream; **b** two UDP streams; **c** a video stream (stream 1) and a TCP stream (stream 2); and **d** a video stream (stream 1) and an UDP stream (stream 2)

Controller to adapt the network service provisioning to each type of access technology.

4.1 Virtualizing the User Equipment

Current smartphones allow accessing online services while moving from one wireless technology to another. UE virtualization implies the virtualization of its network context into the cloud enabling the vMN to perform management decisions on behalf of the UE: the vMN not only allows to anchor the UE when offloading traffic from one wireless PoA, but actually allows to deploy a virtual representation on other devices (Fig. 6). For example, when a user accesses an online video using the UE and is near a TV with a set-top-box connected, the video can be offloaded to it by creating a small virtualized instance of the smartphone in the set-top box. The vMN acts as a proxy for the Internet access and as a mobile anchor, enabling transparent cross-technology handovers, since both transmitter and receiver continue to send/receive traffic to the vMN. Such scenario is further discussed and evaluated in Sect. 6.

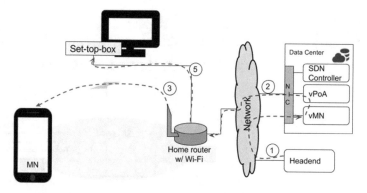

Fig. 6 Virtualization of the MN scenario

4.2 Mobile Node Virtualization Procedures

The interaction between the vMN and its physical counterpart allows the provisioning of information about the connectivity status of the latter. Thus, the network controller (which is involved in the instantiation of the vMN and its traffic flows configuration) gets feedback on the physical link conditions of the MN, allowing optimization. Moreover, the fact that the physical MN is connected to its virtual counterpart, means that all traffic associations between the vMN and other communication entities can be handled inside the data centre, minimizing handover control signalling.

4.2.1 Context Update of the Virtual Mobile Node

During MN attachment (Fig. 7: 1 and 6), the PoA informs the controller of a new connection (Fig. 7: 2 and 7), using an OpenFlow *Packet_in* message. MN attachment information is sent to the SDN controller, registering the necessary information for network (re)configuration. By saving the MN's MAC address and PoA, the controller verifies if the MN is already associated to an existing vMN (Fig. 7: 8) or if a new one is required (Fig. 7: 3 and 4).

After identifying the respective vMN, the controller (re)configures the network to redirect all MN's traffic to the vMN (Fig. 7: 5 and 9). For context updates (e.g. active links and neighbour cells), different approaches can be considered:

i. The PoA/vPoA monitors the different attached MNs and periodically updates the controller (or when triggered), which updates the vMNs (Fig. 7: Case A, 10–11a)

ii. The MN monitors the wireless medium, sending updates to the controller (Fig. 7: Case A, 10–11a and 10–11b).

Fig. 7 Signalling for
instantiation and context
update of the vMN

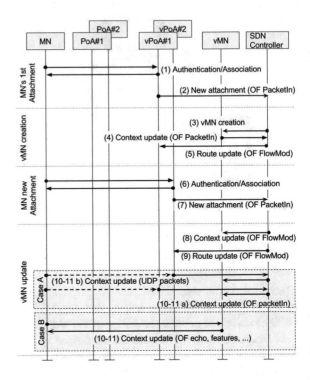

iii. Extend the OpenFlow control path all the way to the MN enabling it (and
the vMN) to directly communicate via OpenFlow messages. This also enables
source offloading mobility mechanisms [28, 29] (Fig. 7: Case B).

In the first approach, the SDN Controller updates vMNs with the respective MN's
context, evidencing a scalability issue when many MNs are connected. As such, the
latter approaches provide a better level of scalability, since each MN directly updates
its context in the cloud, but requires MN modification.

4.2.2 Mobility Procedures

Upon attachment, all MN Internet traffic passes through its vMN. In this way, the
correspondent node will see traffic from both interfaces as being from the same MN,
enabling an offloading mechanism transparent to end nodes. In case of a handover
request from a network entity, depending on the requester (e.g. vPoA or MN) and
the traffic direction (i.e. downstream or upstream) different procedures are taken, as
illustrated in Fig. 8:

i. Downstream: if the trigger is sent from the vPoA to the Controller (1b1), the latter verifies the network state and, after deciding which MN should be offloaded,
updates the flow table (1b2) with the IP address of the MN new receiving

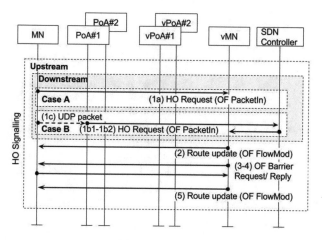

Fig. 8 Handover signalling

interface. For an MN-initiated mobility (1a), the trigger is sent to the vMN, which updates its flow table. Instead of OpenFlow, the MN can send the trigger via an UDP packet (1c), which is converted into an OpenFlow message in the vPoA (1b1), and sent to the Controller (1b2).

ii. Upstream: When the trigger is sent from the vPoA towards the Controller (1b1), the latter verifies the network state, and informs the respective vMN (1b2). If the trigger is sent from the MN (1a), the message is sent to the vMN, responsible for choosing which flow will be offloaded (2–5). In addition, an OpenFlow Barrier message (3–4) is used to verify the correct reception of the first flow-rule, preventing packet loss due to the handover procedure [36].

Table 4 compares the described mechanisms ("virtualization of the PoA and MN") against scenarios "without virtualization" and with "only PoA virtualization".

4.2.3 Network Enhancements

The signalling between a MN and its PoA, can be improved by modifying the wireless driver and allowing the vMN to realize it instead. This enables a deeper exploration of the virtual entities context in the cloud, reducing the signalling wireless impact. For example, in the open system authentication, the intermediary signalling of the MN authentication and association are exchanged between the vPoA and vMN (Fig. 9: 1a/b–2a/b). In this way, for a new L2 attachment, the MN only needs to be involved in the first and last related packets (i.e. authentication request and association response).

Table 5 provides messages details in a L2 and L3 attachment (open system authentication and DHCP, respectively). Section 3.2 shows how the former three approaches are related. For the last approach, it was assumed that the Wi-Fi RTT has a higher delay than Ethernet. For latency, it is expected that the signalling enhanced approach

Table 4 Mobility impact of the MN virtualization

	Mobility impact
Without virtualization [28, 29]	Requires the SDN controller to be aware of all MN wireless state In a handover scenario, the SDN controller needs to be aware MN's path and reconfigure the right nodes in the path to avoid triangular routing
Only PoA virtualization [34, 35, 51]	Requires the SDN controller to be aware of all MN wireless state Easier to reconfigure the path up to the MN (compared to without virtualization approach) Triangular routing with less impact
Virtualization of the PoA and MN [36]	Each vMN keeps the wireless state of the respective MN Easier to reconfigure the path up to MN (compared to former approaches) Avoids triangular routing

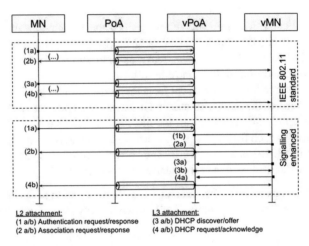

Fig. 9 Access point's (AP) Layer 2 (L2) and Layer 3 (L3) attachment enhanced signalling

has lower delay, since it reduces the total number of over-the-air messages to 50% for the L2 attachment and 25% for DHCP. Notwithstanding, "L3 virtualization", when compared to "without virtualization", increases the same values in the wired signalling (i.e. Ethernet).

In the "without virtualization" procedure, neither the AP or MN were instantiated in the cloud, hence the MN performs eight message interactions (i.e., authentication, association and DHCP). Despite that this approach has only one hop between the MN and the AP, current integration of Wi-Fi and 3GPP (e.g. Hot Spot 2.0) adds new interaction steps where the connectivity procedure also has to be sent towards the operator infrastructure, for AAA, roaming and other procedures, increasing delay.

Table 5 Over-the-air overhead and delay calculation for different AP deployment approaches

AP implementation approaches	No. of messages	No. of bytes	Delay calculation for L2 and L3 attachments
Without virtualization	2×4	$288 + 1368$	4x wifi_rtt + processing
L2 + L3 virtualization	2×4	$288 + 1368$	2x (2x wifi_rtt + 2x eth_rtt) + processing
L3 Virtualization	2×4	$288 + 1368$	2x (2x wifi_rtt + eth_rtt) + processing
L2 + L3 virtualization with signalling enhancement	$2 + 1$	$141 + 342$	1.5x (wifi_rtt + eth_rtt) + processing

Using a "virtualized L2+L3 AP" (presented in Sect. 3) increased delay, due to the Ethernet RTT. To improve the previous case, the "L3 virtualized AP" was presented with a trade-off of losing seamless L2 handovers.

The introduction of a vMN does not negatively affect the MN attachment delay. In fact, besides resulting in similar times when compared to previous cases, the vMN allows the deployment of a mobility mechanism for inter-technology handovers [36]. Moreover, the previous signalling can be enhanced by decoupling the MN from the previously needed authorization and association messages. In this way, the MN sends the first message and receives the last, resulting in a reduction of 50% of the L2 attachment signalling over-the-air. Remaining management messages are exchanged in the cloud between virtualized entities, with lower processing times. Regarding DHCP, this can be directly negotiated with only the vMN, transmitting to the MN only the result, saving 75% of the transmitted bytes.

4.3 Proof-of-Concept Prototype Evaluation

This section provides an experimental assessment of the deployment of a vMN in a data-centre along with vAPs in a physical testbed. All management decisions will be performed in the cloud, assisted by a SDN controller enhanced with mobility management and virtual machine/networking functions. To extend OpenFlow to the MN, one bridge for each wireless interface was created and interconnected with patch ports (Fig. 10) [28].

The framework was tested in two different environments, OpenStack[1] and Docker,[2] in order to evaluate its virtualization impact. OpenStack VMs were instantiated with 1 vCPU and 1 GB RAM, running Ubuntu 64-bit 14.04 LTS. Docker ran in a Virtualbox VM, with 1 vCPU, 4 GB RAM and identical Ubuntu. To evaluate the delay of the on-demand instantiation capability added to the SDN controller, the

---

[1] OpenStack: https://www.openstack.org.
[2] Docker: https://www.docker.com.

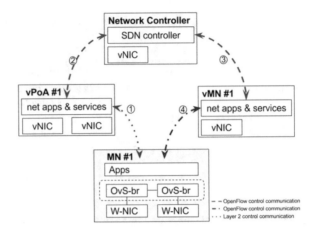

Fig. 10 Conceptual vMN architecture

Table 6 Instantiation and reconfiguration time of the vMN

	Instantiation	Network reconfiguration	MN and vMN control connection
OpenStack (s)	134.34 ± 2.17	134.49 ± 2.27	138 ± 3.75
Docker (ms)	551.49 ± 6.71	551.49 ± 6.71	N/A

time between the initial instantiation request and the first communication performed between the vMN and remaining network entities was registered and is shown in Table 6.

The controller begins communications establishment and has a delay of about 134 s in OpenStack against the 551 ms in Docker. This delay is related to OpenStack VM instantiation which needs to be scheduled, built and initiated. Also, using a regular Ubuntu 64-bit OS, instead of an optimized one, increases the delay. Notwithstanding, the use of Docker significantly reduces the delay, since it does not contain a guest OS, decreasing the boot-time and creation delay [52]. Despite these delay issues, VM creation is not a major issue in this proposal, since it is only created on the first attachment. As such, it can be compared to activating a mobile device, where a user waits several seconds for network connectivity. Moreover, the framework does not aim to be tied to a particular virtualization mechanism, but rather illustrates an approach for 5G realization in the coming future. Studies such as [53] aim to provide new methods for decreasing VM creation delay, along with numerous cloud management solutions under development. These VM instantiation technical issues are beyond the scope of this work.

In [36], the use of a vMN was evaluated in a video upstreaming scenario (Fig. 11), using the same platform described in Sect. 3.3, and assuming that VM creation has already occurred. The scenario evaluates two different triggers in *make-before-break* and *break-before-make* approaches:

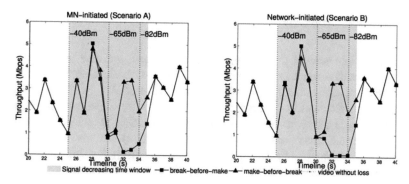

Fig. 11 Video throughput and signal level over time for the live video upstreaming scenario [36]

i. MN-initiated: As the MN moves away from PoA1, link quality decreases and when a better one is detected, it sends a handover request to vMN. The vMN starts the handover, implements the necessary flow rules and the MN starts streaming via AP2.

ii. Network-initiated: The vPoA measures the MN signal level, which is streaming video via AP1. When the signal decreases (past a configurable threshold), vPoA1 sends a handover request to the controller, which analyzes the network state, and spotting the MN, alerts the respective vMN. In turn, the vMN handovers the MN's flow to AP2.

For *make-before-break* approaches the MN sends the handover request as soon as the signal level crosses the threshold (at 30 s with a signal level of −65 dBm), avoiding the loss of 1.16 MB of the video which was registered for *break-before-make* approaches. Handover delay was analysed under the perspective of the MN, showing that MN-initiated scenarios have an extra component from the MN's point of view: the interval between the handover request (i.e. Figure 8: 1a) and the moment that the flow starts to be redirected (i.e. Figure 8: 2). This extra communication increased delay in 25.56 (±8.22) ms and signalling over-the-air in 144 bytes. Notwithstanding, from the network's point of view (i.e. measured in the vMN and defined as the delay between the handover request—Fig. 8:1a—and the flow being actually redirected—Fig. 8: 5) the MN-initiated handover delay had better performance than the network-initiated, since the MN notifies the vMN directly [36].

5 The Impact of Virtualization in 5G Scenarios

This section highlights three major 5G scenarios, namely mobile operators, MEC and smart-cities where the deployment of virtualization-based mechanisms can support mobility management operations.

5.1   Mobile Operator

In current mobile networks, indicators such as interference and load are used to assist in network management. On a virtualized SDN-based, the inherent delay between the controller and network devices needs to be considered. Figure 12 overviews the framework deployment in a mobile operator scenario. By virtualizing the network entities, a shorter path is set between them and the controller, reducing delay, simplifying control and leaving only the actual data to be transported towards end-nodes. The presence of the vMN in the cloud, reduces management decisions, such as PoA attachment and handovers, as well as reducing the control signalling over-the-air. On one hand, the integration of SDN in the MN can require changes to how L3 flow procedures are executed, as well as L2 wireless association. On the other hand, it allows seamless and simpler L2 and L3 handovers. Moreover, it is difficult for a single controller to allocate the right resources in time, due to the dynamics of the channel. As such, the vMN is responsible for managing and adjusting the MN to channel variations. Hence, while the network controller has a global view of the network state, the vMN assists, manages and controls the MN state and handover decisions.

5.2   Mobile-Edge Computing

With network cloudification, MEC places compute and storage resources nearer to the RAN, improving content and applications delivery to end users. This enables operators to better adapt traffic to radio conditions, optimize service quality and improve network efficiency. This allows to explore scenarios where services benefit from being hosted in the distributed cloud close to customers and better adapt content delivery to ensure consistent QoE [54]. Instancing the vMN in a MEC environment

Fig. 12 Mobile operator scenario

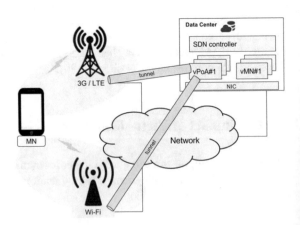

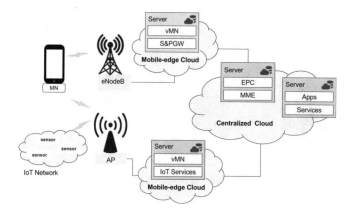

Fig. 13 Mobile-Edge Computing (MEC) scenario

(considering its role as proxy and anchor) allows reducing communication delay between the physical and virtual instances. As such, the framework exploits a RAN-aware video optimization scenario, by using the MEC-enabled vMN to transform the data (i.e. changing a video codec on-the-fly) and adjust traffic rates (i.e. applying QoS). However, due to the MN mobility, virtualization mechanisms procedures might need some form of enhancement to be able to support seamless mobility. Figure 13 presents a MEC scenario framework solution.

5.3 Smart-Cities

In upcoming 5G networks, billions of connected devices with different requirements are expected to be operating simultaneously, divided in three main groups: (i) mobile broadband; (ii) critical and (iii) massive communications. Smart-cities exploit requirements of previous groups. An example can be a city fully equipped with sensors and actuators, not only statically placed (i.e., buildings) but actually moving (i.e. cars and mobile devices), with IoT devices using different protocols and networks. However, users are interested only in services that can be built upon this information. A way of facilitating the development of IoT services is to virtualize such devices. In this way, the information that is acquired by the IoT device is stored in its virtual instance to be further used when necessary, provided from there, and maintained despite the involved mobility.

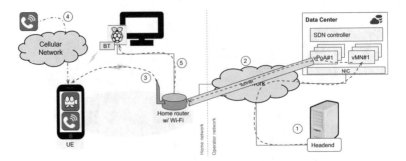

Fig. 14 Deployment scenario for a mobile operator use case

6 Use Case Scenario in Wireless Testbed

This section evaluates the virtualization-based mobile management framework in a mobile operator scenario, introduced in Sect. 4. Here, the user is at home and has a set-top-box[3] connected to the TV. The user is accessing an online live video stream on the smartphone, and receives a voice call via mobile network. The UE is virtualized in the operator's cloud, which seamlessly offloads the video stream from the mobile (or Wi-Fi) network to the TV, upscaling the video quality, and allowing the user to keep watching the video while answering the call. The evaluated scenario is presented in Fig. 14.

The proposed framework was deployed in the wireless testbed described in Sect. 3.3, where the physical PoAs were deployed in the testbed nodes and the virtual instances deployed in Docker containers (as described in Sect. 4.3). The MN was an Android 6.0 Nexus 5. The scenario was evaluated for handover performance (data throughput) and delay.

6.1 Deployment and Signalling

In the scenario (Fig. 14), the user accesses a video in the smartphone using the home Wi-Fi (1–3), when a call arrives (4). An application was developed in the smartphone, allowing it to trigger a handover event when a call is received (4), signalling context to the network (such as indicating that a video is being watched, and that there are other displays nearby (i.e. TV connected to the LAN)) thus triggering a flow handover. By reading Bluetooth beacons,[4] the application is able to identify the TV to the network, which, in turn, can verify if it is a viable receptacle for the video. The controller in the network is able to compose this information into a handover decision, shifting the video flow from the MN towards the TV, while allowing the phone call

---

[3]Deployed in a Raspberry Pi 3.
[4]Bluetooth beacon device: ByteReal TagBeacon 2.0.

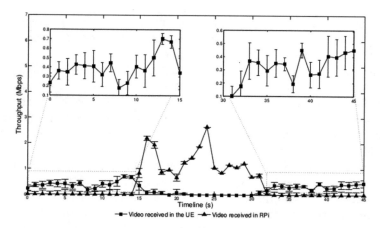

Fig. 15 Video stream throughput for the evaluated scenario

to be processed into the smartphone (5). Security and authentication aspects of this procedure are important, but are outside of the mobility scope of this work.

6.2 Scenario Evaluation

The scenario evaluates the mechanism impact in the MN, when accessing a live video, focusing on the execution of a *make-before-break* handover. Experiments were run 10 times, averaged with a confidence interval of 95%.

When the video is shown in the smartphone, Standard-Definition (SD) quality is used. The handover signalling is related to Fig. 8 (Downstream—Case B). At about $t = 15$ s in Fig. 15, upon reception of the call, the video flow is switched to the TV and upgraded to High-Definition (HD), motivating a higher throughput. When the phone call is over, the video flow is downsized and moved back into the smartphone at $t = 31$ s. The TV accounted for 86% (totalling 2.69 ± 0.07) MB) more video traffic when compared with the smartphone (a total of $1.44(\pm 0.04)$ MB). Despite this flow change, the vMN is always the anchor recipient of the video, and is able to dynamically switch between them. The offloading process had a delay of $31(\pm 3.24)$ ms and $27(\pm 3.26)$ ms (handover delay measured between the handover request and the first video packet towards the TV, as perceived by the vPoA).

7 Conclusions and Future Work

This chapter addressed how virtualization technologies can enhance mobility management for upcoming 5G networks, by exploiting SDN and NFV as key enablers to build systems with greater degree of flexibility and abstraction, proposing to virtualize the RAN (more specifically, the PoAs) and the UEs (or MNs), supported by proof-of-concepts deployed in a wireless testbed.

A state of the art analysis in mobility management was provided, followed by mobile architecture cloudification proposals and virtual APs implementations, pointing out limitations and contributions for the adoption of SDN and NFV in heterogeneous networks. Two different approaches of a virtual Wi-Fi AP were implemented and assessed in terms of signalling, performance and QoS, providing indicators for PoA virtualization in future networks.

The user context was brought to the cloud by virtualizing the MN therein and extending the OpenFlow control path up to the MN, allowing it to become part of the network control, assisting the controller in the mobility management by providing its perspective of the network conditions (such as, detected neighbour cells) and minimizing signalling delay. Such framework was studied and experimentally evaluated, followed by the case study of its deployment in different 5G scenarios. Finally, the proposed framework was deployed in one of the presented scenarios, providing the foundations to improve mobile video QoE in multi-technology networks.

As a final note, virtualization had a minor impact in terms of delay and throughput, and presented advantages for selective flow handover, such as TVs with network capabilities, while decoupling different control challenges of the connectivity at the link layer and management of the mobile service in terms of mobility management, reducing the path between the involved entities, resulting in a handover delay decrease and less information exchanged over the air. As such, promising new prospects for virtualization-assisted MN mobility management can be considered in upcoming generations of mobile networks.

Acknowledgments This work is funded by FCT/MEC through national funds and when applicable co-funded by FEDER PT2020 partnership agreement under the project UID/EEA/50008/2013, by the Integrated Programme of SR&TD SOCA (Ref. CENTRO-01-0145-FEDER-000010), co-funded by Centro 2020 program, Portugal 2020, European Union, through the European Regional Development Fund, and by the FCT Grant SFRH/BD/96553/2013.

References

1. Alliance, NGMN. "5G white paper." *Next generation mobile networks, white paper* (2015).
2. Cisco, Cisco visual networking index: Global mobile data traffic forecast update, 2016–2021 white paper (March 2017).
3. McKeown, Nick, et al. "OpenFlow: enabling innovation in campus networks." *ACM SIGCOMM Computer Communication Review* 38.2 (2008): 69–74.
4. Nguyen, Van-Giang, Truong-Xuan Do, and YoungHan Kim. "SDN and virtualization-based LTE mobile network architectures: A comprehensive survey." *Wireless Personal Communications* 86.3 (2016): 1401–1438.
5. Gudipati, Aditya, et al. "SoftRAN: Software defined radio access network." *Proceedings of the second ACM SIGCOMM workshop on Hot topics in software defined networking*. ACM, 2013.

6. Cai, Yegui, F. Richard Yu, and Shengrong Bu. "Cloud radio access networks (C-RAN) in mobile cloud computing systems." *Computer Communications Workshops (INFOCOM WKSHPS), 2014 IEEE Conference on*. IEEE, 2014.
7. Lasi, Heiner, et al. "Industry 4.0." *Business & Information Systems Engineering* 6.4 (2014): 239–242.
8. Nokia, Dynamic end-to-end network slicing for 5G. White Paper, 2016.
9. Cisco, Cisco Visual Networking Index: Forecast and Methodology 2016–2021, White Paper, June 2017.
10. Perkins, Charles E. "IP mobility support for IPv4, revised." (2010).
11. Rasem, Ahmad, Marc St-Hilaire, and Christian Makaya. "A comparative analysis of predictive and reactive mode of optimized PMIPv6." *Wireless Communications and Mobile Computing Conference (IWCMC), 2012 8th International*. IEEE, 2012.
12. Gundavelli, Sri, et al. *Proxy mobile ipv6*. No. RFC 5213. 2008.
13. Zúniga, Juan Carlos, et al. "Distributed mobility management: a standards landscape." *IEEE Communications Magazine* 51.3 (2013): 80–87.
14. Guo, Hua, et al. "LMA/HA Discovery Mechanism on the interaction between MIPv6 and PMIPv6." *Wireless Communications, Networking and Mobile Computing, 2009. WiCom'09. 5th International Conference on*. IEEE, 2009.
15. Giust, Fabio, Antonio De la Oliva, and Carlos J. Bernardos. "Mobility management in next generation mobile networks." *World of Wireless, Mobile and Multimedia Networks (WoWMoM), 2013 IEEE 14th International Symposium and Workshops on A*. IEEE, 2013.
16. Giust, Fabio, Luca Cominardi, and Carlos J. Bernardos. "Distributed mobility management for future 5G networks: overview and analysis of existing approaches." *IEEE Communications Magazine* 53.1 (2015): 142–149.
17. Nguyen, Tien-Thinh, Christian Bonnet, and Jérôme Harri. "SDN-based distributed mobility management for 5G networks." *Wireless Communications and Networking Conference (WCNC), 2016 IEEE*. IEEE, 2016.
18. Costa-Requena, Jose. "SDN integration in LTE mobile backhaul networks." *Information Networking (ICOIN), 2014 International Conference on*. IEEE, 2014.
19. Wang, Shiwei, et al. "An optimal slicing strategy for SDN based smart home network." *Smart Computing (SMARTCOMP), 2014 International Conference on*. IEEE, 2014.
20. Valtulina, Luca, et al. "Performance evaluation of a SDN/OpenFlow-based Distributed Mobility Management (DMM) approach in virtualized LTE systems." *Globecom Workshops (GC Wkshps), 2014*. IEEE, 2014.
21. Yiakoumis, Y., J. Schulz-Zander, and J. Zhu. "Pantou: OpenFlow 1.0 for OpenWRT (2011)".
22. Yap, Kok-Kiong, et al. "OpenRoads: Empowering research in mobile networks." *ACM SIGCOMM Computer Communication Review* 40.1 (2010): 125–126.
23. Guimaraes, Carlos, et al. "Empowering software defined wireless networks through media independent handover management." *Global Communications Conference (GLOBECOM), 2013 IEEE*. IEEE, 2013.
24. Guimaraes, Carlos, et al. "Enhancing openflow with media independent management capabilities." *Communications (ICC), 2014 IEEE International Conference on*. IEEE, 2014.
25. Bernardos, Carlos J., et al. "An architecture for software defined wireless networking." *IEEE wireless communications* 21.3 (2014): 52–61.
26. Dely, Peter, et al. "Best-ap: Non-intrusive estimation of available bandwidth and its application for dynamic access point selection." *Computer Communications* 39 (2014): 78–91.
27. Lee, Jeongkeun, et al. "meSDN: mobile extension of SDN." *Proceedings of the fifth international workshop on Mobile cloud computing & services*. ACM, 2014.
28. Meneses, Flavio, et al. "Extending sdn to end nodes towards heterogeneous wireless mobility." *Globecom Workshops (GC Wkshps), 2015 IEEE*. IEEE, 2015.
29. Meneses, Flavio, et al. "Multiple flow in extended sdn wireless mobility." *Software Defined Networks (EWSDN), 2015 Fourth European Workshop on*. IEEE, 2015.
30. Makris, Nikos, et al. "Forging Client Mobility with OpenFlow: an experimental study." *Wireless Communications and Networking Conference (WCNC), 2016 IEEE*. IEEE, 2016.

31. Grunenberger, Yan, and Franck Rousseau. "Virtual access points for transparent mobility in wireless LANs." *Wireless Communications and Networking Conference (WCNC), 2010 IEEE*. IEEE, 2010.
32. Berezin, Maria Eugenia, Franck Rousseau, and Andrzej Duda. "Multichannel virtual access points for seamless handoffs in IEEE 802.11 wireless networks." *Vehicular Technology Conference (VTC Spring), 2011 IEEE 73rd*. IEEE, 2011.
33. Lin, You-En, and Ting-Ming Tsai. "Creation, management and migration of virtual access points in software defined WLAN." *Cloud Computing and Big Data (CCBD), 2015 International Conference on*. IEEE, 2015.
34. Dely, Peter, et al. "CloudMAC—An OpenFlow based architecture for 802.11 MAC layer processing in the cloud." *Globecom Workshops (GC Wkshps), 2012 IEEE*. IEEE, 2012.
35. Suresh, Lalith, et al. "Towards programmable enterprise WLANS with Odin." *Proceedings of the first workshop on Hot topics in software defined networks*. ACM, 2012.
36. Meneses, Flavio, et al. "An abstraction framework for flow mobility in multi-technology 5G environments using virtualization and SDN." *Network Softwarization (NetSoft), 2017 IEEE Conference on*. IEEE, 2017.
37. Sherwood, Rob, et al. "Carving research slices out of your production networks with Open-Flow." *ACM SIGCOMM Computer Communication Review* 40.1 (2010): 129–130.
38. ONF, Open Networking Foundation. TR-256 Applying SDN Architecture to 5G Slicing, Apr 2016.
39. Ericsson. 5G Systems White Paper, Jan 2017.
40. 3GPP. Study on Architecture for Next Generation System, Jun 2016.
41. Jiang, Menglan, Massimo Condoluci, and Toktam Mahmoodi. "Network slicing management & prioritization in 5G mobile systems." *European Wireless 2016; 22th European Wireless Conference; Proceedings of*. VDE, 2016.
42. Samdanis, Konstantinos, Xavier Costa-Perez, and Vincenzo Sciancalepore. "From network sharing to multi-tenancy: The 5G network slice broker." *IEEE Communications Magazine* 54.7 (2016): 32–39.
43. Zhou, Xuan, et al. "Network slicing as a service: enabling enterprises' own software-defined cellular networks." *IEEE Communications Magazine* 54.7 (2016): 146–153.
44. Xun Hu, Rong Chai, Guixiang Jiang, and Haipeng Li. A joint utility optimization based virtual ap and network slice selection scheme for sdwns. In 2015 10th International Conference on Communications and Networking in China (ChinaCom), pages 448–453, Aug 2015.
45. Ericsson, Cloud RAN: the benefits of virtualization, centralization and coordination - White Paper (Sep 2015).
46. Li, Li Erran, Z. Morley Mao, and Jennifer Rexford. "Toward software-defined cellular networks." *Software Defined Networking (EWSDN), 2012 European Workshop on*. IEEE, 2012.
47. Akyildiz, Ian F., Pu Wang, and Shih-Chun Lin. "SoftAir: A software defined networking architecture for 5G wireless systems." *Computer Networks* 85 (2015): 1–18.
48. Govindan, S., et al. *Objectives for control and provisioning of wireless access points (capwap)*. No. RFC 4564. 2006.
49. Li, Hongxing, et al. "WiCloud: Innovative uses of network data on smart campus." *Computer Science & Education (ICCSE), 2016 11th International Conference on*. IEEE, 2016.
50. Martins, Joao, et al. "Experimentation made easy with the AMazING panel." *Proceedings of the seventh ACM international workshop on Wireless network testbeds, experimental evaluation and characterization*. ACM, 2012.
51. Pfaff, Ben, et al. "The Design and Implementation of Open vSwitch." *NSDI*. 2015.
52. Seo, Kyoung-Taek, et al. "Performance comparison analysis of linux container and virtual machine for building cloud." *Advanced Science and Technology Letters* 66 (2014): 105–111.
53. Sotiriadis, Stelios, et al. "Cloud virtual machine scheduling: Modelling the cloud virtual machine instantiation." *Complex, Intelligent and Software Intensive Systems (CISIS), 2012 Sixth International Conference on*. IEEE, 2012.
54. Juniper Networks, Mobile Edge Computing Use Cases & Deployment Options, White Paper, (July 2016).

Printed in the United States
By Bookmasters